# Recent Developments in Separation Science

## Volume IX

Editors

**Norman N. Li, Sc.D.**
Director
Separation Science and Technology
Signal Research Center, Inc.
Des Plaines, Illinois

**Joseph M. Calo, Ph.D.**
Associate Professor of Engineering
Division of Engineering
Brown University
Providence, Rhode Island

CRC Press, Inc.
Boca Raton, Florida

**Library of Congress Cataloging-in-Publication Data**
(Revised for volume 9)

Recent developments in separation science.

(CRC uniscience series)
Vol.     editors, Norman N. Li, Joseph M. Calo.
Vols.     lack series statement.
Includes bibliographical references and indexes.
1. Separation (Technology)   I. Li, Norman N.
II. Calo, J. M.   III. Series.
TP156.S45R44     660.2'842     72-88417
ISBN 0-8493-5031-X (v. 1)

This book represents information obtained from authentic and highly regarded sources. Reprinted material is quoted with permission, and sources are indicated. A wide variety of references are listed. Every reasonable effort has been made to give reliable data and information, but the author and the publisher cannot assume responsibility for the validity of all materials or for the consequences of their use.

Direct all inquiries to CRC Press, Inc., 2000 Corporate Blvd., N.W., Boca Raton, Florida, 33431.

© 1986 by CRC Press, Inc.

International Standard Book Number 0-8493-5490-0

Library of Congress Card Number 72-88417
Printed in the United States

# PREFACE

Ninth in this popular series, this volume continues the tradition of its predecessors in presenting a unique and useful blend of recent developments in theory, applications, and process technology in the important field of separation science. The chapters comprising the current volume are a combination of comprehensive reviews and the latest work covering a wide spectrum of separation applications. Among the comprehensive reviews concerned with new and developing technologies are chapters on supercritical fluid extraction, the application of coordination complexes for gas separations, multicomponent adsorption and cyclic processes, rotating-disk, thin-layer chromatography, and laser-induced separations. Important new applications in membrane technology include separations of olefins, barbiturates, and europium, as well as an analysis of the role of recycle in process permeators. New applications/analyses of adsorption and extraction include chapters on separation of heavy metal chelates by activated carbon, removal of nitrogenous heterocyclic compounds in shale-derived oils via solvent extraction, and interfacial charge and mass transfer in liquid-liquid extraction of metals. Some new process technologies are reviewed in chapters on Monsanto PRISM℗ technology, acid gas removal via the CNG process, and PEDS (Peltier effect diffusion separation) process technology for evaporation/concentration applications for aqueous streams. In summary, this volume provides a sampler of the very latest state of technology in the continuously developing area of separation science.

We wish to thank the CRC Press staff for their editorial assistance. We would like also to extend our sincere gratitude to all the chapter contributors who made this volume possible.

Norman N. Li
Joseph M. Calo

# THE EDITORS

**Norman N. Li, Sc.D.** is a Research Director at Signal Research Center, Inc. Des Plaines, Illinois. Previously, he was with Exxon Research and Engineering Co. He has about 100 scientific publications and U.S. patents in the areas of crystallization, extraction, blood oxygenation, enzyme membranes, water treatment, liquid membranes, and polymeric membranes. Of the 40 or so patents he holds, 35 alone deal with the basic invention and various applications of liquid membranes.

Dr. Li was a consultant on gas diffusion for the U.S. Apollo project and was the Chairman of the Industrial and Engineering Chemistry Division of American Chemical Society. He has given lectures at many universities and industrial research laboratories. Dr. Li teaches two short courses for the American Institute of Chemical Engineers — "New Separation Processes" and "Surface Chemistry and Emulsion Technology". He served as the Chairman of two Gordon Research Conferences — "Separation and Purification" and "Transport Phenomena in Synthetic and Biological Membranes" and of the recent Engineering Foundation Conference on "New Directions in Separations Technology". He was also the chairman of several symposia on separations at the American Chemical Society and American Institute of Chemical Engineers meetings.

Dr. Li is currently serving on National Research Council's Committee on Separation Science and Technology and Committee on Biotechnology.

**Joseph M. Calo, Ph.D.,** received his B.S.Ch.E. degree from Newark College of Engineering and his Ph.D. in chemical engineering from Princeton University. Upon graduation from Princeton, he served on active duty as a Captain in the United States Air Force at the Air Force Cambridge Research Laboratories (AFCRL, now the Air Force Geophysics Laboratory), Hanscom AFB, Massachusetts. At AFCRL, he conducted basic research in the areas of atmospheric composition and chemical kinetics. After his four-year term, Dr. Calo joined Exxon Research and Engineering Co. in Florham Park, New Jersey, where he worked in the areas of reactor fundamentals and fluid dynamics. Dr. Calo returned to Princeton University as Assistant Professor of Chemical Engineering in 1976. In 1981, he became Associate Professor of Engineering at Brown University, where he is currently participating in the establishment of a new chemical engineering program in the Division of Engineering.

Professor Calo has taught courses in chemical reaction engineering, separation processes, fluid dynamics, and associated laboratories. He is an associate editor of the Stagewise and Mass Transfer Operations area of the Chemical Engineering Modular Instruction (CHEMI) Project.

Professor Calo is a licensed professional engineer in the states of Rhode Island and New Jersey. He is also a member of the American Institute of Chemical Engineers, the American Chemical Society, the American Association for the Advancement of Science, and Sigma Xi.

Dr. Calo's current research interests are applied chemical kinetics in chemical engineering and atmospheric phenomena. In addition to publications in these areas, Dr. Calo has presented papers at national and international meetings on these subjects. He is also currently an active consultant to industry and government agencies in these and related areas.

# CONTRIBUTORS

**E. W. Albaugh, Ph.D.**
Research Associate, Scientific Staff
Petroleum and Alternate Energy Research
 Division
Gulf Research and Development
 Company
Pittsburgh, Pennsylvania

**Dibakar Bhattacharyya, Ph.D.**
Ashland Professor
Department of Chemical Engineering
College of Engineering
University of Kentucky
Lexington, Kentucky

**W. A. Bollinger**
Engineered Products Division
Monsanto Company

**G. Bouboukas**
Centre Réacteurs et Processus
Ecole Nâtionale Superiéure des Mines de
 Paris
Paris, France

**William R. Brown, Ph.D.**
President
Helipump Corporation
Cleveland, Ohio

**Joseph M. Calo, Ph.D.**
Associate Professor of Engineering
Division of Engineering
Brown University
Providence, Rhode Island

**Jiang Changyin**
Institute of Nuclear Energy Technology
Qinghua University
Beijing, People's Republic of China

**Wei-Niu Chen**
Associate Professor of Chemical
 Engineering
Zhejiang University
Hangzhou, Zhejiang, People's Republic
 of China

**P. Colinart**
Centre Réacteurs et Processus
Ecole Nâtionale Supérieure des Mines des
 Paris
Paris, France

**Robert W. Farmer**
Research Associate
Department of Chemical and Bio
 Engineering
Arizona State University
Tempe, Arizona

**Richard T. Hallen, M.S.**
Research Scientist
Department of Chemical Technology
Battelle Pacific Northwest Laboratory
Richland, Washington

**Robert D. Hughes, Ph.D.**
Senior Research Chemist
Corporate Research Department
Amoco Corporation
Naperville, Illinois

**Brian R. James, Ph.D.**
Professor
Department of Chemistry
University of British Columbia
Vancouver, Canada

**Yu Jianhan**
Institute of Nuclear Energy Technology
Qinghua University
Beijing, People's Republic of China

**D. E. King**
Engineered Products Division
Monsanto Company

**M. S. Kuk, Ph.D.**
Project Engineer, Scientific Staff
Petroleum and Alternate Energy Research
 Division
Gulf Research and Development
 Company
Pittsburgh, Pennsylvania

**R. J. Laub, Ph.D.**
Professor
Department of Chemistry
San Diego State University
San Diego, California

**Nabil M. Lawandy**
Assistant Professor
Division of Engineering
Brown University
Providence, Rhode Island

**Chung-Li Lee, Ph.D.**
Postdoctoral Research Fellow
Department of Chemistry
University of British Columbia
Vancouver, Canada

**Kwang-Lung Lin, Ph.D.**
Department of Metallurgy and Materials
  Engineering
National Cheng Kung University
Tainan, Taiwan

**Donald L. MacLean**
Monsanto Development Center, Inc.
Research Triangle Park, North Carolina

**John A. Mahoney, Ph.D.**
Director
Synthetic Fuel Program Department
Amoco Corporation
Chicago, Illinois

**Lester G. Massey, Ph.D.**
Manager, Process Development (Ret.)
Research Department
Consolidated Natural Gas Company
Cleveland, Ohio

**Mark A. McHugh**
Assistant Professor
Department of Chemical Engineering
University of Notre Dame
Notre Dame, Indiana

**Milton Meckler**
President
The Meckler Group
Encino, California

**J. C. Montagna, Ph.D.**
Staff Engineer
Petroleum and Alternate Energy Research
  Division
Gulf Research and Development
  Company
Pittsburgh, Pennsylvania

**R. S. Narayan**
Engineered Products Division
Monsanto Company

**David A. Nelson, Ph.D.**
Senior Research Scientist
Department of Chemical Technology
Battelle Pacific Northwest Laboratory
Richland, Washington

**K. Osseo-Asare, Ph.D.**
Professor
Department of Materials Science and
  Engineering
Pennsylvania State University
University Park, Pennsylvania

**H. Renon, Ph.D.**
Centre Réacteurs et Processus
Ecole Nâtionale Supérieure des Mines des
  Paris
Paris, France

**Kamalesh K. Sirkar, Ph.D.**
Professor of Chemical Engineering
Department of Chemistry and Chemical
  Engineering
Stevens Institute of Technology
Hoboken, New Jersey

**Edward F. Steigelmann, Ph.D.**
Research Supervisor
Research and Development
Amoco Oil Company
Naperville, Illinois

**Steven Teslik**
Process Engineer
Department of Process Engineering
GA Technologies
San Diego, California

**G. Trouvé**
Centre Réacteurs et Processus
Ecole Nationale Supérieure des Mines des
 Paris
Paris, France

**Ralph T. Yang, Ph.D.**
Professor
Department of Chemical Engineering
State University of New York at Buffalo
Amherst, New York

**Zhu Yongjun**
Institute of Nuclear Energy Technology
Qinghua University
Beijing, People's Republic of China

**D. L. Zink**
Merck Institute
Rahway, New Jersey

# TABLE OF CONTENTS

Chapter 1

# APPLICATION OF COORDINATION COMPLEXES FOR SEPARATION OF GASEOUS MIXTURES

**David A. Nelson, Richard T. Hallen,**
**Chung-Li Lee, and Brian R. James**

## TABLE OF CONTENTS

# I. INTRODUCTION

Several methods have been proposed for separating individual components from gas mixtures instead of cryogenic rectification. These include pressure-swing adsorption, permeable hollow fiber technology (e.g., Prism®), and bimetallic hydrides.[1] The separation of oxygen from air has received the most attention in the search for alternative procedures to cryogenic techniques. Both pressure-swing adsorption and high temperature inorganic oxide techniques offer varying degrees of success in separating oxygen and nitrogen. However, another procedure, the use of coordination complexes, offers technical and economic potential for gas separation which yet may prove competitive with cryogenic separation. Indeed, coordination complexes have a history of being experimentally employed to provide oxygen aboard military ships[2] and aircraft.[3]

Chemical abbreviations for many of the ligands and their groups discussed below are presented in Table 1.

The initial research on coordination complexes which can reversibly bind oxygen centered about salcomine, a complex of cobalt(II) with bis(salicylidene)-diaminoethane.[4,5] Exposing the complex to air between 0 and 25°C allows oxygen to be bound preferentially to the cobalt center as a cobalt(III) superoxide structure.[6] Heating or reducing pressure facilitates release of oxygen from the complex. Unfortunately, several difficulties plagued commercialization of salcomine including a rigorous activation stage of the crystalline complex,[7,8] slow oxygenation rates above 10°C, and sensitivity to water. Further, irreversible oxidation to a bridge peroxo species was cited as the most important cause of cobalt(II) complex deterioration.[8,9] Many of these difficulties, except the activation stage, were minimized with the use of the cobalt(II) complex of bis(3-fluorosalicylidene)diaminoethane or "fluomine".[3] Considerable research has been carried out with these and other cobalt(II) complexes as well as iron(II) and manganese(II) complexes in an attempt to understand natural $O_2$ systems such as hemoglobin. Several reviews have compiled the results concerning oxygen coordination and separation.[9-12]

The efforts involving the separation of air with coordination complexes have stimulated application of this technique to other gases. In particular, the separation of individual components from mixtures of carbon monoxide, carbon dioxide, hydrogen, nitrogen, and methane is of considerable interest. Such mixtures of two or more of these gases are produced by coal gasification, the water-gas shift reaction, steam reforming, and partial oxidation of petroleum. Refinery off-gas is another source. In order for coordination complexes to be applicable to industrial gas separation they must possess rapid sorption, easy gas desorption, and a long recycle life (low degradation over time). Selectivity for specific gases is also an important requirement.

Although methane and other alkanes can add oxidatively to coordination complexes, to yield hydridoalkyl metal species, the interaction is extremely weak and subject to competition with solvents and other gases.[13-15] On the other hand, carbon monoxide, carbon dioxide, and hydrogen are superb candidates for separation from gaseous mixtures due to their ability to act as stronger coordinating ligands. For instance, several mono- and bimetallic complexes have been shown to be reversible carbon dioxide carriers. These include the bifunctional $Co^I$(salen)M · THF complex, where M is Li, Na, K, or Cs.[16-18] In order to activate $CO_2$ for coordination, the basic cobalt center to which the C atom binds requires an alkali cation to interact with the O atoms and complete the acid-base interaction. Carbon dioxide desorption can be achieved by warming the complexes to 30°C or by decreasing the pressure over the mixture of the $CO_2$ complex and solvent. The three tertiary phosphine-rhodium complexes $Rh(P^nBu_3)_3Cl$, $Rh(PhPMe_2)_3Cl$, and $Rh(Ph_2PEt)_3Cl$ have been reported to bind $CO_2$.[19] Only the latter two complexes desorbed $CO_2$ under vacuum or by heating above 80°C. $Ni(PCy_3)_3$[20] and $(NiL_4)$, where L = $PEt_3$ or $PBu_3$,[21] also coordinate $CO_2$. Heating the $CO_2$ adducts of

## Table 1
### CHEMICAL ABBREVIATIONS FOR LIGANDS, ASSOCIATED GROUPS, AND SOLVENTS

| Abbreviation | Structural name |
|---|---|
| Me | Methyl |
| Et | Ethyl |
| Pr | Propyl |
| Bu | Butyl |
| Cy | Cyclohexyl |
| Ph | Phenyl |
| THF | Tetrahydrofuran |
| salen | Bis(salicylidene)diaminoethane |
| dmpe | Bis(dimethylphosphino)ethane |
| depe | 1,2-Bis(diethylphosphino)ethane |
| dpm | Bis(diphenylphosphino)methane |
| diars | *o*-Phenylenebis(dimethylarsine) |
| dma | Dimethylacetamide |

the nickel complexes *in vacuo* resulted in phosphine decomposition and incomplete $CO_2$ recovery, while treatment with iodine or sulfuric acid did release $CO_2$. Several iridium complexes have been reported to reversibly bind $CO_2$. The earliest work was by Flynn and Vaska[22] who investigated the reactivity of *trans*-Ir(PPh$_3$)$_2$(CO)(OH). Complete decarboxylation of the $CO_2$ complex required 10 hr at 25°C under vacuum. A companion $CO_2$-rhodium complex underwent desorption only after several weeks of vacuum exposure. The other iridium systems were claimed in a patent by Herskovitz and Parshall.[23] Two of the more successful $CO_2$ carriers were [Ir(dmpe)$_2$]Cl and [Ir(PMe$_3$)$_4$]Cl, both absorbing $CO_2$ while in hydrocarbon suspension. Desorption was achieved by heat, vacuum, or treatment with dichloromethane. The [Rh(diars)$_2$]Cl complex has also been observed to bind $CO_2$.[24] The $CO_2$ bonding within this complex is considerably different from that in most others since no metal-oxygen interactions occur; i.e., bonding occurs only between rhodium and the carbon of $CO_2$. Carbon dioxide coordination with metallo-systems has also been reviewed recently.[25,26]

A considerable number of metal carbonyls and metal cluster carbonyls[27] are known to activate small gas molecules, such as CO and $H_2$, and act as homogeneous catalysts for carbonylation and hydrogenation processes, for example. Many of the mononuclear complexes serve also as CO or $H_2$ carriers under the same conditions. Several complexes have been investigated for their ability to reversibly bind these gases. For instance, CH$_3$Mn(CO)$_5$ was studied as a CO carrier in various solvents, with equilibrium and the forward and reverse kinetics being determined at several temperatures.[28] Vaska[29] reported the reversible carbonylation of *trans*-IrCl(CO)(PPh$_3$)$_2$ in solution, a system that also reversibly binds oxygen. A series of complexes of structure Pd$_2$(dpm)$_2$X$_2$ (where dpm bridges the single-bonded Pd–Pd moiety and X = halogen, SCN, NCO, N$_3$, or NO$_2$) has been observed to reversibly bind carbon monoxide,[30] sulfur dioxide,[31] and acetylene.[32] Carbon monoxide inserts directly between the Pd-Pd bond forming a complex described as a molecular A-frame.[30] The reversible reaction between carbon monoxide and phthalocyanatoiron(II) in dimethylsulfoxide has been studied.[33] The equilibrium and rate constants have been determined. Iron(II) porphyrin systems have been similarly investigated due to their reversible affinity for carbon monoxide.[34-36]

The reversible binding of hydrogen by *trans*-IrCl(CO)(PPh$_3$)$_2$[37] and Rh(PPh$_3$)$_3$Cl[38] are noteworthy, since the oxidative addition ($+H_2$) and reductive elimination ($-H_2$) processes are common in hydrogenation and homogeneous catalysis, generally. Many other complexes add hydrogen at a single metal center to form a dihydride. Decamethyltitanocene reversibly binds hydrogen and nitrogen, but reacts irreversibly with carbon monoxide.[39] Several hours under vacuum at 25°C are required to desorb hydrogen. Hydrogen adds to platinum(O) trialkylphosphine complexes to give five-coordinate dihydrides at 25°C.[40] However, the hydrides were unstable unless stored below −50°C. Hydrogen has also been added to IrCl(PCy$_3$)$_3$ at 20°C, but a cyclohexyl group of the ligand has been observed to partially dehydrogenate upon desorption.[41] The complexes W(CO)$_3$(PCy$_3$)$_2$ and Mo(CO)$_3$(PCy$_3$)$_2$ both reversibly coordinate hydrogen, nitrogen, and ethylene.[42] The hydrogen complexes are of considerable interest, since H$_2$ binds as a distinct molecule (i.e., in a $\eta^2$-mode) and not as two hydride ligands. A further example of hydrogen binding involves Ru(CO)$_2$(PPh$_3$)$_3$.[43] The hydrogen can be removed with either O$_2$ or CO with concomitant generation of the dioxygen or carbon monoxide complexes. Of further interest, a reversible exchange can be achieved between O$_2$ and H$_2$ on the metal center under slight hydrogen pressures. A recent review covers many other examples of hydrogen binding by mono- and polynuclear metallic complexes.[44]

Thus, it is evident that a considerable number of coordination complexes are capable of reversibly binding CO, CO$_2$, and H$_2$. Since our long-term interest involves the separation of specific components from medium-BTU gas mixtures, the present investigation was undertaken to determine the selectivity and gas-uptake kinetics of a few of the complexes mentioned above.

## II. EXPERIMENTAL

### A. Materials

Carbon monoxide (CP grade), carbon dioxide, hydrogen, ethylene, and acetylene were used as received from Matheson. Bis(diphenylphosphino)methane, 1,2-bis(dimethyl-phosphino)ethane, and trimethylphosphine were obtained from Strem Chemicals or Alfa Products. Solvents such as tetrahydrofuran and dimethylacetamide were dried over LiAlH$_4$ and CaH$_2$, respectively, prior to distillation. The Co(salen)Na · THF,[18] Rh(PBu$_3$)$_3$Cl,[19] Rh(EtPPh$_2$)$_3$Cl,[19] and Pd$_2$(dpm)$_2$X$_2$[30] complexes were prepared according to the literature procedures. The constant pressure apparatus for measuring gas uptakes has been described elsewhere.[45]

### B. Sodium [*N,N'*-Ethylenebis(*o*-Hydroxybutyrophenylideneimato)] Cobaltate(I)

This complex was prepared in a multistep process presented below.[18] Spectroscopic and elemental analyses confirmed the structure of each isolated material.

The ligand Pr-salen was prepared by combining 54 g (0.33 mol) *o*-hydroxybutyrophenone and 10 g (0.17 mol) diaminoethane in ethanol. The reagents were refluxed 8 hr, then cooled. No crystalline product formed. The solvent was removed, and the yellow oil was placed in a freezer. Crystals which formed under these conditions were recrystallized from 8% aqueous methanol to give yellow platelets (mp 119 to 122°C), 54.3 g for 95% yield.

Preparation of Co(Pr-salen) was performed in a helium-filled glovebox. All solvents were degassed using a vacuum freeze-thaw process. To 10 g (29.7 mmol) of Pr-salen dissolved in 130 m$\ell$ of 95% ethanol was added 7.4 g (29.7 mmol) of cobalt(II) acetate in 30 m$\ell$ of water. Then, 2 g (49.3 mmol) of sodium hydroxide in 15 m$\ell$ of water was added. The solution was stirred for 4 hr and refluxed for an additional 45 min. The red-orange solid was filtered and washed with water followed by ethanol. The solid complex was recrystallized from 450 m$\ell$ THF. A microcrystalline solid was recovered: 3.6 g for 30% yield.

Preparation of the bimetallic complex was performed in Schlenk glassware to avoid air exposure. To 0.5 g of Co(Pr-salen) was added 0.1 g sodium metal. Degassed THF (20 m$\ell$) was transferred to the solid Co(Pr-salen) and Na. The mixture was magnetically stirred for

18 hr, then filtered (sintered glass) to remove excess Na metal. In a number of cases, excess Na metal was necessary to induce the reaction (color change red to green). The green solution was transferred to a vacuum vessel to remove the solvent. The remaining dark green solid could be transferred to storage vessels within the He glovebox. Transfer of THF solutions of the bimetallic complex within the glovebox generally involved immediate oxidation (a color change of green to brown) due to the presence of minute amounts of oxygen. The crystalline yield was 0.4 g (50%). The green THF solution, when exposed to $CO_2$, produced a quantitative amount of reddish brown adduct.

## C. Chlorobis(dimethylphosphino)ethane Iridium(I)

This and the following iridium complex were prepared in a manner similar to that of Herskovitz.[46] Degassed solvents were essential to both syntheses. To 2.0 g (2.33 mmol) $Ir_2Cl_2(C_8H_{14})_4$[47] in 20 m$\ell$ of benzene was added 1.3 g (8.9 mmol) of bis(dimethylphosphino)ethane. The mixture (darkening to deep orange) was stirred for 18 hr, then filtered. The orange precipitate (2.3 g) was dried *in vacuo*, yield 99%. Exposure of the powder to air yielded a white to pink solid. Exposure to $CO_2$ gave a white solid within 5 to 10 sec.

## D. Chlorotetrakis(trimethylphosphino) Iridium(I)

To 2.0 g (2.2 mmol) $Ir_2Cl_2(C_8H_{14})_4$ in 50 m$\ell$ of toluene at $-25°C$ was added 1.4 g (18 mmol) trimethylphosphine in 10 m$\ell$ toluene. The yellow solution darkened to orange and was held at $-25°C$ for 3 days. The product was filtered with a medium, sintered glass filter and washed with 50 m$\ell$ of hexane. The yellow solid was dried *in vacuo* for 18 hr, 2.3 g for 77.5% yield.

## E. Dicarbonyltris(triphenylphosphine)ruthenium(O)

This complex was prepared in a multistep process presented below.[48] The structure of each isolated material was confirmed by spectroscopic and elemental analysis; $RuCl_2(PPh_3)_3$, $Ru(CO)_2(PPh_3)_2Cl_2$, and $Ru(CO)_2(PPh_3)_3$ have all been reported previously.

To $RuCl_2$ (1.0 g, 3.18 mmol), dissolved in 500 m$\ell$ hot methanol and cooled, was added triphenylphosphine (5.0 g, 19 mmol). The solution was refluxed 6 hr, cooled, and filtered. The precipitate was washed three times with methanol, then vacuum dried. The brown crystals of $RuCl_2(PPh_3)_3$ were air stable. A quantitative yield was achieved (3.0 g).

The $RuCl_2(PPh_3)_3$ crystals from the previous reaction were dissolved in 50 m$\ell$ dichloromethane and stirred 20 hr under 1 atm of CO. The solvent volume was reduced to 20 m$\ell$ under a slow flow of CO. To this solution was added 30 m$\ell$ of methanol to precipitate the white crystals of $Ru(CO)_2(PPh_3)_2Cl_2$. The crystals were filtered and dried *in vacuo* for 24 hr, 1.9 g for 80% yield.

To $Ru(CO)_2(PPh_3)_2Cl_2$ (2.0 g, 2.65 mmol) dissolved in 400 m$\ell$ of dry, degassed tetrahydrofuran was added triphenylphosphine (1.39 g, 2.65 mmol). After stirring the solution for a few minutes, an excess of sodium amalgam was added. The colorless mixture turned orange. After 2 days the mixture was filtered through Celite®. The solvent was removed from the filtrate under vacuum, and the residue extracted with degassed benzene and filtered through Celite®. After solvent removal, the residue was dissolved in a minimum amount of degassed tetrahydrofuran. The solvent volume was reduced until yellow-orange crystals appeared. Hexane was added to facilitate crystallation. The crystals of $Ru(CO)_2(PPh_3)_3$ were filtered and dried *in vacuo*, 2.0 g for 80% yield.

## III. RESULTS AND DISCUSSION

The coordination complexes examined during this work appeared to be among the most

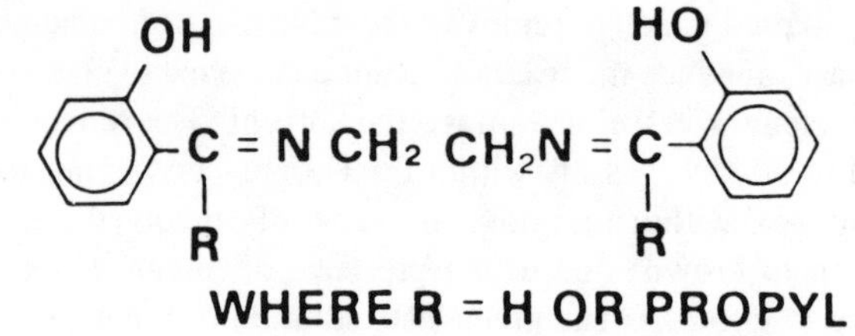

FIGURE 1.    $H_2$(salen) ligands in bimetallic complex.

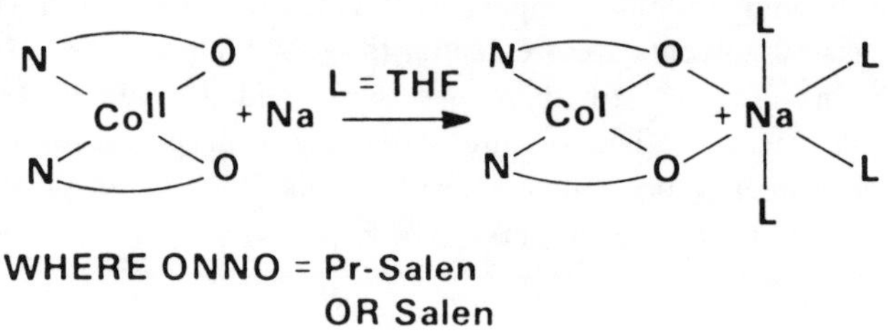

FIGURE 2.    Incorporation of ligands into bimetallic cobalt-sodium complex.

promising for reversibly binding carbon dioxide, carbon monoxide, and hydrogen. Criteria for selection of the complexes were based upon known solution kinetics, gas selectivities, and potential crystalline state kinetics.

## A. Carbon Dioxide Coordination

The bimetallic complexes Co(Pr-salen)Na · THF and Co(salen)Na · THF were prepared by Floriani's group[17,18] using the salen ligands shown in Figure 1. The salen and Pr-salen ligands were reacted with cobalt(II) to form [Co(salen)] and [Co(Pr-salen)]. The red and orange complexes, respectively, were dissolved in tetrahydrofuran (THF) and reacted with sodium metal, which reduced the oxidation state of cobalt while becoming bonded as a cation to the oxygen ends of the salen ligand (see Figure 2). The reduction of the Co(II) complexes was accompanied by a color change (red or orange to green). The isolated green bimetallic complexes in THF rapidly absorbed carbon dioxide dissolved in the solution. In each case, the $CO_2$ uptake was stoichiometric with Co. Samples of 0.025 $M$ of complex in THF absorbed a full complement of $CO_2$ within 5 to 10 sec at 25°C using 1 atm of constant pressure of the gas over the solution. Stopped-flow equipment will be needed to examine this rapid reaction more quantitatively.

Upon exposure to $CO_2$, THF solutions of $Co^I$(Pr-salen)Na · THF formed a red-brown precipitate while $Co^I$(salen)Na · THF formed red crystals. These crystals released $CO_2$ *in vacuo* and also redissolved in THF to form the original green solution. Carbon dioxide can also be removed from the $CO_2$ complexes by suspending them in other solvents and partially vacuum distilling the solvent at 30°C.[18]

These bimetallic complexes are extremely sensitive to oxygen while in THF solution. Our preliminary experiments suggest that irreversible oxidation occurs even in atmospheres containing less than 50 ppm oxygen. For this reason all syntheses and manipulations had to be performed in Schlenk (vacuum) glassware. Due to this extreme oxygen sensitivity further work with these complexes was terminated.

The report that solutions of $Rh(PBu_3)_3Cl$ and $Rh(EtPPh_2)_3Cl$ absorbed $CO_2$[19] was not confirmed by later work.[49] In the Richland laboratory neither complex in solution was found to react with $CO_2$ even after 24 hr exposure. The reversibility of the reported $Rh(PBu_3)_3(CO_2)Cl$ system is also doubtful in that upon standing, the $CO_2$ complex is claimed to rearrange to form a species containing both coordinated phosphine oxide and CO.[19] These complexes were not considered further.

The yellow complexes $[Ir(PMe_3)]Cl$ and $[Ir(dmpe)_2]Cl$ were reported to reversibly bind $CO_2$ in benzene suspension to yield white adducts, and the bound $CO_2$ of $Ir(dmpe)_2(CO_2)Cl$ could be displaced by CO or $H_2$.[23,50] We find that a white complex was formed when either of the precursors was exposed to oxygen.

The $[Ir(dmpe)_2]Cl$ complex was examined in the present work while dissolved in dimethylacetamide saturated with $CO_2$ at 1 atm; a 0.2-g sample was dissolved in 25 m$\ell$ dma with a constant $CO_2$ pressure which allowed for pseudo-first-order conditions. Gas uptake measurements revealed standard first-order absorption curves from which the $k_{obs}$ was found to be $5.9 \times 10^{-3}$ $sec^{-1}$ at 25°C and $4.2 \times 10^{-3}$ $sec^{-1}$ at 0.9°C. A full complement of $CO_2$ was absorbed with a $t_{1/2}$ of ~3 min at 0.9°C. Carbon dioxide absorption by $[Ir(PMe_3)_4]Cl$ in dma was complete in less than 1 min at 0.9°C.

Both $[Ir(depe)_2]Cl$ and $[Ir(PMe_3)_4]Cl$ have been shown to metalate and, in the presence of $CO_2$, carboxylate activated hydrocarbon solvents.[51] Acetonitrile, acrylonitrile, and phenylacetylene all undergo these reactions. The processes and accompanying NMR data were consistent with the following reactions:

$$[Ir(depe)_2]Cl + CH_3CN \overset{K_1}{\rightleftarrows} [Ir(H)(CH_2CN)(depe)_2]Cl \qquad (1)$$

$$[Ir(depe)_2]Cl + CH_3CN + CO_2 \overset{K_2}{\rightleftarrows} [Ir(H)(O_2CCH_2CN)(depe)_2]Cl \qquad (2)$$

where $K_2 >> K_1$. The reaction of acetonitrile with $[Ir(PMe_3)_4]Cl$ at 25°C was rapid ($t_{1/2} <$ 10 min). It should be noted that the $CO_2$ carboxylation reaction with both these two iridium complexes was reversible, and $CO_2$ could be removed by gas evacuation at 80°C. It seemed plausible that dma might also react with $[Ir(dmpe)_2]Cl$ and $[Ir(PMe_3)_4]Cl$ via reactions such as Reactions 1 and 2, and, thus, the $CO_2$ reaction with both complexes was examined in the absence of solvent. The solid phase uptake of $CO_2$ by $[Ir(PMe_3)_4]Cl$ was quite slow; only 41% of the stoichiometric amount of $CO_2$ was absorbed in 48 hr. However, the solid phase 1:1 uptake of $CO_2$ by $[Ir(dmpe)_2]Cl$ occurred in less than 10 sec; on heating to 100°C under vacuum for 30 min, the white $CO_2$ complex returned to the orange color of the original complex. Reexposure of this material to $CO_2$ instantly (5 to 10 sec) turned the complex white again. Thus, $[Ir(dmpe)_2]Cl$ retains the rapid $CO_2$ binding rate even after a single recycle.

The slow $CO_2$ absorption by $[Ir(PMe_3)_4]Cl$ in the absence of solvent suggests that it possesses a highly ordered crystalline structure that allows for $CO_2$ binding by molecules on the crystalline surface. This should disrupt the ordered surface and provide a pseudo-diffusion controlled process. The reported rapid $CO_2$ sorption in benzene or toluene slurries by this complex may be explained by solvation of the crystalline complex which enhances surface structural disorder. The rapid $CO_2$ sorption by $[Ir(dmpe)_2]Cl$ in the solid state suggests that the dmpe ligand confers a definite degree of disorder in the crystalline lattice. It is interesting that the solid-phase $CO_2$ absorption was more rapid than that performed in dma solvent; this might be a function of diffusion control in the solution reaction or, perhaps, the solution species is different than that in solid state.

## B. Carbon Monoxide Coordination

The phosphine-bridged Pd(I) dimers $Pd_2(dpm)_2X_2$ ($\underline{1}$, where X = Cl, Br, or NCO) were

**Table 2**

**VALUES OF $K_{obs}$ OBTAINED WITH FIXED CARBON MONOXIDE CONCENTRATION ($1.74 \times 10^{-3} M$, 150 TORR) AND VARIATION OF COMPLEX CONCENTRATION IN DIMETHYLACETAMIDE AT 24°C**

| $Pd_2(dpm)_2Cl_2$ ($M \times 10^{-3}$) | $k_{obs}$ (sec$^{-1}$) |
|:---:|:---:|
| 3.33 | 5.21 |
| 6.46 | 5.42 |
| 8.69 | 5.58 |
| 11.7 | 5.49 |
| 13.7 | 5.58 |
| 15.4 | 5.22 |
| 17.6 | 5.03 |

examined for the reversible sorption of carbon monoxide in dimethylacetamide by a stopped-flow spectrophotometric method.

$$Pd_2(dpm)_2X_2 + CO \underset{k_{-1}}{\overset{k_1}{\rightleftarrows}} Pd_2(dpm)_2(\mu\text{-CO})X_2 \tag{3}$$

$$\underline{1} \qquad\qquad\qquad \underline{2}$$

The rate law for Reaction 3 can be expressed by Equation 4 in accordance with a first-order dependence upon both the dimeric palladium complex and the carbon monoxide concentrations for the forward reaction, and a simple first-order process for decarboxylation.

$$\frac{-d[\underline{1}]}{dt} = \frac{d[\underline{2}]}{dt} = k_1[\underline{1}][CO] - k_{-1}[\underline{2}] \tag{4}$$

The integrated form of Equation 4 becomes

$$\ln \frac{[A_o - A_e]}{[A_t - A_e]} = (k_1[CO] + k_{-1})t \tag{5}$$

where $A_o$ is the initial absorbance due to $\underline{1}$, $A_e$ is the absorbance at equilibrium, and $A_t$ is the absorbance at any time t.

At constant [CO], the observed rate constant $k_{obs}$ (which equals $k_1[CO] + k_{-1}$) did not change with variations of [$\underline{1}$] (see Table 2), thereby showing a first-order dependence on the dimeric palladium complex. The linear relationship of $k_{obs}$ vs. the carbon monoxide concentration (Figure 3) demonstrated a first-order dependence on carbon monoxide, the slope of the line yielding $k_1$. Under these experimental conditions, $k_{-1}$ is seen to be negligible since the lines intersect the graph at the origin. The 1:1 stoichiometry was confirmed by gas uptake measurements and was further demonstrated by the equilibrium spectrophotometric experiment shown in Figure 4. Decarbonylation is readily accomplished at temperatures around 40°C under vacuum.

The reactivities of $Pd_2(dpm)_2Cl_2$ and $Pd_2(dpm)_2Br_2$ toward CO ($k_1$ parameters) were the same within experimental error, both having an activation enthalpy of ~15.0 kJM$^{-1}$ and

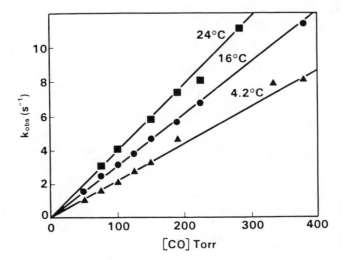

FIGURE 3.    Plot of observed rate against partial pressure of CO at a fixed
Pd$_2$(dpm)$_2$Cl$_2$ concentration (1.23 × 10$^{-4}$ M) in dma at various temperatures.

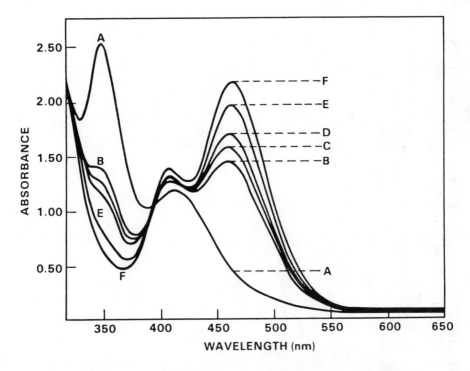

FIGURE 4.   Equilibrium constants measured by varying CO partial pressure in dma at 24°C; A =
100% Pd$_2$(dpm)$_2$Cl$_2$ and F = 100% Pd$_2$(dpm)$_2$(μ-CO)Cl$_2$ at (Pd$_2$) = 1.64 × 10$^{-4}$ M, while B = 2.72
× 10$^5$ M$^{-1}$ (0.92 torr), C = 2.52 × 10$^5$ M$^{-1}$ (1.62 torr), D = 2.24 × 10$^5$ M$^{-1}$ (2.32 torr), and E
= 2.54 × 10$^5$ M$^{-1}$ (5.87 torr).

an entropy of activation of − 120 JK$^{-1}$ (Table 3 and Figure 5). However, a somewhat lower
rate constant, reflected by a greater activation enthalpy, was observed in the case of
Pd$_2$(dpm)$_2$(NCO)$_2$; the stronger electron withdrawing nature of the isocyanato ligand may
reduce the reactivity of the Pd − Pd bond toward carbon monoxide.

Table 3

RATE CONSTANTS, ACTIVATION ENTHALPY, AND ENTROPIES FOR THE INSERTION OF CARBON MONOXIDE INTO $Pd_2(dpm)_2Cl_2$, $Pd_2(dpm)_2Br_2$, and $Pd_2(dpm)_2(NCO)_2$

| $Pd_2(dpm)_2Cl_2$ | | $Pd_2(dpm)_2Br_2$ | | $Pd_2(dpm)_2(NCO)_2$ | |
|---|---|---|---|---|---|
| Temperature (°C) | $k_1(M^{-1}sec^{-1})$ | Temperature (°C) | $k_1(M^{-1}sec^{-1})$ | Temperature (°C) | $k_1(M^{-1}sec^{-1})$ |
| 4.2 | $3.95 \times 10^3$ | 4.0 | $3.62 \times 10^3$ | 10 | $1.80 \times 10^3$ |
| 16.0 | $5.13 \times 10^3$ | 10.0 | $4.11 \times 10^3$ | 16 | $2.46 \times 10^3$ |
| 24.0 | $6.60 \times 10^3$ | 16.0 | $4.96 \times 10^3$ | 24 | $3.15 \times 10^3$ |
| | | 24.0 | $6.00 \times 10^3$ | | |
| $\Delta H^* = 15.4$ kJ    $\Delta S^* = -120$ JK$^{-1}$ | | $\Delta H^* = 14.9$ kJ    $\Delta S^* = -120$ JK$^{-1}$ | | $\Delta H^* = 25.5$ kJ    $\Delta S^* = -92$ JK$^{-1}$ | |

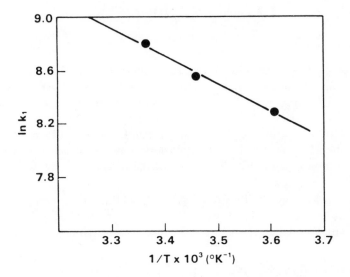

FIGURE 5.    Arrhenius plot of the reaction of Pd$_2$(dpm)$_2$Cl$_2$ with CO.

A series of experiments was performed with Pd$_2$(dpm)$_2$Cl$_2$ dissolved in dma to determine the effect of various gases on the rate of CO sorption. This solution was exposed to a gas mixture of 150 torr CO and 230 torr of each of the gases argon, air, hydrogen, acetylene, carbon dioxide, and ethylene. The k$_{obs}$ was 5.4 ± 0.1 sec$^{-1}$ at 24°C in each experiment. Thus, in the presence of these gases, the specificity of Pd$_2$(dpm)$_2$Cl$_2$ for CO was not altered. A complete account of the kinetics and thermodynamics of the Pd$_2$(dpm)$_2$X$_2$/CO systems will be reported subsequently.[52]

## C. Hydrogen Coordination

Reversible hydrogen and carbon monoxide binding has been previously reported with Ru(CO)$_2$(PPh$_3$)$_3$.[43] The absorption of CO by dma solutions of this complex was initially investigated here. The two steps considered to be required for CO coordination are shown in Reactions 6 and 7.

$$Ru(CO)_2(PPh_3)_3 \; \underset{k_{-1}}{\overset{k_1}{\rightleftarrows}} \; Ru(CO)_2(PPh_3)_2 + PPh_3 \tag{6}$$

$$Ru(CO)_2(PPh_3)_2 + CO \overset{k_2}{\longrightarrow} Ru(CO)_3(PPh_3)_2 \tag{7}$$

The overall rate expression for these reactions is

$$k_{obs} = \frac{k_1 k_2 [Ru(CO)_2(PPh_3)_3][CO]}{k_{-1}[PPh_3] + k_2[CO]} \tag{8}$$

Pseudo-first-order kinetics were achieved by maintaining both the CO and PPh$_3$ concentrations constant during any one experiment. Thus, the following relationships are obtained:

$$k_{obs} = \frac{k_1 k_2 [CO]}{k_{-1}[PPh_3] + k_2[CO]} \tag{9}$$

$$k_{obs}^{-1} = \frac{k_{-1}[PPh_3]}{k_1 k_2 [CO]} + \frac{1}{k_1} \tag{10}$$

These relationships were utilized by measuring the observed rate while varying the $[PPh_3]$. A straight line was obtained by plotting $k_{obs}^{-1}$ vs. $[PPh_3]$, the slope of which is $k_{-1}/k_1 k_2$ with an intercept of $1/k_1$. Using stopped-flow spectrophotometry at 24°C with dma solutions, the $k_1$ value $= 0.12$ sec$^{-1}$ and $k_{-1}/k_2 = 4.3 \times 10^{-2}$. The data indicate that under the conditions employed the $k_2[CO]$ term is considerably larger than $k_{-1}[PPh_3]$ in Equation 9. Thus, the overall rate is largely determined by the $PPh_3$ loss shown in Reaction 6. Using vacuum techniques, the desorption of CO which involved the reverse $k_{-2}$ step of Reaction 7 was extremely slow in dma at 25°C.

Coordination of hydrogen by $Ru(CO)_2(PPh_3)_3$ to give $RuH_2(CO)_2(PPh_3)_2$ occurs at about the same rate as CO binding, and a two-step reaction corresponding to Reactions 6 and 7 is assumed. Consequently, Reactions 6 and 11 represent the hydrogenation process.

$$Ru(CO)_2(PPh_3)_2 + H_2 \xrightarrow{k_2} Ru(CO)_2(PPh_3)_2(H)_2 \tag{11}$$

Under pseudo-first-order conditions, as described by equations similar to Equations 9 and 10, some very preliminary values have been obtained at 24°C: $k_1 \sim 0.10$ sec$^{-1}$ and $k_{-1}/k_2 \sim 1.5 \times 10^{-1}$. These values indicate that reaction of CO with the $Ru(CO)_2(PPh_3)_2$ intermediate is about 3.5 times faster than the reaction with $H_2$. Although the overall reaction rates are very similar (governed by the common $k_1$ step), selectivity can still result from differences in $k_2$ (and also $k_{-2}$). Further data, including equilibrium constants for the reactions, are required to establish the selectivity of $Ru(CO)_2(PPh_3)_3$ in a mixture of CO and $H_2$.

## IV. CONCLUSIONS

Several coordination complexes were examined for their ability to reversibly bind $CO_2$, CO, or $H_2$. Three of these complexes appear worthy of further investigation. Carbon dioxide at 1 atm was rapidly absorbed by crystalline $[Ir(dmpe)_2]Cl$ within 10 sec, while desorption of $CO_2$ was achieved with heat and decreased pressure. The series $Pd_2(dpm)_2X_2$, where X = Cl, Br, or NCO, was found to be of considerable interest for CO coordination. Absorption of CO was rapid, and the kinetics, which appeared to be quite favorable, were not influenced by the presence of $CO_2$, $N_2$, $O_2$, $H_2$, ethylene, or acetylene. Thus, $Pd_2(dpm)_2X_2$ showed considerable specificity for CO. Hydrogen and carbon monoxide both formed adducts with $Ru(CO)_2(PPh_3)_3$, the complex releasing a phosphine ligand prior to gas coordination. The kinetics of both gas uptakes were examined, but definite conclusions concerning selectivity require further experimentation. Further research on the kinetics and the selectivity of these complexes is in progress.

## ACKNOWLEDGMENTS

This investigation was supported by the U.S. Department of Energy, Morgantown Energy Technology Center, Coal Projects Management Division.

# REFERENCES

1. **Santangelo, J. G. and Chen, G. T.,** Use metal hydrides to recover hydrogen, *Chem. Technol.,* 13, 621, 1983.
2. **Fogler, B. B.,** Regenerating unit for generating oxygen, *Ind. Eng. Chem.,* 39, 1353, 1947.
3. **Adduci, A. J.,** The case of aircraft $O_2$ system based on metal chelates, *Chem. Technol.,* 6, 575, 1976.
4. **Calvin, M., Bailes, R. H., and Wilmarth, W. K.,** The oxygen-carrying synthetic chelates compounds. I, *J. Am. Chem. Soc.,* 68, 2254, 1946.
5. **Barkelew, C. H. and Calvin, M.,** Oxygen-carrying synthetic chelate compounds. II, *J. Am. Chem. Soc.,* 68, 2257, 1946.
6. **Getz, D., Melamud, E., Silver, B. L., and Dori, Z.,** Electronic structure of dioxygen in cobalt(II) oxygen carriers, singlet oxygen or $O_2^-$?, *J. Am. Chem. Soc.,* 97, 3846, 1975.
7. **Vogt, L. H., Jr.,** Synthetic reversible oxygen-carrying chelates, *Chem. Rev.,* 63, 269, 1963.
8. **Stewart, R. F., Eslep, P. A., and Sebastian, J. J. S.,** Investigation of oxygen production by metal chelates, *U.S. Bur. Mines Inf. Circ.,* No. 7906, 1959.
9. **Henrici-Olivé, G. and Olivé, S.,** Activation of molecular oxygen, *Angew. Chem. Int. Ed. Engl.,* 13, 29, 1974.
10. **Erskine, R. W. and Field, B. O.,** Reversible oxygenation, *Struct. Bonding,* 28, 1, 1976.
11. **McLendon, G. and Martell, A. E.,** Inorganic oxygen carriers as models for biological systems, *Coord. Chem. Rev.,* 19, 1, 1976.
12. **Jones, R. D., Summerville, D. A., and Basolo, F.,** Synthetic oxygen carriers related to biological systems, *Chem. Rev.,* 79, 139, 1979.
13. **Crabtree, R. H., Mellea, M. F., Mihelcic, J. M., and Quirk, J. M.,** Alkane dehydrogenation by iridium complexes, *J. Am. Chem. Soc.,* 104, 107, 1982.
14. **Shilov, A. E.,** *Activation of Saturated Hydrocarbons by Transition Metal Complexes,* Reidel, Dordrecht, 1984.
15. Methane adds oxidatively to iridium complexes, *Chem. Eng. News,* 61, 33, 1983.
16. **Fachinetti, G., Floriani, C., and Zanazzi, P. F.,** Bifunctional activation of carbon dioxide, *J. Am. Chem. Soc.,* 100, 7405, 1978.
17. **Fachinetti, G., Floriani, C., Zanazzi, P. F., and Zanzari, A. R.,** Bifunctional model complexes active in carbon dioxide fixation, *Inorg. Chem.,* 18, 3469, 1979.
18. **Gambarotta, S., Francesco, A., Floriani, C., and Zanazzi, P. F.,** Carbon dioxide fixation: bifunctional complexes containing acidic and basic sites working as reversible carriers, *J. Am. Chem. Soc.,* 104, 5082, 1982.
19. **Aresta, M. and Nobile, C. F.,** Carbon dioxide-transition metal complexes. III, *Inorg. Chim. Acta,* 24, L49, 1977.
20. **Aresta, M., Nobile, C. F., Albano, V. G., Forni, E., and Manassero, M.,** New nickel-carbon dioxide complex: synthesis, properties, and crystallographic characterization of (carbon dioxide)-bis(tricyclohexylphosphine)nickel, *J. Chem. Soc. Chem. Commun.,* 636, 1975.
21. **Aresta, M. and Nobile, C. F.,** (Carbon dioxide)-bis(trialkylphosphine)-nickel complexes, *J. Chem. Soc. Dalton Trans.,* p. 708, 1977.
22. **Flynn, B. R. and Vaska, L.,** Reversible addition of carbon dioxide to rhodium and iridium complexes, *J. Chem. Soc. Chem. Commun.,* 703, 1974.
23. **Herskovitz, T. and Parshall, G. W.,** Carbon Dioxide Complexes of Rh, Ir, Ni, Pd, and Pt, U.S. Patent 3,954,821, 1976.
24. **Calabrese, J. C., Herskovitz, T., and Kinney, J. B.,** Carbon dioxide coordination chemistry. V. The preparation and structure of $Rh(CO_2)(Cl)(diars)_2$, *J. Am. Chem. Soc.,* 105, 5914, 1983.
25. **Eisenberg, R. and Hendriksen, D. E.,** Binding and activation of CO, $CO_2$ and NO, in *Advances in Catalysis,* Vol. 28, Eley, D. D., Pines, H., and Weisz, P. B., Eds., Academic Press, New York, 1979, 121.
26. **Sneeden, R. P. A.,** Reactions of $CO_2$ with organotransition metal systems, in *Comprehensive Organometallic Chemistry,* Vol. 8, Wilkinson, G., Stone, F. G. A., and Abel, E. W., Eds., Pergamon Press, Oxford, 1982, 229.
27. **Pittman, C. U. and Ryan, R. C.,** Metal cluster catalysts, *Chem. Technol.,* 8, 170, 1978.
28. **Calderazzo, F. and Cotton, F. A.,** Carbon monoxide insertion reactions. I, *Inorg. Chem.,* 1, 30, 1962.
29. **Vaska, L.,** Reversible combination of carbon monoxide with a synthetic oxygen carrier complex, *Science,* 152, 769, 1966.
30. **Benner, L. S. and Balch, A. L.,** Novel reactions of metal-metal bonds. Insertion of isocyanides and carbon monoxide into the palladium-palladium bond of some Pd(I) dimers, *J. Am. Chem. Soc.,* 100, 6099, 1978.
31. **Balch, A. L., Benner, L. S., and Olmstead, M. M.,** Novel reactions of metal-metal bonds. Addition of sulfur dioxide and sulfur to $Pd_2(dpm)_2Cl_2$ and the oxidation of coordinated sulfur, *Inorg. Chem.,* 18, 2996, 1979.

32. **Balch, A. L., Lee, C.-L., Lindsay, C. H., and Olmstead, M. M.,** Insertion of acetylenes in the Pd–Pd bond of Pd₂(dpm)₂Cl₂, *J. Organomet. Chem.,* 177, C22, 1979.

33. **Ercolani, C., Monacelli, F., Pennesi, G., Rossi, G., Antonini, E., Ascenzi, P., and Brunori, M.,** Equilibrium and kinetic study of the reaction between phthalocyaninatoiron(II) and carbon monoxide in dimethyl sulfoxide, *J. Chem. Soc. Dalton Trans.,* 1120, 1981.

34. **James, B. R., Reimer, K. J., and Wong, T. C. T.,** Reaction of carbon monoxide with ferrous porphyrins, *J. Am. Chem. Soc.,* 99, 4815, 1977.

35. **Hashimoto, T., Dyer, R. L., Crossley, M. J., Baldwin, J. E., and Basolo, F.,** Ligand, oxygen and carbon monoxide affinities of iron(II) modified capped porphyrins, *J. Am. Chem. Soc.,* 104, 2101, 1982.

36. **Collman, J. P., Brauman, J. I., Iverson, B. L., Sessler, J. L., Morris, J. M., and Gibson, Q. H.,** O₂ and CO bonding to iron(II) porphyrins, *J. Am. Chem. Soc.,* 105, 3052, 1983.

37. **Vaska, L. and DiLuzio, J. W.,** Activation of hydrogen by a transition metal complex at normal conditions leading to a stable molecular dihydride, *J. Am. Chem. Soc.,* 84, 679, 1962.

38. **Halpern, J. and Wong, C. S.,** Hydrogenation of tris(triphenylphosphine)-chlororhodium(I), *J. Chem. Soc. Chem. Commun.,* 629, 1973.

39. **Bercaw, E., Marvich, R. H., Bell, L. G., and Brintzinger, H. H.,** Titanocene as an intermediate in reactions involving molecular hydrogen and nitrogen, *J. Am. Chem. Soc.,* 94, 1219, 1972.

40. **Gerlach, D. H., Kane, A. R., Parshall, G. W., Jesson, J. P., and Muetterties, E. L.,** Reactivity of trialkylphosphine complexes of platinum(O), *J. Am. Chem. Soc.,* 93, 3543, 1971.

41. **Hietkamp, S., Stufkens, D. J., and Vrieze, K.,** Activation of C–H bonds by transition metals. IV, *J. Organomet. Chem.,* 152, 347, 1978.

42. **Kubas, G. J.,** Five co-ordinate molybdenum and tungsten complexes, M(CO)₃(PCy₃)₂, which reversibly add dinitrogen, dihydrogen, and other small molecules, *J. Chem. Soc. Chem. Commun.,* 61, 1980.

43. **Porta, F., Cenini, S., Giordano, S. and Pizzotti, M.,** Low oxidation state ruthenium chemistry. IV, *J. Organomet. Chem.,* 150, 261, 1978.

44. **James, B. R.,** Addition of hydrogen and hydrogen cyanide to carbon-carbon double and triple bonds, in *Comprehensive Organometallic Chemistry,* Vol. 8, Walkinson, G., Stone, F. G. A., and Abel, E. W., Eds., Pergamon Press, Oxford, 1982, 285.

45. **Dekleva, T. W.,** The Solution Behavior and Reactivity of Some Trialkylphorphine Complexes of Ruthenium, Ph.D. dissertation, University of British Columbia, Vancouver, 1983.

46. **Herskovitz, T.,** Carbon dioxide complexes of rhodium and iridium, *Inorg. Synth.,* 21, 99, 1982.

47. **Herdé, J. L. and Senoff, C. V.,** μ-Dichlorotetrakis (cyclooctene) diiridium(I), *Inorg. Nucl. Chem. Lett.,* 7, 1029, 1971.

48. **James, B. R. and Mahajan, D.,** Catalytic asymmetric hydrogenation using hydridorhodium(I) diop complexes, *Isr. J. Chem.,* 15, 214, 1977.

49. **James, B. R., Preece, M., and Robinson, S. D.,** Tricyclohexylphosphine complexes of ruthenium, rhodium, and iridium and their reactivity toward gas molecules, in *Catalytic Aspects of Metal Phosphine Complexes, Advances in Chemistry Series,* No. 196, Alyea, E. C. and Meek, D. W., Eds., American Chemical Society, Washington, D.C., 1982, 145.

50. **Herskovitz, T.,** Carbon dioxide coordination chemistry. III, *J. Am. Chem. Soc.,* 99, 2391, 1977.

51. **English, A. D. and Herskovitz, T.,** Metalation and carboxylation of activated carbon-hydrogen bonds by complexes of iridium and rhodium, *J. Am. Chem. Soc.,* 99, 1648, 1977.

52. **Lee, C.-L., James, B. R., Nelson, D. A., and Hallen, R. T.,** Kinetics and thermodynamics of the reversible reaction between carbon monoxide and palladium(I) dimers containing bis(diphenylphosphino)methane, *Organometallics,* 3, 1360, 1984.

Chapter 2

## ACID GAS REMOVAL: THE NEW CNG PROCESS

**Lester G. Massey and William R. Brown**

### TABLE OF CONTENTS

# I. OVERVIEW

## A. General Considerations

Raw gases from coal gasifiers, in general, are hot and dusty, present a surprisingly broad mixture of components, and may be at pressures ranging from nearly atmospheric to 1000 psi or more. The capital cost for cleaning and purifying this gas frequently exceeds the corresponding cost of any section of a coal gasification plant, amounting to 15 to 20 + % of total plant cost. Preliminary conditioning usually amounts to cooling (quenching) and removal of condensables and particulates, so that the residual mixture of gases can be treated for removal of $CO_2$ and $H_2S$ (the acid gases) by one of a number of possible established processes. All of the known processes are aimed at the selective removal of $H_2S$ and $CO_2$ in the presence of a mixture of other "permanent" gases with minimal loss of the permanent constituents.

Most gas purification plants consist (see Figure 1) of an absorption tower (absorber) and regenerator (stripper). In the absorber, the solvent flows countercurrently to the ascending gas, absorbing the removable components physically or chemically. Suitable internals in the form of packing or trays promote mass and heat transfer. The rich (or "fat") solvent is withdrawn from the bottom of the absorber and sent to the regenerator where absorbed components are removed by thermal or chemical treatment. The lean solvent is returned to the top of the absorber for further duty.

Absorption generally occurs at low temperature and under elevated pressure, while regeneration is accomplished at high temperature and low pressure. The temperature gradient is utilized by heat exchange between the cool fat solution and the hot regenerated solution, and the pressure gradient energy is recovered with expansion turbines. All absorption-regeneration processes operate continuously.

The majority of gas purification known processes were developed to meet the needs for natural gas sweetening, for petroleum refining, and for manufacture of hydrogen and ammonia. These gas streams typically are relatively "pure" when compared with the quenched raw gas from a coal gasifier, as they lack some of the numerous so-called "trace" components found in raw coal gas. Adaptation of these processes to purify coal-derived gas leads to economic penalties to overcome the disparity of feed gas composition. Such penalties are encountered in increased capital cost and increased operating expense. The gas compositions shown in Table 1 illustrate the large differences in coal gas composition between two different coal gasification processes and the large difference between both of these coal gases and a typical refinery sour gas stream. Clearly, purification of raw coal gas streams requires a most flexible and adaptable process to meet the technical and economic requirements of the various coal gasification processes.

Coal gas differs markedly from the refinery gas of Table 1. For example, (1) the level of $CO_2$ in the refinery gas is roughly 70% of that for Koppers-Totzek (K-T) and about 16% of that for Lurgi; (2) the $H_2S$ level in the refinery gas is from 40 to 57 times that for the coal gases; (3) the acid gas ratio, $CO_2/H_2S$, in the coal gas ranges from about 55 to about 340 times that for the refinery gas; and (4) the coal gas streams contain a large variety of contaminants not present in the refinery gas, some of them highly toxic. Additional "trace" substances can and do appear in some coal gasifier raw gases, examples of which are benzene, toluene, xylene, phenols, thiocyanates, and mercury. These trace contaminants in coal gas have the potential to generate serious gas purification problems through recycle build-up in the liquid absorbent used for gas absorption. Toxic contaminants such as hydrogen cyanide, carbonyl sulfide, and mercury are not tolerable in the product gas.

The importance of the ratio $CO_2/H_2S$ is related to the recovery of sulfur from the acid gas after its removal from the main stream, usually by means of the Claus process which requires a feed containing at least 20% $H_2S$. Thus, the $CO_2/H_2S$ ratio should be less than

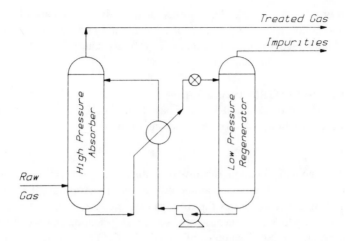

FIGURE 1.    Simple gas purification.

**Table 1**
**TYPICAL COMPOSITIONS, DRY GAS TO ACID GAS**
**REMOVAL (MOL%)[1]**

| | Quenched raw gas | | Refinery |
| Component | Lurgi | K-T | sour gas |
|---|---|---|---|
| Hydrogen | 37.6 | 32.74 | |
| Oxygen | 0.6 | | |
| Carbon monoxide | 16.7 | 57.35 | |
| Methane | 9.1 | | 8.4 |
| Ethane | 0.7 | | 5.2 |
| | | | |
| Propane | | | 4.6 |
| Isobutane | | | 2.5 |
| Normal butane | | | 7.5 |
| Normal pentane | | | 3.4 |
| Hexane | | | 1.0 |
| | | | |
| Naphthalene | 0.7 | | |
| Nitrogen + argon | 0.6 | 1.16 | |
| Ammonia | 0.2 | | |
| Hydrogen cyanide | 2.7 | | |
| | | | |
| Carbonyl sulfide | | 0.114 | |
| Hydrogen sulfide | 1.1 | 1.59 | 62.5 |
| Carbon dioxide | 30.0 | 7.05 | 4.9 |
| | 100.0 | 100.00 | 100.00 |
| | | | |
| $CO_2/H_2S$ | 27.3 | 4.43 | 0.08 |
| Temp, °F | 850 | 2700 | 122 |
| Press, psia | 300 | 15 | 22 |

4:1 for acceptable sulfur recovery economics. If all the acid gas of the K-T process of Table 1 were absorbed simultaneously and produced as feed to a Claus process, the feed would contain only about 18% $H_2S$; for the Lurgi process it would be only 3.5% $H_2S$. A desirable gas purification process would split the absorbed acid gas into a stream of high purity $CO_2$ for venting to the atmosphere and a stream of concentrated $H_2S$ (at least 40%) in $CO_2$ for feed to a Claus sulfur recovery unit. In the case of refinery sour gas (Table 1), the acid gas feed to the Claus unit would contain nearly 93% $H_2S$.

## B. Acid Gas Removal (AGR) Process Types

Processes for removal of $CO_2$ and $H_2S$ are based on (1) chemical reaction, (2) physical absorption, (3) condensation, and (4) combinations of these.

### 1. Chemical Reaction

In general, the chemical reaction cases involve substantial exothermic heats of absorption at elevated pressure and substantial endothermic heats of desorption at lower pressures. Absorption via chemical reaction has the disadvantage of low thermodynamic process reversibility. Other difficulties arise because of absorbent degradation and because of corrosion of vessels, pumps, and piping. Examples of chemical AGR processes using thermal regeneration are the proprietary Hot Carbonate process, the Alkazid Wash, Adip Wash, Sulfinol Wash, and the generally nonselective, low pressure amine-based processes commonly found in the petroleum and natural gas industries. Some success has been achieved with selective absorption of $H_2S$ and $CO_2$ employing membrane technology with potassium carbonate solution as the absorption medium.

### 2. Physical Absorption

Physical absorption processes depend solely on pressure and temperature of the gas-liquid equilibrium, which, to a limited degree, is dependent upon the choice of liquid absorbent. Absorptive capacity is generally increased by absorption at lower temperature and at higher pressure. Energies of absorption and desorption are normally much smaller than in chemical absorption, giving rise to greatly reduced thermal effects per mole of gas absorbed. For example, absorption of a pound-mole of $CO_2$ in the Hot Carbonate process releases about 13,500 Btu, but physical absorption releases about 3500 Btu, or one quarter of the thermal effect of chemical absorption. Nevertheless, the enthalpy decrease of carbon dioxide which accompanies physical absorption is not small. It causes a significant rise in temperature of the liquid absorbent, with a consequent lowering of equilibrium concentration caused by the rise of temperature. This effect frequently is so great that cooling of the absorbent liquid is required, normally by withdrawing the hot liquid to an external heat exchanger and returning the cooled liquid to the absorption tower on the next lower tray or packed section. An alternative of limited value is to increase the flow rate of liquid absorbent to provide a larger heat sink and, thus, a smaller temperature rise. Economics of increased tower size and internals vs. external interstage cooling determines the choice between these alternatives.

Theoretically, any liquid can be used for physical absorption of gaseous components. Practical commercial considerations limit severely the choices available on the basis of (1) availability, (2) cost, (3) solvent loss via vapor pressure, (4) solvent degradation with time, (5) corrosion of equipment, (6) specificity for absorption of a particular component, (7) molecular weight, (8) density, (9) viscosity, and, to a lesser degree, (10) specific heat. Items 7 to 9 are principal determinants of absorption efficiency, usually referrred to as tray efficiency or, for packed towers, HETP, height equivalent to a theoretical plate. Absorption efficiency is enhanced by an absorbent liquid of low molecular weight, high density, and low viscosity. Since gases are physically absorbed on a molar basis, a liquid absorbent of low molecular weight promotes high gas solubility per unit mass of absorbent. Thus, liquid circulation rates are reduced with consequent reduction of tower size and pumping requirements. These lead to reduced capital and operating costs.

Water has been employed for absorption of $CO_2$ and $H_2S$ from pressured coal gasification gases, natural gas, and refinery gas, but with limited success. Organic sulfur compounds, gum-formers, and hydrogen cyanide are not removed completely. Relatively low gas solubility requires large water circulation rates, thereby entailing large power consumption costs. Much more successful physical absorption processes are the well-known proprietary processes shown in Table 2. Rectisol is used commercially to purify coal-derived gases. Selexol® has been specified for two coal gasification pilot plants but has not yet operated in a commercial coal gasification process. Both processes have seen duty in petroleum refineries. Purisol and Fluor processes have not been applied to pilot plant or commercial coal gas purification.

Physical absorption processes have potential for selective absorption of one component over the one or more others to be removed. However, ability to remove a component completely (say, to less than 1 ppm) depends heavily upon complete absence of the component in the regenerated solvent. Thorough regeneration of solvent is essential, since unremoved components exert their equilibrium effects in the absorber and tend to inhibit complete absorption. The myriad "trace" components found in raw coal gas virtually assures the presence of components difficult to remove in regeneration. Such components accumulate in the recycling solvent until reaching a concentration such that removal by regeneration is exactly in balance with accumulation by absorption. If the rate of removal by regeneration is less than the rate of feed to the absorber, absorption will be incomplete and the treated gas will be impure. Unsatisfactory gas purification can result from difficulty in absorption (poor solubility) or from difficulty in adequate regeneration with respect to the offending component. If the offending component is toxic, corrosive, or tends to promote formation of gums, deposits, and the like, achievement of complete removal is essential.

Physical absorption processes generally operate with substantial pressure differences between absorber and regenerator, accompanied by a thermal difference to aid in the stripping function of the regenerator. While conservation of energy is practiced via heat exchange and power recovery, it is done at the expense of capital equipment and the inevitable loss of thermodynamic efficiency. Physical absorption processes would approach perfection with: (1) minimum pressure difference between absorber and regenerator; (2) minimum temperature difference between absorber and regenerator; (3) complete regeneration to produce solvent containing none of the feed gas components; and (4) a solvent with good specificity for desired components, low molecular weight, low viscosity, and very low vapor pressure.

### 3. Condensation

When a component to be removed is present at adequate partial pressure, cooling of the gas to its dew point temperature causes condensation to begin. Figure 2 shows the dew point of a pressurized synthesis gas containing 30% $CO_2$. Continued gradual cooling causes condensation to occur in a close approach to thermodynamic reversibility, a process of maximum thermodynamic efficiency. Cocondensation of other components, according to their individual solubilities, occurs simultaneously. Further cooling yields further condensation, but extended cooling provides diminishing condensate per unit temperature drop and is limited by the condensate freezing temperature. The temperature at which cooling is halted is an economic choice, related to specification for residual $CO_2$ in the treated gas and technology available for further treatment, if necessary, to achieve treated gas specification. Condensation can remove bulk $CO_2$ (at adequate partial pressure) from a syngas at adequate total pressure and, in so doing, contributes significantly to good process efficiency and improved process economics.

Bulk condensation of $CO_2$ leads to potential departure from standard absorption-regeneration technology in that separation of the condensate components may be performed by distillation, with or without preliminary flash separations. Distillation to produce pure $CO_2$, however, is made difficult by the approach to equal volatility for $CO_2$ and $H_2S$ at high $CO_2$

Table 2

FEATURES OF PHYSICAL ABSORPTION ACID GAS
REMOVAL PROCESSES

| Process | Solvent | T/P | Selectivity | | Solvent loss | Util. |
| | | | $H_2S/CO_2$ | $CO_2/HC$ | | |
| --- | --- | --- | --- | --- | --- | --- |
| Rectisol | Methanol | Low/high | Good | Poor | Hi | Mod/low |
| Selexol® | DMEPG[a] | Mod/high | Good | Mod | Low | Low |
| Purisol | NMP[b] | Mod/high | Good | Mod | Low | Low |
| Fluor | Propylene carbonate | Mod/high | Mod | Mod | Low | Low |

[a]  Dimethyl ether of polyethylene glycol.
[b]  N-Methyl-2-pyrrolidone.

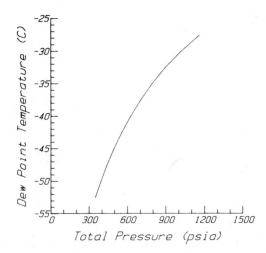

FIGURE 2.   Dew point vs. pressure for a synthesis gas with 30% carbon dioxide.

concentration and by the formation of a $CO_2$-ethane azeotrope (67% $CO_2$, 33% $C_2H_6$). Separation of methane from $CO_2$ by distillation, which appears easy, encounters severe solid $CO_2$ deposition at pressures below 715 psia. All schemes to avoid this problem have led to unacceptably high capital and energy cost when compared with available alternatives. An exception is found in the Ryan-Holmes process, in which addition of butane or natural gas liquids at the condenser prevents solid $CO_2$ formation in the demethanizer, prevents ethane-$CO_2$ azeotrope formation in the deethanizer, and enhances the separation of $H_2S$ from $CO_2$ by distillation.[2] Gazzi et al.[3] employ similar technology for upgrading natural gases.

In the CNG Acid Gas Removal Process, distillation has been abandoned in favor of triple-point crystallization to produce virtually 100% pure $CO_2$ from a condensate containing $H_2S$ and all other trace components thus far identified. Triple-point crystallization of carbon dioxide is an efficient, effective means to: (1) produce pure liquid $CO_2$ for complete absorption of all trace components; (2) produce pure $CO_2$ as a product or for venting to the atmosphere; (3) completely remove all contaminants from the treated gas; (4) eliminate all trace contaminants from the recycle $CO_2$ used for absorption; (5) provide for rejection of contaminants with the small acid gas stream sent to sulfur recovery; and (6) produce an acid gas stream rich in $H_2S$ for substantial improvement of sulfur recovery economics. High thermodynamic efficiency results from the nearly reversible process of triple-point crystallization, as only slight pressure changes cause melting or freezing of carbon dioxide without heat exchange surfaces.

## 4. Combinations

Process types can be combined in ways that are advantageous for specific applications. When the concentration of acid gas is low, chemical or physical absorption alone is likely to be adequate. As the carbon dioxide concentration and total pressure of the raw gas are increased, the potential for simple condensation becomes attractive, as is evident in Figure 2. Cooling for condensation of $CO_2$ is limited to about $-56.6°C$ at partial pressures of at least 5.112 atm (75 psia) as indicated by the triple point in the $CO_2$ phase diagram of Figure 3. Carbon dioxide remaining in the vapor phase at the above condition depends upon total pressure. Thus, at 1000 psia the gas contains $(75/1000)(100) = 7.5\%$ $CO_2$; at 300 psia the concentration is 25% $CO_2$. These residual levels can be further reduced to 1.0 or even 0.1% $CO_2$ by absorption-regeneration processes such as Selexol®, Rectisol, Hot Carbonate, etc.

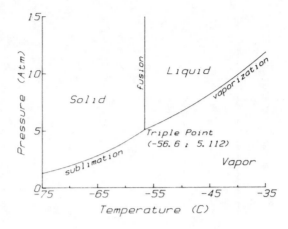

FIGURE 3.    Carbon dioxide phase diagram.

For gases of high $CO_2$ partial pressure, condensation offers improved economics and avoids forcing the absorption-regeneration system to function beyond its intended range of economic usefulness. Carbon dioxide condensation should be considered for $CO_2$ partial pressures exceeding 75 psia, and the process configuration decided on the basis of economics. Examples of combined process types are found in Ryan-Holmes[2] and in the Cryofac[3] processes for natural gas treating.

## C. Summary

The foregoing discussion of acid gas removal technology has identified the principal characteristics that are desired, if not actually needed, in an AGR process applied to coal gasification. The process should:

1.    Produce high purity $CO_2$, preferably with sulfur content well below 1 ppm, as an item of commerce or for environmentally acceptable venting to the atmosphere
2.    Produce an acid gas stream for sulfur recovery that is enriched in hydrogen sulfide (40 to 75 + %) for maximum favorable combined acid gas removal and sulfur recovery economics
3.    Remove completely all "trace" contaminants from the treated gas
4.    Reject completely all removed "trace" contaminants into the small acid gas stream for sulfur recovery to minimize or eliminate recycle absorbent build-up; if special treatment is required for any specific one or more of these contaminants, treatment cost is minimized because the acid gas stream is small and contaminant concentrations are maximized
5.    Regenerate the absorbent to a purity approaching 100% to permit essentially complete removal of all contaminants from the main gas stream
6.    Employ an absorbent of near-optimal properties at process conditions for maximum absorption efficiency and desirable economics: low molecular weight, low vapor pressure, high liquid density, low liquid viscosity, high specific heat, good chemical stability, noncorrosive to equipment, readily available at reasonable cost
7.    Achieve goals 1 through 6 within the constraints of acceptable capital and operating costs

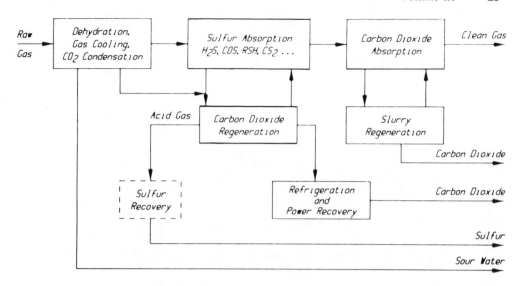

FIGURE 4. CNG process overview.

## II. CNG ACID GAS REMOVAL PROCESS

### A. Major Applications

The major applications of acid gas removal appear to be (1) treating natural gas containing sulfurous compounds and carbon dioxide, (2) treating shift gases in substitute natural gas production from coal, (3) treating synthesis gas made by partial oxidation of sulfur-contaminated feedstocks, e.g., gasoline production from coal via synthesis gas (Fischer-Tropsch reaction or through the intermediates methanol or dimethyl ether), (4) acid gas removal in hydrogen production from coal or residual feedstocks for uses such as shale oil refining, and (5) treating synthesis gas made from methane and steam (production of ammonia, hydrogen, and oxo compounds). In short, acid gas removal is a key step in the upgrading of natural gas and synthesis gas made from natural gas, petroleum, or coal, and acid gas removal is an essential step in most envisioned fossil fuel energy conversion processes.

### B. Process Description and Features

The CNG Acid Gas Removal Process is distinguished from existing acid gas removal processes by three features. The first is the use of pure liquid carbon dioxide as absorbent for sulfurous compounds; the second is the use of triple-point crystallization to separate pure carbon dioxide from sulfurous compounds; and the third feature is the use of a liquid-solid slurry to absorb carbon dioxide below the triple-point temperature of carbon dioxide. Pure liquid carbon dioxide is an effective scavenger for sulfurous compounds and trace contaminants; triple-point crystallization economically produces pure carbon dioxide and concentrated hydrogen sulfide; for bulk carbon dioxide absorption the slurry absorbent diminishes absorbent flow and controls the carbon dioxide absorber temperature rise without external refrigeration. The sequence of gas treatment is shown in Figure 4, an overview of the CNG acid gas removal process.

Carbon dioxide plays a central role in the CNG process both as a pure component and in mixture with other compounds. The triple point of carbon dioxide is referred to frequently in the following discussion; it is the unique temperature and pressure at which solid, liquid, and vapor phases of carbon dioxide can exist at equilibrium ($-56°C$, 5.1 atm). The carbon dioxide triple point is shown in Figure 3, a phase diagram for carbon dioxide.

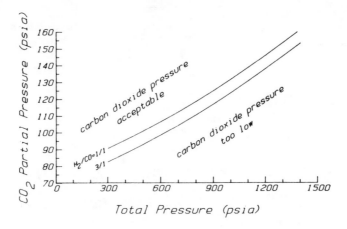

FIGURE 5.    Carbon dioxide partial pressures required for sulfurous compound absorption with liquid carbon dioxide.

### 1. Precooling, Water Removal

The raw gas is cooled and residual water vapor is removed to prevent subsequent icing. The method of water removal depends on the crude gas composition. If the crude gas is relatively free of $C_2 +$ hydrocarbons, a regenerable molecular sieve can be used; otherwise, water can be removed by solvent washing, e.g., dimethyl formamide, with dry solvent regenerated by distillation. The economics of the two methods appear competitive. The water-free crude gas is further cooled to its carbon dioxide dew point by countercurrent heat exchange with return clean gas and separated carbon dioxide.

The dew point of the gas depends on the gas pressure and the carbon dioxide content. At fixed carbon dioxide composition, the dew point temperature is lowered as total pressure decreases; at fixed total pressure, the dew point temperature is lowered as carbon dioxide content decreases. Calculated dew points for a synthesis gas with 30 mol % carbon dioxide are shown in Figure 2 for synthesis gas pressures up to 1500 psia.

The dew point temperature must be warmer than $-56.6°C$ to permit use of liquid carbon dioxide absorbent, because pure liquid carbon dioxide cannot exist below the triple point. The carbon dioxide partial pressure of two synthesis gas mixtures with $-56.6°C$ dew points is plotted vs. synthesis gas pressure in Figure 5. Increasing the $H_2/CO$ ratio of the synthesis gas decreases the carbon dioxide partial pressure required for a $-56.6°C$ dew point. Liquid carbon dioxide can be used to absorb sulfur molecules for any combination of synthesis gas pressure and carbon dioxide partial pressure which lies above the curve of Figure 5.

### 2. Carbon Dioxide Condensation, Sulfurous Compounds Absorption

Carbon dioxide is condensed by cooling the gas from its dew point to about $-55°C$; the condensate is contaminated with sulfurous compounds and other "trace" constituents. Substantial amounts of carbon dioxide can be condensed if the dew point is relatively warm. A synthesis gas at 1000 psia with 30 mol % carbon dioxide has a dew point of about $-30°C$, and approximately 65% of the carbon dioxide condenses to a liquid in cooling to $-55°C$.

The gas at $-55°C$ is scrubbed clean of sulfurous compounds and remaining trace contaminants using pure liquid carbon dioxide. Liquid carbon dioxide has several desirable properties as an absorbing medium for sulfurous compounds. Surprisingly, liquid carbon dioxide absorbs carbonyl sulfide (COS) more effectively than it absorbs hydrogen sulfide as shown by their relative vaporization coefficients, i.e., $K_{COS} < K_{H_2S}$ ($K_i = Y_i/x_i$, $y_i$ = vapor mole fraction component i, $x_i$ = liquid mole fraction component i). Thus, an absorber for sulfurous compounds designed to remove hydrogen sulfide using liquid carbon dioxide

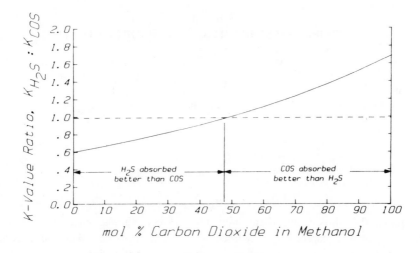

FIGURE 6. Equilibrium vaporization constant ratios.

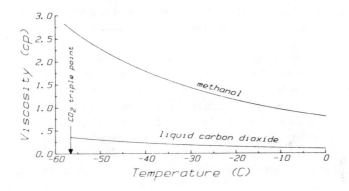

FIGURE 7. Viscosity of methanol and liquid carbon dioxide.

will also remove COS and all other less volatile sulfurous compounds in the gas. The opposite is true concerning COS for both cold methanol (Rectisol) and the dimethyl ether of polyethylene glycol (Selexol® solvent). For comparison, the ratio $K_{H_2S}/K_{COS}$ is plotted in Figure 6 vs. $H_2S$-free liquid phase composition from 100 mol % methanol to 100 mol % carbon dioxide. The ratio $K_{H_2S}/K_{COS}$ for Selexol® solvent is even lower than for methanol.

The physical properties of liquid carbon dioxide also enhance its use as an absorbent in general. The low viscosity (about 0.35 cP at −55°C) and high density of liquid carbon dioxide promote high stage efficiency. For comparison, the viscosity of methanol is 7 to 8 times greater over the temperature range 0 to −55°C (Figure 7), and Selexol® solvent viscosity at process (room) temperature is 15 to 30 times that of liquid carbon dioxide (the viscosity of water at 25°C is about 1 cP). Liquid carbon dioxide is denser (sp gr 1.17 at −55°C) than most absorbents, which, together with its low viscosity, enables high liquid and gas flow rates in absorption and stripping towers. For comparison, methanol has a specific gravity of about 0.85 (−40°C), and Selexol® solvent about 1.03 at room temperature. Carbon dioxide has a relatively low molecular weight of 44 compared to Selexol® solvent molecular weight of over 200.[4] Low molecular weight favors high gas solubility per unit volume of solvent. A comparison of physical properties for methanol, Selexol® solvent, and liquid carbon dioxide is summarized in Table 3. Finally, fresh solvent supply is not required because carbon dioxide is obtained from the crude gas being treated.

**Table 3**
**PROPERTIES OF PHYSICAL ABSORBENTS**

| Solvent | MW | Sp gr$^{T°C}$ | $\mu^{T°C}$ | bp (°C) | fp (°C) |
|---|---|---|---|---|---|
| Methanol | 32 | $0.85^{-40}$ | $1.8^{-40}$ | 64.8 | −97.5 |
| Dimethyl ether of polyethylene glycol | 200 | 1.03 | 6.25 | — | −20 to −30 |
| Liquid carbon dioxide | 44 | $1.17^{-55}$ | $0.35^{-55}$ | −78.5[a] | −56.6[b] |

[a]   Solid sublimation temperature.
[b]   Triple point, 75.1 psia.

The liquid carbon dioxide absorbent, with sulfurous compounds, other trace components, and, perhaps, some coabsorbed light hydrocarbons such as methane and ethane, is combined with the contaminated liquid carbon dioxide condensed in precooling to −55°C. The combined carbon dioxide stream, typically 3 to 5 mol % hydrogen sulfide, is stripped of light hydrocarbons, if necessary, and sent to the carbon dioxide regenerator. The treated gas, containing less than 1 ppm $H_2S$, leaves the sulfur absorber at essentially −55°C, with carbon dioxide the only significant impurity remaining to be removed.

### 3. Carbon Dioxide Regeneration by Triple-Point Crystallization

Crystallization of carbon dioxide produces pure carbon dioxide (less than 1 ppm sulfur) and concentrated hydrogen sulfide (up to 75 mol %). The great advantage of crystallization is that pure solid carbon dioxide crystals are formed from mother liquor containing sulfurous and various other trace compounds. In contrast, separation of pure carbon dioxide from hydrogen sulfide by distillation is difficult because of the very small relative volatility between carbon dioxide and hydrogen sulfide at low hydrogen sulfide concentrations.[5] Solid carbon dioxide is not known to form solid solutions with any of the trace compounds likely to be absorbed by liquid carbon dioxide in the sulfurous compound absorber. Consequently, the CNG process sharply rejects trace contaminants with the hydrogen sulfide-rich acid gas stream.

The crystallization process employed is direct-contact triple-point crystallization with vapor compression and is a continuous separation cascade analogous to continuous distillation. The cascade operates at temperatures and pressures near the triple point of carbon dioxide such that vapor, liquid, and solid phases coexist nearly in equilibrium. Solid carbon dioxide is formed by flashing; solid carbon dioxide is melted by direct contact with condensing carbon dioxide vapor. No heat-exchange surfaces are necessary to transfer the latent heat involved. Crystal formation occurs at pressures slightly below the triple point; crystal melting occurs at pressures slightly above the triple point (several psi in each case). Carbon dioxide vapor is compressed from the crystal formation pressure to the crystal melting pressure. Only a few stages of recrystallization are required to achieve extremely pure carbon dioxide. Pure carbon dioxide produced by triple-point crystallization is split into two streams, one of which becomes absorbent and is recycled to the sulfurous compound absorber; the other is returned back through the process for refrigeration and power recovery and is delivered as a product stream or is vented to the atmosphere. The amount of excess liquid carbon dioxide produced at triple-point conditions can be substantial. For example, in treating a 1000-psia synthesis gas with 30 mol % carbon dioxide, nearly 40% of the total carbon dioxide content of the synthesis gas is separated as pure triple-point liquid available for refrigeration and power recovery. The other product of triple-point crystallization, a concentrated hydrogen sulfide stream, also is recycled back through the process for refrigeration recovery before being sent to the sulfur recovery unit.

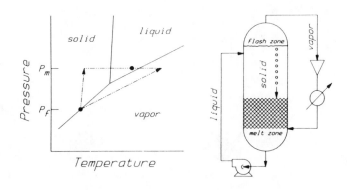

FIGURE 8.    Carbon dioxide triple-point crystallization.

Triple-point crystallization of carbon dioxide is illustrated in Figure 8, which shows a schematic carbon dioxide phase diagram expanded about the triple point and a closed-cycle triple-point crystallizer operating with pure carbon dioxide. The operation of this closed-cycle unit is identical to that of a unit in the stripping section of a continuous crystallizer cascade, except that in the cascade, vapor would pass to the unit above and liquid would pass to the unit below.

Adiabatic flash pressure $P_f$, maintained slightly below the triple-point pressure, spontaneously causes some liquid to vaporize and some to solidify. The ratio of solid to vapor is determined by the heats of fusion and vaporization; for carbon dioxide about 1.7 mol of solid are formed for each mole vaporized. The solid, more dense than the liquid, falls through a liquid head and forms a loosely packed crystal bed at the bottom. The liquid head is about 10 to 12 ft and increases the hydrostatic pressure on the solid to melter pressure $P_m$. The crystal bed depth is about 2 ft. In actual operation with sulfur compounds present, a small backwash flow is maintained upward through the bed to wash the crystals and prevents mother liquor from penetrating into the bed. Vapor is withdrawn from the flash zone, compressed to melter pressure $P_m$, sensibly cooled to near saturation, and dispersed under the solid bed where it condenses and causes solid to melt. The liquid is withdrawn and pumped to the flash zone. The process cycle is traced with dashed lines on the phase diagram of Figure 8.

Carbon dioxide triple-point crystallization is economically attractive because of the high triple-point pressure (75.1 psia) of carbon dioxide. The relatively high density of carbon dioxide vapor at the triple point permits use of conventional compression machinery. The above ambient triple-point pressure guarantees prevention of air leaks into the crystallizer which would degrade the efficiency of direct contact heat transfer.

Desalination of sea water by triple-point crystallization was extensively developed during the 1960s[11,12] and is similar to triple-point crystallization of carbon dioxide, with the major exception of triple-point pressure: 0.089 psia for water, 75.1 psia for carbon dioxide. Efficient compression of triple-point water vapor, about 2800 times more voluminous than triple-point carbon dioxide vapor on a mass basis, requires custom compressor design, and the high efficiency of melting by direct vapor-solid contact is degraded by air leakage into the subambient pressure sea water crystallizer.

Triple-point crystallization of carbon dioxide is best explained by describing a basic crystallization unit consisting of a vessel with external liquid and vapor flows. Within the basic unit there are liquid and solid internal flows. Basic units are combined in series to form a complete crystallization cascade.

### a. Basic Unit

Consider the vessel shown in Figure 9a. The bulk of the vessel is filled with liquid and

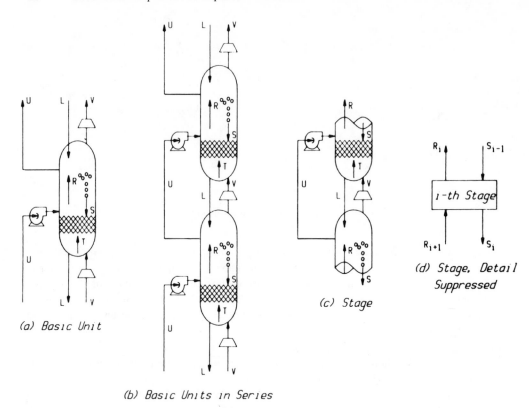

(a) Basic Unit

(b) Basic Units in Series

(c) Stage

(d) Stage, Detail Suppressed

FIGURE 9.     Continuous triple-point crystallization of carbon dioxide.

solid; a small vapor space exists at the top of the vessel above the liquid and solid. Liquid streams L enter the vessel top and leave the vessel bottom. Vapor streams V enter the vessel bottom and leave from the vapor space at the top. The liquid head in the vessel provides a pressure differential from top to bottom of a basic unit sufficient to effect flashing at the top and melting at the bottom.

The flash pressure at the vessel top is maintained slightly below the triple-point pressure, causing liquid carbon dioxide to spontaneously vaporize and freeze in the upper portion of the basic unit, forming about 1.7 mol of solid for each mole vaporized. Crystals formed at the top fall by gravity and form a loosely packed bed at the bottom. A small upward backwash flow T is maintained through the bed to wash crystals and prevent downward penetration of the crystal bed by mother liquor. Carbon dioxide vapor V from an adjacent unit below is compressed, sensibly cooled to near saturation, and dispersed under the solid bed where it condenses and melts solid by direct contact. Melt liquid L is withdrawn and sent to the flash zone of the unit below; about 10% of the melt liquid is sufficient to backwash the crystal bed.

Modest liquid flows R through the basic crystallizer vessel carry mother liquor upward. The upward flow R is the sum of the small backwash stream T and reflux stream U. Streams U serve to carry hydrogen sulfide and other trace contaminants upward through the cascade; hence, they originate from the location with the highest concentration of hydrogen sulfide.

Figure 9b shows two basic crystallizer units joined in series. Any number of basic units can be cascaded to achieve a specified degree of separation.

### b. Stage Definition

The concept of an equilibrium crystallization stage is useful for calculating the number

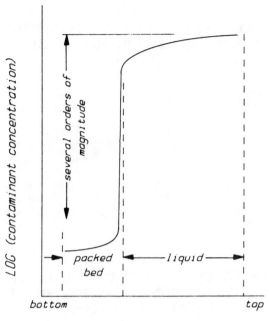

FIGURE 10.    Qualitative concentration profile in packed crystal bed
of carbon dioxide triple-point crystallizer.

of basic units required in a cascade and the magnitude of streams such as U, L, V, and the
internal solids flow S. Figure 9c defines an "equilibrium" crystallization stage. The stage
comprises the upper and lower halves of two adjacent basic units shown in Figure 9b. Stream
R, which exits the top, is the sum of backwash T and reflux U. Reflux U is generally much
greater than backwash T, and, therefore, stream R is nearly as concentrated in hydrogen
sulfide and trace contaminants as mother liquor reflux U, which is in equilibrium with solid
S leaving the bottom of the stage. Thus, the crystallization stage, shown in Figure 9d with
details suppressed, is very nearly a "conventional" equilibrium stage, i.e., exit streams $R_i$
and $S_i$ are nearly in equilibrium, and the differences $R_i - S_{i-1}$ or $R_{i+1} - S_i$ give the net
flow through the stage.

### c. Experimental and Design Considerations

Streams U, L, and V at the top of a basic unit are very much richer in hydrogen sulfide
and trace contaminants than streams U, L, and V at the bottom of a basic unit. Concentration
changes observed between mother liquor in the flash zone (top) and liquid product L in the
melt zone (bottom) of an experimental triple-point crystallizer have been dramatic. A qualitative concentration profile typical of those observed in an experimental basic unit is shown
in Figure 10. Mother liquor concentration is relatively uniform above the packed bed, but
a sharp drop in contaminant concentration occurs within the top several inches of the loosely
packed crystal bed. Concentration changes of the order 500- to 5000-fold have been observed
for representative sulfurous compounds and trace contaminants contained in raw coal gasifier
gas, including hydrogen sulfide, carbonyl sulfide, methyl mercaptan, ethane, and ethylene.
Concentration profiles calculated for the packed bed of solid carbon dioxide using a conventional packed bed axial dispersion model predict well the observed experimental profiles.

### Table 4
### RELATIVE CRYSTALLIZER INTERNAL
### FLOWS VS. CARBON DIOXIDE PRODUCT
### PURITY

| CO₂ purity (ppm H₂S) | S, V | U | S/U |
|---|---|---|---|
| 1.00 | 1.00 | 1.00 | 38.2 |
| 0.10 | 1.05 | 2.86 | 13.8 |
| 0.01 | 1.18 | 7.41 | 6.0 |

*Note:*  Feed = 100, 4% $H_2S$; acid gases = 16, 25% $H_2S$ to S recovery; $CO_2$ = 84; S = internal solid flow, V = vapor flow, U = upward reflux, separation factor = 250, three crystallization stages.

### Table 5
### RELATIVE CRYSTALLIZER INTERNAL FLOWS
### VS. ACID GAS CONCENTRATION (AGC) TO
### SULFUR RECOVERY

| AGC mol% H₂S | CO₂ | Acid gases | S, V | U | S/U |
|---|---|---|---|---|---|
| 25 | 84.0 | 16.0 | 1.00 | 1.00 | 38.2 |
| 50 | 92.0 | 8.0 | 1.10 | 1.24 | 33.2 |
| 75 | 94.7 | 5.3 | 1.14 | 1.42 | 29.9 |

*Note*:  Feed = 100, 4% $H_2S$; $CO_2$ purity = 1 ppm $H_2S$; separation factor = 250; three crystallization stages; S = internal solid flow, V = vapor flow, U = upward reflux.

The separation of carbon dioxide and hydrogen sulfide achieved in one experimental basic unit indicates that three basic crystallization units (two below the feed, one above the feed) can easily produce pure carbon dioxide bottom product (<1 ppm $H_2S$) and a 25% hydrogen sulfide top product from a feed containing 3 to 5 mol % hydrogen sulfide. More stringent product specifications (down to 0.01 ppm $H_2S$ bottom, up to 75% $H_2S$ top) also can be met in three basic crystallization units by modest increase of internal crystallizer flows. Relative changes in crystallizer internal flows to produce increasingly pure carbon dioxide (fixed overhead acid gas composition) are shown in Table 4. Relative changes in crystallizer internal flows to produce an increasingly concentrated hydrogen sulfide acid gas stream (fixed carbon dioxide purity) are shown in Table 5. The eutectic of the $H_2S$-$CO_2$ system, 85% $H_2S$, is the maximum acid gas concentration which can be produced by triple-point crystallization of carbon dioxide. The relative crystallizer internal flows of Table 5 are calculated assuming a very conservative separation factor of 250 for the key components hydrogen sulfide and carbon dioxide.

A triple-point crystallizer comprising three crystallization stages is shown in flow sheet Figure 11. For comparison with distillation, the crystallizer of Figure 13 has been designed to produce pure carbon dioxide (1 ppm $H_2S$) and 25% $H_2S$ content acid gas from a 4% hydrogen sulfide-carbon dioxide feed mixture. Minor process equipment is suppressed from Figure 11 for clarity.

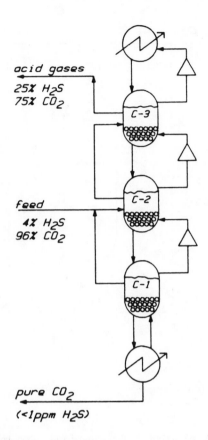

FIGURE 11.    Staged carbon dioxide triple-point crystallizer.

### d. Distillation-Based Processes

**General basis** — Simple distillation and extractive distillation are examined as alternative processes for separating carbon dioxide and hydrogen sulfide. The feed to each process is identical with the triple-point crystallizer feed, a mixture of hydrogen sulfide and carbon dioxide containing 4 mol % hydrogen sulfide. Each distillation-based separation process is designed to produce pure carbon dioxide (1 ppm $H_2S$) and a carbon dioxide-hydrogen sulfide acid gas stream containing 25% $H_2S$. The design pressure, 100 psia, prevents solid carbon dioxide formation while enhancing the relative volatility of the hydrogen sulfide-carbon dioxide system ($^K H_2S = 0.67$) and minimizing the temperature differential between condenser and reboiler. The small temperature differential, about 1.2°C, permits use of vapor recompression to condense carbon dioxide overhead vapor, after slight compression, in the boiler.

**Simple distillation** — A simple distillation process to separate carbon dioxide and hydrogen sulfide is shown in flow sheet Figure 12. With the exception of vapor recompression, the process is conventional. Overhead carbon dioxide vapor is compressed to provide a temperature driving force of 3°C for condensation in the reboiler. Trade-offs in compressor-reboiler capital cost and compressor operating cost were considered in selecting the 3°C reboiler temperature approach. Condensed carbon dioxide is cooled in exchanger R-1 and split into reflux and product streams. The desired separation is achieved with 64 equilibrium stages using a reflux ratio of 3.4, about 1.7 times the minimum reflux.

**Extractive distillation** — Extractive distillation employs an extractive agent which serves to lower the volatility of hydrogen sulfide and ease its separation from carbon dioxide. For

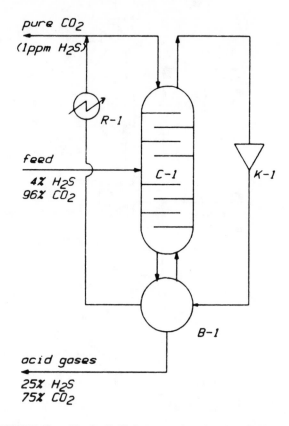

FIGURE 12.     Simple distillation separation of carbon dioxide and
hydrogen sulfide.

development of an extractive distillation flow sheet to compare with triple-point crystallization, an extractive agent with the following properties is assumed:

- A mixture of 25% agent in liquid carbon dioxide is used as reflux absorbent.
- The agent is assigned the physical properties of dimethyl ether.
- The agent lowers the equilibrium vaporization constant of the binary $H_2S-CO_2$ from
  0.67 to 0.25.

The equilibrium vaporization constant K = 0.25 is a lower bound determined experimentally
for various candidate-extractive agents, primarily oxygen-containing organic compounds
such as alcohols, ethers, and ketones. The use of K = 0.25 is a best-case assumption for
the preliminary analysis of extractive distillation. Ryan and Holmes[2] report improved separation of $H_2S$ and $CO_2$ by addition of a hydrocarbon extractive agent, normal butane, but
do not reveal modified equilibrium constants for this system. The extractive distillation
process is shown in flow sheet Figure 13. The reflux absorbent in the rectification section
of C-1 is a mixture of 25% extractive agent in carbon dioxide. A small liquid reflux of pure
carbon dioxide is introduced at the top of C-1 to reabsorb vaporized extractive agent. The
overhead product of column C-1 is pure carbon dioxide (1 ppm $H_2S$). The bottom product
is a ternary mixture of extractive agent and 25 mol % hydrogen sulfide in carbon dioxide.
This ternary mixture is separated in column C-2 into extractive agents (bottoms) and acid
gases (overhead). The desired separation is achieved with 37 equilibrium stages using a
reflux ratio of 0.45, about 1.4 times the minimum reflux.

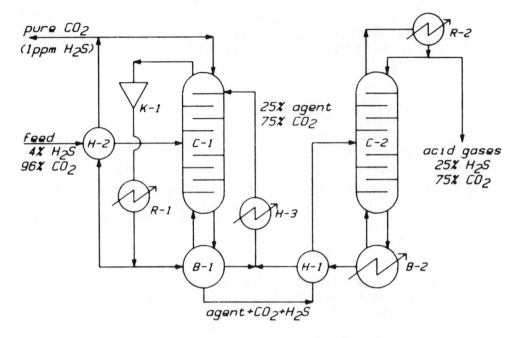

FIGURE 13.    Extractive distillation separation of carbon dioxide and hydrogen sulfide.

### Table 6
### RELATIVE ECONOMICS DISTILLATION VS.
### TRIPLE-POINT CRYSTALLIZATION

| Process | Capital cost | Operating cost (annual) | Combined cost |
|---|---|---|---|
| Continuous TP crystallization | 1.0 | 1.0 | 1.0 |
| Simple distillation | 3.1 | 1.9 | 2.7 |
| Extractive distillation | 2.0 | 1.5 | 1.8 |

*Note:* Combined cost = capital + 5-year operating costs; carbon dioxide purity = 1 ppm $H_2S$; acid gas concentration = 25% $H_2S$ to sulfur recovery.

### e. Economic Comparison

Presented here is a preliminary economic analysis of triple-point crystallization and of distillation for production of pure carbon dioxide from dilute mixtures of hydrogen sulfide in liquid carbon dioxide. Installed capital costs are estimated using the methods and general cost correlations of Guthrie[13] and are believed accurate for relative ranking of the processes.

Estimates of installed capital cost, annual operating cost, and a combined capital plus 5-year operating cost are given in Table 6. All estimates of Table 6 are relative to the triple-point crystallizer base case. The combined cost comparison assumes an average plant payout period of 5 years.

Vapor flow is the primary determinant of process capital and operating costs for both triple-point crystallization and the distillation processes. The negative impact of the poor equilibrium vaporization constant and consequent large vapor flows on simple distillation economics is evident in Table 6. The minimum vapor flow (minimum reflux, infinite stages)

## Table 7
### CARBON DIOXIDE AND HYDROGEN SULFIDE
### LATENT HEATS (TEMPERATURE = −55°C)

|                                | BTU/lb | BTU/lb mol |
|--------------------------------|--------|------------|
| Hydrogen sulfide (vaporization) | 235    | 8010       |
| Carbon dioxide (vaporization)   | 150    | 6600       |
| Carbon dioxide (fusion)         | 86     | 3780       |

for simple distillation is about six times the vapor flow per stage of crystallization, or about double the combined vapor flow through the compressors in a three-stage triple-point crystallizer. Although vapor flows decrease and economics become more attractive for hypothetical extractive distillation with the agent defined above, a suitable agent remains to be identified.

### f. Conclusions

Triple-point crystallization of carbon dioxide can easily produce pure carbon dioxide (0.01 ppm $H_2S$ and concentrated hydrogen sulfide (up to 75% $H_2S$) from dilute mixtures of hydrogen sulfide in carbon dioxide. Crystallization offers reduced capital and operating costs compared to distillation processes which must rely on vapor-liquid equilibrium to effect the separation. Crystallization also assures pure carbon dioxide with respect to trace sulfur-containing compounds and other contaminants associated with crude coal gasifier gas, a feat difficult to achieve with conventional vapor-liquid contacting processes.

### 4. Carbon Dioxide Absorption; Refrigerant Absorbents
### a. General Comments

We define a refrigerant absorbent as one which contains materials that undergo a phase change during the absorption and/or regeneration steps. The latent heat of phase change enhances the thermal capacity of the absorbent over the desired absorption temperature range. A refrigerant absorbent is distinguished from an absorbent that is dependent solely on its specific heat to moderate absorber temperature rise. One class of refrigerant absorbents comprises liquids which vaporize while absorbing acid gases; a second class is slurries wherein the solid phase melts upon acid gas absorption.

An example of a liquid refrigerant absorbent is the liquid carbon dioxide used in the CNG process to absorb hydrogen sulfide and trace contaminants. Liquid $CO_2$ contacts crude gas in the sulfurous compound absorber under $CO_2$ dew point conditions, a small fraction of the liquid $CO_2$ absorbent is vaporized by the heat of absorption of hydrogen sulfide, thus maintaining nearly isothermal conditions in the absorber.

Latent heats for carbon dioxide and hydrogen sulfide are shown in Table 7. Each mole of $H_2S$ absorbed vaporizes about 1.2 mol of $CO_2$ (8010/6600). Carbon dioxide absorbent losses in the absorber are made up from excess carbon dioxide condensed in the initial crude gas cooling step and subsequently purified in the triple-point crystallizer.

An example of a slurry refrigerant absorbent is solid carbon dioxide slurried in $C_4$ to $C_6$ ketones or ethers as used in the $CO_2$ absorber of the CNG process. Figure 14 shows an equilibrium stage of the carbon dioxide absorber. As slurry S + L + K descends through the absorption tower, solid carbon dioxide S in the slurry melts and carbon dioxide V in the crude gas is absorbed (condensed). The liquid carbon dioxide L formed becomes part of the downward flowing slurry. The solubility of carbon dioxide in carrier liquid K increases due to a small temperature rise on the equilibrium stage, and permits carrier liquid K to accommodate the melted solid and condensed vapor. All solid $CO_2$ is melted in the absorber. The bulk of the slurry refrigerant is regenerated by staged flashing down to 1 atm.

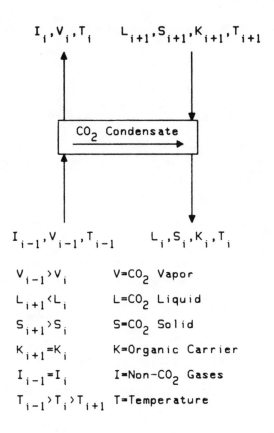

$V_{i-1} > V_i$      $V = CO_2$ Vapor

$L_{i+1} < L_i$      $L = CO_2$ Liquid

$S_{i+1} > S_i$      $S = CO_2$ Solid

$K_{i+1} = K_i$      $K =$ Organic Carrier

$I_{i-1} = I_i$      $I =$ Non-$CO_2$ Gases

$T_{i-1} > T_i > T_{i+1}$   $T =$ Temperature

FIGURE 14.    Slurry absorber equilibrium stage.

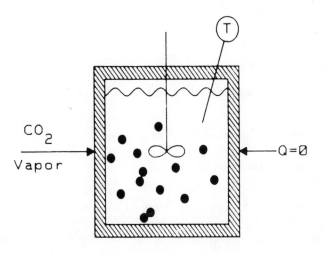

FIGURE 15.    Insulated adiabatic batch cell.

The ability of refrigerant slurry to moderate absorber temperature rise is illustrated in the following example. Consider an insulated batch cell which contains a 1-lb mass of refrigerant slurry, e.g., solid carbon dioxide slurried in ketone. The slurry is initially at $-80°C$ and contains 43 wt % solid $CO_2$. A schematic diagram of the cell is shown in Figure 15. Carbon dioxide vapor is admitted to the cell with pressure sufficient to cause condensation. Under

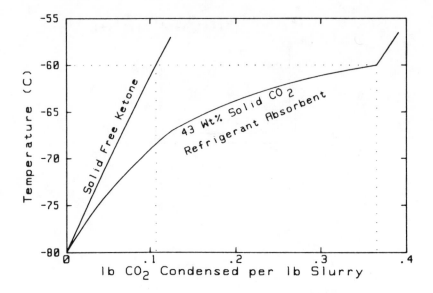

FIGURE 16.     Slurry temperature rise vs. amount of carbon dioxide condensed.

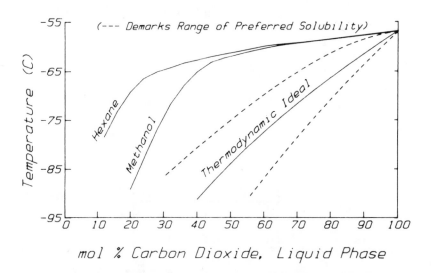

FIGURE 17.     Solubility of solid carbon dioxide in selected solvents.

adiabatic conditions, the condensation of $CO_2$ releases heat and causes the slurry temperature to rise. Figure 16 shows the slurry temperature rise plotted vs. pounds mass of $CO_2$ condensed. The break in the refrigerant absorbent curve at $-60°C$ is the point where all solid carbon dioxide has melted. Also shown in Figure 16 is the temperature rise experienced in a batch cell initially charged with pure ketone at $-80°C$. For the same temperature rise, $-80$ to $-60°C$, the refrigerant slurry absorbs well over three times as much carbon dioxide vapor as does the solid-free ketone absorbent.

Not all organic liquids are acceptable candidates for use in refrigerant slurry absorption of carbon dioxide.[7] The solubility of solid carbon dioxide in hexane, methanol, and a thermodynamically ideal solvent is shown in Figure 17. Neither hexane nor methanol has acceptable solubility for solid carbon dioxide to function efficiently as a carrier of solid

**Table 8**
**CRUDE GAS TO ACID GAS REMOVAL**

| Component | lb mol/hr | mol % |
|---|---|---|
| $CO_2$ | 32,406 | 32.45 |
| CO | 13,074 | 13.09 |
| $CH_4$ | 12,192 | 12.21 |
| $H_2$ | 40,536 | 40.60 |
| $N_2$ | 516 | 0.52 |
| $H_2S$ | 1,125 | 1.13 |
| | 99,849 | 100.00 |

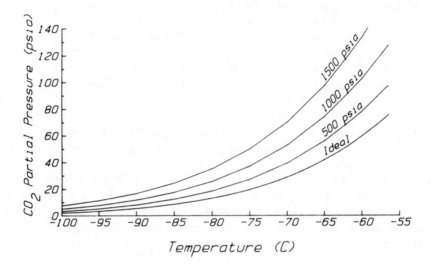

FIGURE 18.     Solubility of solid carbon dioxide in synthesis gas.

carbon dioxide for refrigerant absorption. The preferred range of carbon dioxide solubility for slurry absorption is defined qualitatively by the dashed lines of Figure 17, and is derived from computer simulation of refrigerant slurry absorption based on experimental solubility data for various solvents, and of hypothetical slurry absorption based on constant carbon dioxide partial pressure driving force per stage. Various $C_4$ to $C_6$ ketones and ethers fall within the preferred solubility range and have other desirable characteristics for use as a carrier of solid $CO_2$: (1) a nearly linear change in solubility with temperature (as contrasted with hexane or methanol), (2) low viscosity when saturated with carbon dioxide, (3) low melting point, and (4) low vapor pressure.[14]

Carbon dioxide removal by slurry absorption is attractive down to about $-75°C$, a temperature easily achieved by slurry regeneration, to slightly above 1 atm carbon dioxide pressure. The level to which carbon dioxide can be removed from a treated crude gas by refrigerant slurry absorption depends on the solubility of solid carbon dioxide in the treated gas. The solubility of solid carbon dioxide in synthesis gas (for composition, see Table 8) is illustrated in Figure 18 for several crude gas pressures. For example, with a 1000-psia crude gas pressure and a $-75°C$ exit gas temperature from the $CO_2$ absorber, refrigerant absorption reduces the carbon dioxide content from 13 to about 4 mol %. Fine removal of $CO_2$ to lower levels is accomplished by conventional absorption into a slip stream of the refrigerant absorbent regenerated to meet particular product gas $CO_2$ specifications.

### b. Slurry Absorption in the CNG Process

In the CNG process, carbon dioxide remaining in the gas after sulfurous compound absorption at $-55°C$ is removed at temperatures below the carbon dioxide triple point. Carbon dioxide is absorbed into a slurry consisting of a liquid solvent saturated with carbon dioxide and containing suspended solid carbon dioxide particles. As carbon dioxide vapor is absorbed, the released latent heat melts solid carbon dioxide. Thus, the solid carbon dioxide circulated with the liquid solvent provides direct refrigeration for the condensation and absorption of carbon dioxide vapor.

The large effective heat capacity of liquid-solid slurry absorbent enables relatively small slurry flows to absorb the heat released with only mild temperature rise. This contrasts with other acid gas removal processes in which solvent flow to the carbon dioxide absorber is often determined by the large absorber heat effects rather than vapor-liquid equilibrium considerations. Small slurry absorbent flows permit smaller tower diameters because allowable vapor velocities generally increase with reduced liquid loading.[6]

The liquid solvent of the slurry absorbent is the sink for absorbed carbon dioxide vapor and melted solid carbon dioxide. The carbon dioxide-rich solvent exits the absorber near the triple-point temperature, but contains no solid carbon dioxide and is stripped of methane and other light molecules if necessary. The carbon dioxide-rich absorbent is next flashed in stages at successively lower pressures to generate, simultaneously, pure carbon dioxide gas and a progressively colder slurry of liquid solvent and solid carbon dioxide crystals. The carbon dioxide gas is recycled back through the process for refrigeration and power recovery, and is delivered as a product stream or is vented to the atmosphere. The regenerated slurry absorbent is recirculated to the carbon dioxide absorber.

The absorption of carbon dioxide with slurry and the regeneration of slurry by flashing carbon dioxide require small temperature and pressure driving forces. Consequently, the CNG process is nearly reversible and consumes less energy than conventional carbon dioxide removal processes. Compared with other subambient temperature carbon dioxide removal processes, the CNG process requires less refrigeration even though process temperatures are often lower.

### c. Crude Gas Specifications

Crude manufactured or natural gas mixtures with moderate to high carbon dioxide partial pressure are candidate feed gases for treatment by the CNG acid gas removal process. The minimum acceptable crude gas carbon dioxide partial pressure to permit use of liquid $CO_2$ absorbent for hydrogen sulfide absorption is about 75 psia, the carbon dioxide triple-point pressure. Below the $CO_2$ triple-point pressure, liquid carbon dioxide cannot exist. As crude gas pressure increases, the minimum $CO_2$ partial pressure needed to insure the existence of liquid $CO_2$ absorbent also increases. The relation between crude gas (total) pressure and the required $CO_2$ partial pressure is shown in Figure 5. The boundary curve shifts slightly up or down depending on the hydrogen/carbon monoxide ratio of the crude synthesis gas.

The CNG process removes carbon dioxide in two steps: (1) by condensation above the triple point of $CO_2$, and (2) by absorption with a refrigerant slurry below the triple point of $CO_2$. The relative amounts of $CO_2$ removed in the two steps vary with the $CO_2$ partial pressure of the crude gas. At low $CO_2$ partial pressures, most of the $CO_2$ is removed by absorption with refrigerant slurry below the triple point. As the $CO_2$ partial pressure increases, progressively more of the $CO_2$ can be removed in the initial condensation step. For example, the fraction of $CO_2$ in a crude gas initially containing 30 mol % $CO_2$ and which is removed in the CNG process by absorption with refrigerant slurry is shown in Figure 19 as a function of crude gas pressure. Clearly, the advantage of refrigerant slurry absorption is greatest at moderate to low crude gas pressures where little carbon dioxide is condensed in cooling to the triple point, and the bulk of carbon dioxide must be removed by absorption.

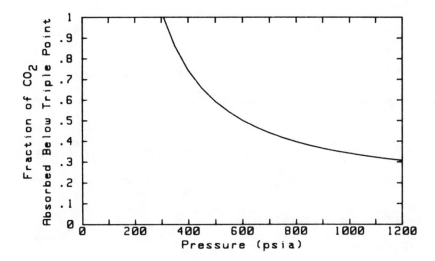

FIGURE 19. Fraction of total carbon dioxide absorbed below triple-point vs. total gas pressure for a crude gas containing 30% carbon dioxide.

### d. Conclusions

The flow rates of the refrigerant absorbents are smaller than the flow rates of customary physical absorbents, and the use of refrigerant absorbents is economically attractive in the integrated low temperature CNG acid gas removal process. The attractive economics result from reduced absorbent flow rates and a closer approach to thermodynamic reversibility, both of which contribute to smaller equipment and reduced energy consumption.

## III. CNG PROCESS CHARACTERISTICS

Several attractive characteristics are inherent in the CNG acid gas removal process.

**Removal of all sulfurous compounds and other trace impurities** — Pure liquid $CO_2$ absorbs all sulfurous compounds and all other trace impurities at reasonable absorbent flow rates. Liquid $CO_2$ has the unique property, among all the known physical and chemical absorbents, that it absorbs COS better than it absorbs $H_2S$, as discussed previously in Section II. Thus, an absorber designed to remove $H_2S$ automatically removes all COS and other trace impurities.

**Production of concentrated hydrogen sulfide** — Triple-point crystallization can easily produce up to 75% $H_2S$ concentrations for sulfur recovery, and with minimal energy consumption.

**Production of pure carbon dioxide** — The triple-point crystallizer has exceptional ability to purify $CO_2$, since each stage of crystallization reduces impurities by factors ranging from 500 to 5000. Present designs call for 1 ppm impurities, but 10 ppb or even less can be achieved with little additional capital or operating cost.

**Low flow rates of absorption liquids** — The flow rate of liquid carbon dioxide used to absorb sulfurous compounds and other trace impurities is less than the absorbent flow rate in any other physical absorption acid gas removal process. The low flow rate is partly due to low molecular weight and to the unique ability of liquid carbon dioxide to absorb carbonyl sulfide (and other trace impurities) better than it absorbs hydrogen sulfide. Low absorption temperature also contributes to the low flow rate; hydrogen sulfide is absorbed at $-56°C$, just slightly above the lowest temperature at which carbon dioxide can exist in the liquid state (triple-point temperature, $-56.6°C$). Even at the triple point, liquid carbon dioxide

**Table 9**
## CHARACTERISTICS OF ACID GAS REMOVAL PROCESSES[a]

| Process | Type | T(°C) | Absorbent | MW | Gal/hr to remove | | Regen steam (lb/hr) | Claus feed (% H$_2$S) | CO$_2$ to atm (ppm S) |
| | | | | | H$_2$S | CO$_2$ | | | |
|---|---|---|---|---|---|---|---|---|---|
| CNG | Phys | −56 | CO$_2$L,S | 44[b] | 60,000[c] | 50,000[d] | 0 | 70 | <1 |
| Benfield[e] | Chem | 100+ | K$_2$CO$_3$[f] | | — | — | 1,300,000 | 15—20 | 3,400 |
| Selexol®[g] | Phys | 25 | DEPG | 200[h] | 120,000 480,000[i] | 1,200,000 | 140,000 | 25 | 100 |
| Rectisol[g] | Phys | −40 | CH$_3$OH | 32[j] | 100,000 150,000[i] | 1,500,000 | 100,000 | 25 | 100 |

[a]  Based on BIGAS process, 250 MMSCFD of SNG.[9]
[b]  CO$_2$ liquid; CO$_2$ solid in liquid.
[c]  Liquid CO$_2$ absorbent.
[d]  CO$_2$ solid in liquid absorbent.
[e]  See Reference 9.
[f]  Hot solution in water.
[g]  Estimated from Reference 10.
[h]  Dimethyl ether of polyethylene glycol.
[i]  Upper figure to remove COS.
[j]  Refrigerated methanol.

physical properties are attractive for an absorbent liquid: low viscosity, low surface tension, and relatively high density (see Section II). Similarly, because of its enhanced heat capacity, the slurry used for carbon dioxide removal has a lower flow rate than conventional physical absorbents. Condensation by absorption of carbon dioxide vapor into a liquid releases considerable latent heat. Dissipating this latent heat without excessive absorbent temperature rise in most absorbers requires interstage cooling with heat exchangers or very large liquid absorbent flow rates to provide sufficient heat capacity. The slurry absorbent provides the necessary heat capacity *in situ* and in a concentrated form through the nearly isothermal phase change from solid carbon dioxide to liquid carbon dioxide.

**Low energy consumption** — The CNG acid gas removal process utilizes relatively small temperature and pressure driving forces to effect the desired separations: (1) bulk carbon dioxide from crude gas by condensation; (2) minor acid gases from crude gas by physical absorption; (3) carbon dioxide from hydrogen sulfide by triple-point crystallization; (4) carbon dioxide from crude gas by absorption with slurry absorbent; and (5) carbon dioxide from solvent by adiabatic flashing. To remove sulfurous compounds and trace contaminants, the CNG process uses no absorption agent foreign to the crude gas, thus, easing the subsequent recovery of absorption agent (carbon dioxide) from acid gases (carbon dioxide, hydrogen sulfide, trace contaminants). The CNG process is more nearly reversible in the thermodynamic sense than are other known acid gas removal processes, and the CNG process energy requirements are only a fraction of those normally experienced in acid gas removal.

**Noncorrosive system** — Because of the low temperatures and poor solubility of water in liquid carbon dioxide, little or no corrosion of equipment can occur. Carbon steel construction is adequate at temperatures warmer than −40°C; mild nickel steels are adequate for CNG process temperatures below −40°C.

## IV. PROCESS COMPARISONS

Table 9 compares some of the major characteristics of the CNG process with commercially available acid gas removal processes: Benfield (hot carbonate), Selexol® (dimethyl ether of polyethylene glycol), and Rectisol (refrigerated methanol). Benfield is a chemical absorption process; Selexol® and Rectisol are physical absorption processes. The comparisons are for the crude gas stream defined in Table 8, and are based on information available in the open literature[4,10] (and recommended reading [Hockgesand]).

The volumetric absorbent flow rate in Selexol® and Rectisol for hydrogen sulfide removal is about twice that in the CNG process. If Selexol® is required to remove all COS in addition to hydrogen sulfide, the flow rate of Selexol® solvent is about eight times the flow rate in the CNG process. The lower CNG absorbent flow rate is a result of: (1) unique COS absorption capability of liquid carbon dioxide; (2) lower molecular weight of carbon dioxide (44) as compared to Selexol® solvent (about 200); and (3) lower absorption temperature possible in the CNG process. When compared with Rectisol, the lower temperature of absorption in the CNG process is beneficial.

The volumetric absorbent flow rate for carbon dioxide removal in the CNG process is 1/6 to 1/30 of that for Selexol® and Rectisol. This lower flow rate is a result of: (1) lower temperature of absorption and (2) prior bulk condensation of carbon dioxide. The high effective heat capacity of the slurry containing solid carbon dioxide (an *in situ* phase-change refrigerant) is highly beneficial.

Thermal energy needed to regenerate the Benfield hot carbonate solution is high because it is a chemical absorbent; that needed by Selexol® and Rectisol is large because the absorbent flow rates are large. In the CNG process, liquid carbon dioxide absorbent used for hydrogen sulfide and other trace impurity removal is regenerated by energy-efficient triple-point crystallization. The slurry absorbent is regenerated by simple adiabatic flashing.

High hydrogen sulfide concentration (up to 75% $H_2S$), achieved by triple-point crystallization in the CNG process, has most desirable economic consequences in sulfur recovery via the Claus process. Other available acid gas removal processes, when applied to gases of high $CO_2/H_2S$ ratio, have difficulty producing sulfur recovery feed streams with more than 25% $H_2S$, thus, incurring added sulfur recovery costs for handling the diluent gases.

The carbon dioxide discharged to the atmosphere is extremely pure in the CNG process, again a consequence primarily of triple-point crystallization. If desired, food-grade carbon dioxide can be produced with virtually no change in process economics.

In summary, the CNG Acid Gas Removal Process achieves to a great extent the desired process characteristics listed at the end of Section I of this chapter:

- High purity $CO_2$ (with impurities well below 1 ppm) is produced as an item of commerce or for environmentally acceptable venting to the atmosphere.
- A sulfur-rich (to 75% $H_2S$) acid gas stream is produced for sulfur recovery, thereby enhancing significantly the economics of sulfur recovery.
- All trace contaminants are removed from the treated gas.
- All trace contaminants removed from the treated gas are rejected completely into the small stream of acid gas to sulfur recovery. Thus, concentrated, trace contaminants may be readily removed or modified if necessary, and at minimal cost.
- Pure liquid $CO_2$ absorbent is readily produced by means of especially effective and economical triple-point crystallization. Impurities can easily be held below 1 ppm.
- Pure liquid $CO_2$ possesses nearly ideal absorbent characteristics for efficient gas absorption and desirable economics: low molecular weight, low vapor pressure, high liquid density, low liquid viscosity, high specific heat, good chemical stability, is noncorrosive to equipment, and is at no cost as a feed gas component.
- All six characteristics above are achieved with reduced capital and operating expense when compared with alternative processes.

The CNG Acid Gas Removal Process thus holds considerable promise for economic and effective treatment of low grade natural gases and synthesis gases manufactured for production of chemicals and fuels. Additionally, the triple-point crystallizer will find considerable application in the manufacture of high-purity carbon dioxide for the food, beverage, and chemical industries.

## STATUS OF CNG PROCESS (1984); ACKNOWLEDGMENTS

This new acid gas removal process has been under development by CNG Research Company (a unit of Consolidated Natural Gas Company) since 1973. Late in 1980, the Morgantown Energy Technology Center of the U.S. Department of Energy joined the project under Contract Number DE-AC21-80MC14399. A new contract (DE-AC21-83MC20230) began preparation late in 1983 to provide for 2 years of joint support by CNG, DOE, and the Gas Research Institute. Most of the research and development has been done by Helipump Corporation at Case Institute of Technology, Case Western Reserve University, Cleveland, Ohio. The authors wish to express their appreciation for the excellent and patient typing service provided by Miss Sakina Jaffer.

## RECOMMENDED READING

**Adler, R. J., Lun, A., Brosilow, C. B., Brown, W. R., Cook, W. J., Gardner, N. C., Liu, Y. C., Hise, R. E., Massey, L. G.,** The CNG Process, a New Low Temperature, Energy Efficient Acid Gas Removal Process, presented at the Symp. on Industrial Gas Separations, American Chemical Society, Washington, D.C., June 14 and 15, 1982.

**Adler, R. J. et al.,** Triple-point crystallization separates and concentrates acid gases, paper 52A, presented at AIChE Spring Natl. Meet., Houston, Tex., March 30, 1983.

**Beavon, D. K., Kouzel, B., and Ward, J. W.,** Claus processing for novel acid gas streams, Symp. Sulfur Recovery and Utilization, presented before the Division of Petroleum Chemistry, Inc., American Chemical Society, Atlanta, March 29 to April 3, 1981.

**Christensen, K. G. and Stupin, W. J.,** Merits of acid gas removal processes, *Hydrocarbon Process.,* 57, 125, 1978.

**Christensen, K. G. and Stupin, W. J.,** Comparison of Acid Gas Removal Processes, U.S. Department of Energy Report Fe-2240-49, prepared by C. F. Braun & Co. under contract No. EX-76-C-01-2240, April 1978.

Direct methanation promises lower gas costs, *Chem. Eng. News,* 60, 30, 1982.

Conf. on Gas Cleaning in the Production of Synfuels, sponsored by the Chemical Engineering Department, North Carolina State University, Raleigh, N.C., April 29 to 30, 1982.

**Edwards, M. C.,** $H_2S$-Removal Processes for Low-BTU Coal Gas, U.S. Department of Energy Report ORNL/TM-6077, Distribution Category UC-90c, January 1979.

**Eickmeyer, A. G. and Gangriwala, H. A.,** The role of acid gas removal in synfuels production, presented at the Symp. Gas Purification at the AIChE National Meeting, Houston, Tex., April 5 to 9, 1981.

Factored Estimates for Western Coal Commercial Concepts, C. F. Braun & Co., Alhambra, Calif. Project 4568-NW, ERDA-AGA, 1976.

Gas processing handbook, *Hydrocarbon Process.,* 61(4), 90, 1982.

**Hise, R. E., Massey, L. G. et al.,** The CNG Process, a New Approach to Physical Absorption Acid Gas Removal, paper 112G presented at AIChE Annual Meeting, Los Angeles, November 14 to 18, 1982.

**Hochgesand, G.,** Rectisol and purisol, *Ind. Eng. Chem.,* 62(7), 37, 1970.

**Kaplan, L. J.,** Methane from coal aided by use of potassium catalyst, *Chem. Eng.,* 89, 64, 1982.

**Rosseau, R. W., Kelly, R. M., and Ferrell, J. K.,** Evaluation of Methanol as a Solvent for Acid Gas Removal in Coal Gasification Processes, Symp. Gas Purification, AIChE Spring National Meeting, Houston, Tex., April 1981.

Lurgi Gesellschaften, Purification of Industrial Gases, Waste Gases and Exhaust Air, 1969 ed.

**Schianni, G. C.,** Cryogenic removal of carbon dioxide from natural gas, *Inst. Chem. Eng. Symp.,* Ser. No. 44, 1976.

**Cook, W. J. et al.,** Refrigerant Absorbents for Use in Acid Gas Removal, paper No. 65B presented at the AIChE Summer Natl. Meet., Denver, Colo., August 29, 1983.

# REFERENCES

1. **Ghassemi, M., Strehler, D., Crawford, K., and Quinlivan, S.,** Applicability of Petroleum Refinery Control Technologies to Coal Conversion, EPA Report No. EPA-600/7-78-190, NTIS No. PB-288 630, October 1978.
2. **Holmes, A. S. and Ryan, J. M.,** Process improves acid gas separation, *Hydrocarbon Process.,* 61(5), 131, 1982.
3. **Gazzi, L. et al.,** New process makes production of highly acid gas economical, *World Oil,* August 1, 1982, 73.
4. **Sweny, J. W.,** The Selexol® Solvent Process in Fuel Gas Treating, paper presented at 81st Natl. AIChE Meet., Kansas City, Mo., April 11 to 14, 1976.
5. **Sobocinski, D. P. and Kurata, F.,** Heterogeneous phase equilibria of the hydrogen sulfide-carbon dioxide system, *AIChE J.,* 5(4), 545, 1959.
6. **King, C. J.,** *Separation Processes,* McGraw-Hill, New York, 1971, 547.
7. **Kurata, F.,** Solubility of Solid Carbon Dioxide in Pure Light Hydrocarbons and Mixtures of Pure Light Hydrocarbons, GPA Research Report RR-10, February 1974.
8. **Adler, R. J., Brosilow, C. B., Brown, W. R., and Gardner, N. C.,** Gas Separation Process, U.S. Patent 4,270,937, June 2, 1981.
9. **Engineering Study and Technical Evaluation of the Bituminous Coal Research, Inc., Two Stage Super Pressure Gasification Process, Air Products and Chemicals, Inc., Contract No. 14-32-0001-1204,** Research and Development Report No. 60, prepared for the Office of Coal Research.
10. **Kohl, A. L. and Riesenfeld, F. C.,** *Gas Purification,* 2nd ed., Gulf Pub., Houston, 1974.
11. **Sherwood, T. K., Brian, P. L. T., and Sarafim, A. F.,** Desalination by Freezing — An Investigation of Ice Production from Brine in a Continuous Crystallizer, PB 203 284, September 1969.
12. **Fraser, J. H. and Ammerman, K. K.,** Vacuum Freezing Vapor Compression Process: 60,000 GPD Pilot Plant Evaluation, PB 200 593, July 1970.
13. **Guthrie, K. M.,** *Process Plant Estimating, Evaluation and Control,* Craftsman Book Company of America, Solana Beach, Calif., 1974.

Chapter 3

# REMOVAL OF NITROGENOUS HETEROCYCLICS FROM SHALE-DERIVED OILS

**M. S. Kuk, E. W. Albaugh, and J. C. Montagna**

## TABLE OF CONTENTS

# I. INTRODUCTION

Shale oil has been known for decades to have a potential as a substitute petroleum crude. Compared to the other potential substitutes for petroleum products, such as oil derived from coal and tar sands, shale oil has the properties most similar to petroleum crude. Nonetheless, raw shale oil contains higher amounts of nitrogen, arsenic, and iron than petroleum crude and cannot be processed in unmodified existing conventional refineries. Nitrogen content amounting to as much as 2.5 wt % of shale oil has been noted. These nitrogenous compounds, which are mostly heterocyclics such as alkyl or cycloalkyl-substituted pyridines, quinolines, acridines, pyrroles, indoles, carbazoles, etc. are not only deleterious for the catalytic refining process, but also, residual amounts of certain types of these nitrogen compounds in shale-derived fuels are believed to promote instability.[1]

The removal or reduction of the nitrogen compounds contained in shale oil is technically feasible via catalytic hydrogenation.[2] However, the hydrotreatment of shale oil crude is an expensive technology, almost amounting to 25% of the syncrude cost.[3] To reduce the nitrogen compounds to a level where shale-derived fuels are as stable as the petroleum counterparts, the amount of hydrogen consumption is quite high and severe conditions for hydrodenitrogenation (HDN) are required. For example, shale oil-derived jet fuels or catalytic reforming feedstock require hydrogenation to a few ppm or less level of the nitrogen compounds in order to protect the catalysts from poisoning and insure the stability of the products. The incentives to explore more economical and technically viable alternatives to the straight HDN of shale oils exist.

# II. APPROACHES TO SHALE OIL DENITROGENATION

Nitrogen compounds in shale oil can be typed as basic, weakly basic, and nonbasic compounds.[4-6] The basic nitrogen compounds include various alkylpyridines, alkylquinolines, alkylacridines, hydroquinolines, and hydroxypyridines. The nonbasic constituents are pyrroles, indoles, carbazoles, and their various alkyl substitutes. Some of the potential denitrogenation methods are (1) catalytic hydrogenation, (2) acid extraction, (3) adsorption/ion exchange, and (4) solvent extraction with and without enhancing agents. The basic features of these methods are described below.

## A. Catalytic Denitrogenation

It is not within the scope of this paper to evaluate or describe the details of various catalytic denitrogenation methods, but to introduce briefly some of the highlights of the methods available in the literature. A typical hydrotreatment which is performed in two stages, pretreatment and main hydrotreatment, reduces the nitrogen content from about 2.0 wt % to about 500 ppm, consuming between 1500 and 2000 SCF per barrel of shale syncrude. The main hydrotreatment is reportedly performed at about 2000 psia and a temperature of about 750°F. Reported catalysts for HDN are Ni-Mo on alumina, Co–Mo on $\gamma$-$Al_2O_3$, Ni–W on silica-alumina, and other combinations of these metals on various supporting materials. Further details are available in the literature.[7-12]

## B. Acid Extraction

Use of various acids such as hydrochloric,[14] acetic,[15] and phosphoric[16] has been reported in the literature. In the case of anhydrous hydrochloric acid extraction, insoluble hydrochloride salts are formed which are separated from the hydrocarbon stream by conventional methods. In the case of organic acids, the operation is essentially similar to solvent extraction. As can be expected, the acid-extracted shale oils still contain a trace quantity of the nonbasic nitrogen compounds, as reported by Nowack.[13]

**Table 1**
**SOLVENT SCREENING BY EXTRACTION OF MODEL NITROGEN COMPOUNDS[a]**

| | | | Raffinate | |
|---|---|---|---|---|
| Solvent | Additives (salt/nitrogen) | Yield (%)[b] | Denitrogenation (%) | Selectivity[c] |
| Furfural (5% H$_2$O) | None | 67 | 57 | 1.7 |
| DMF[d] (5% H$_2$O) | None | 95 | 35 | 7.0 |
| MeOH (15% H$_2$O) | None | 93 | 72 | 10.3 |
| DMSO[d] (5% H$_2$O) | None | 85 | 60 | 4 |
| γ-Butyrolactone (5% H$_2$O) | None | 91 | 80 | 9 |
| Ethylene carbonate (5% H$_2$O) | None | 92 | 64 | 10 |
| Sulfolane (5% H$_2$O) | None | 88 | 74 | 6 |
| Sulfolane (5% H$_2$O) | FeCl$_3$ (0.5:1) | 89 | 88 | 8 |
| Sulfolane (5% H$_2$O) | AlCl$_3$ (0.5:1) | 89 | 81 | 8 |
| Sulfolane (5% H$_2$O) | Cu(ClO$_4$)·6H$_2$O (0.2:1) | 93 | 99 | 15 |

[a] Model compounds mixture included pyridine, quinoline, pyrrole, and indole.
[b] Yield = oil weight in raffinate/oil weight in feed.
[c] Selectivity index $\simeq$ denitrogenation efficiency (%)/(100-yield).
[d] DMF = dimethylformamide; DMSO = dimethylsulfoxide.

## C. Adsorption/Ion Exchange

In this approach, the shale-derived distillate or partially hydrogenated shale syncrude is passed through a bed of solid adsorbents such as silica-alumina, bentonite, kaolin,[17] or an ion exchange resin.[18]

## D. Enhanced Solvent Extraction

As an alternative to the above-mentioned methods, a process characterized by mild hydrotreatment followed by solvent extraction or enhanced solvent extraction is being considered. Expected benefits are twofold: (1) reduction in hydrogen consumption in hydrotreatment and (2) capability of extracting the basic and nonbasic nitrogen species.

Some of the results obtained in our laboratory are presented in the subsequent section.

## III. EXPERIMENTAL RESULTS AND DISCUSSIONS

Since the major portion of the nitrogen compounds in shale oil syncrude are basic, highly polar solvents were selected for screening. The solvent screening was performed in two steps. In the first step, polar solvents were screened with a model compounds system. The model compounds system contained carrier fluid, 50/50 vol % of n-heptane and toluene, and model nitrogen compounds: pyridine, quinoline, pyrrole, and indole, which represent the basic and nonbasic compounds in shale oil. In the second screening step, whole boiling range raw Paraho shale crude and partially hydrotreated shale oils were used. The basic L–L equilibrium data between the screened solvent and the nitrogen compounds were obtained by batch experiments using a bench-scale mixer-settler in a thermal bath. In Table 1, a summary of the solvent screening results with the model compounds is presented and the physical properties of the screened solvents are given in Table 2. The elemental analysis and physical properties of raw Paraho shale oil are presented in Table 3.

### Table 2
### PHYSICAL PROPERTIES OF SCREENED SOLVENTS

| Solvent | Dipole moment (Debye at 25°C) | Normal boiling point (°C) | Sp gr |
|---|---|---|---|
| MeOH | 1.7 | 64.7 | 0.79 |
| Furfural | 3.6 | 161 | 1.15 |
| DMF | 3.9 | 153 | 0.94 |
| DMSO | 4.0 | 189 | 1.09 |
| γ-Butyrolactone | 4.1 | 204 | 1.12 |
| Sulfolane | 4.8 | 287 | 1.26 |
| Ethylene carbonate | 4.9 | 238 | 1.32 |

### Table 3
### ELEMENTAL ANALYSIS AND OTHER PHYSICAL PROPERTIES OF PARAHO SHALE CRUDE

| Wt % | Element |
|---|---|
| 2.05 | Nitrogen |
| 0.66 | Sulfur |
| 1.5 | Oxygen |
| 11.45 | Hydrogen |
| 84.34 | Carbon |

Gravity, °API: 21.7
Pour point, °F: +85
Distillation, 50% °F: 790
  EP, °F: 1060

In the tests with the model compounds system, furfural, sulfolane, dimethyl sulfoxide, γ-butyrolactone, and ethylene carbonate were effective solvents for the denitrogenation (Table 1). Polar solvents such as methanol, ethylene carbonate, and dimethylformamide, which were highly selective for removing the model nitrogen compounds, did not exhibit similar selectivity in the extraction with the actual shale oil (Table 4). This was attributed to the differences between the model compounds system and the actual shale oil. As shown in Table 4, it was found that γ-butyrolactone is the most selective of the reported solvents in the whole-range Paraho shale. However, for middle distillate from partially hydrogenated shale oil, furfural was more selective, as shown in Table 5. Overall, furfural and γ-butyrolactone were significantly more effective in the denitrogenation than other screened solvents in both applications — partially hydrotreated and raw shale oil.

A distinct advantage of these solvents was the negligible solubility in the shale oils. Methanol and DMF were highly soluble in the unhydrogenated whole-range shale crude. A high intensity light or UV light was used to read the interface in the experiments. It was, in general, difficult to identify the interface in the raw shale oil crude experiments.

A typical isothermal L–L equilibrium relationship between the nitrogen compounds contained in the Paraho shale crude and the screened solvent is given in Figure 1. The nitrogen analyses were performed either by gas chromatography equipped with a chemiluminescence detector and an Antek® pyrolyzer or by a micro Kjeldahl method.

As reported in the literature, the selectivity of certain nitrogen compounds can be enhanced by the addition of transition metal salts.[19] One such system that has been applied to hydro-

**Table 4**
**EXTRACTION OF PARAHO SHALE OIL[a] NITROGEN**
**COMPOUNDS WITH POLAR SOLVENTS**

| Solvent | Yield (%) | Raffinate Denitrogenation (%) | Selectivity index |
|---|---|---|---|
| MeOH (5% $H_2O$) | 59 | 11 | <1 |
| Furfural (5% $H_2O$) | 82 | 40 | 2.2 |
| Sulfolane | 90 | 20 | 2 |
| γ-Butyrolactone | 92 | 25 | 3 |
| Ethylene carbonate | 95 | 8 | 1.6 |
| DMSO | 85 | 25 | 1.6 |

[a]   Raw Paraho shale oil crude containing 2.05 wt % nitrogen. Extraction temperature 40°C.

**Table 5**
**EXTRACTION OF PARTIALLY HYDROTREATED**
**PARAHO SHALE OIL WITH POLAR SOLVENTS AT 40°C**

| Solvent | Feed | Raffinate analysis Yield (%) | Denitrogenation (%) |
|---|---|---|---|
| Furfural | Full range[a] | 98 | 32 |
| γ-Butyrolactone | Full range[a] | 98 | 31 |
| TMSO | Full range[a] | 98 | 32 |
| Furfural | Middle distillate[b] | 95 | 70 |
| γ-Butyrolactone | Middle distillate[b] | 95 | 55 |

[a]   Total nitrogen content = 0.71 wt %.
[b]   Total nitrogen content = 0.20 wt %.

genated distillate from petroleum and coal liquid is furfural and ferric chloride hexahydrate. The apparent mechanism of this enhancement is the formation of a Werner-type complex between the iron and nitrogen atoms with hydrogen bonding between the water molecules of the complex and the solvent. In this process, the oil is extracted with solutions of furfural containing from 0.001 to 10 wt % ferric chloride hexahydrate.

The furfural-ferric chloride hexahydrate system was used in extractions of full range shale oil that contained 2.05% total nitrogen. The results are given in Table 6. These extractions were carried out successively at a 3:1 solvent/feed ratio. These results show that this system will remove 91% of the nitrogen in three extractions to a level below 0.2%. In each extraction with a fresh solvent mixture, the estimated recovery is about 70%. To increase the efficiency of the system, the amount of ferric chloride hexahydrate was increased to 40%. As shown in Table 7, 96% of the nitrogen was removed in two extractions, producing a raffinate with 0.1% total nitrogen at 40% recovery of the shale oil. This appears to be the lower level of nitrogen achievable with this system.

In view of these low recoveries, further experiments were performed on lightly hydrogenated shale oil that contained 0.71% nitrogen, of which 0.51% was basic. Table 8 shows the results from extracting this oil with γ-butyrolactone containing several metal salts. This

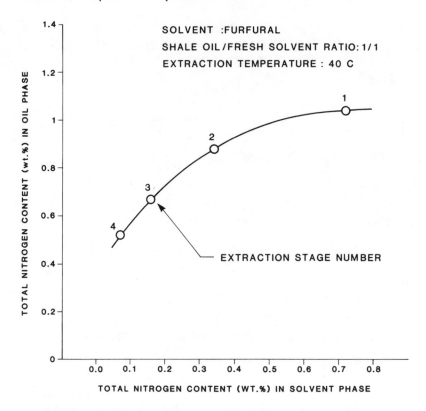

FIGURE 1.    Typical cross extraction L-L equilibrium relationship between nitrogen compounds in full boiling range Paraho shale oil and polar solvents.

### Table 6
### EXTRACTION OF FULL RANGE SHALE OIL WITH FURFURAL CONTAINING 2% FERRIC CHLORIDE HEXAHYDRATE

| Extraction no. | Raffinate nitrogen | | Yield | |
| :---: | :---: | :---: | :---: | :---: |
| | Content (%) | Extracted (%) | Per batch | Total |
| 1 | 0.88 | 57 | 68 | 68 |
| 2 | 0.56 | 73 | 66 | 45 |
| 3 | 0.17 | 91 | 78 | 35 |

*Note*:  Solvent-to-shale oil ratio = 3:1; temperature = 40°C; nitrogen content of original shale oil = 2.05%.

### Table 7
### EXTRACTION OF FULL RANGE SHALE OIL WITH FURFURAL CONTAINING 40% FERRIC CHLORIDE HEXAHYDRATE

| Extraction no. | Raffinate nitrogen | | Yield | |
| :---: | :---: | :---: | :---: | :---: |
| | Content (%) | Extracted (%) | Per batch | Total |
| 1 | 0.64 | 69 | 60 | 60 |
| 2 | 0.10 | 96 | 67 | 40 |

*Note*:  Solvent-to-shale oil ratio = 1:1; temperature = 40°C; nitrogen content of original shale oil = 2.05 wt %.

## Table 8
## SOLVENT EXTRACTION OF LIGHTLY
## HYDROGENATED FULL RANGE SHALE OIL WITH
## γ-BUTYROLACTONE CONTAINING 10% OF
## VARIOUS METAL SALTS

| Metal salts | Nitrogen content of raffinate (%) | | Recovery (%) |
|---|---|---|---|
| | Total | Basic | |
| None | 0.49 | | 98 |
| $FeCl_3 \cdot 6 H_2O$ | 0.15 | 0.073 | 89 |
| $Fe(NO_3)_3 \cdot 3 H_2O$ | 0.14 | 0.044 | 89 |
| $Fe(NO_3)_3 \cdot 9 H_2O$ | 0.13 | 0.043 | 88 |
| $Cu(No_3)_2 \cdot 3 H_2O$ | 0.14 | 0.052 | 89 |
| $CuCl_2 \cdot 2 H_2O$ | 0.16 | 0.071 | 89 |

*Note*: Total nitrogen content in feed = 0.71 wt %.

solvent was chosen because at the expected process condition, it is relatively inert, insoluble in the raffinate phase, and has a high selectivity for nitrogen removal. It was found that the hydrates of the chloride and nitrates are sufficiently soluble to be effective while the anhydrous salts and sulfates are not. Extraction with the solvent alone reduces the total nitrogen content to 0.49% (Table 8). Addition of the salts listed in Table 8 increases the efficiency of the solvent, yielding a product on extraction with 0.15% total nitrogen (0.073% basic nitrogen) with a 89% recovery. The enhancing ability of the various salts is essentially equal, showing little cationic, anionic, or hydration effects. Processes to recover the metal from the extract, such as treatment with caustic, extraction of the liberated nitrogen compounds, and then regeneration of the metal hydroxide with acids, may be feasible.

A simple multistage, countercurrent, bench-scale extraction was performed with furfural and hydrotreated shale oil, which contained 0.32% total nitrogen. The basic schematic of the continuous extraction is given in Figure 2. The continuous extraction contained about 12 mixing and settling zones. It produced raffinate containing about 400 ppm total nitrogen and 350 ppm basic nitrogen, with about 85% yield. The solvent-treated product had significantly improved color characteristics and stability. The results of the continuous extraction are presented in Table 9. It was found that the raffinate contained a negligible amount of the solvent. In a subsequent investigation, which is not reported here, a partially hydrogenated shale distillate was denitrogenated to lower than 10 ppm with an enhanced solvent system and had stability similar to the petroleum-derived counterpart. In view of these investigations, it appears that enhanced solvent extraction improves the shale-derived oil quality. By combining a mild hydrogenation step with enhanced solvent extraction, it will be possible to reduce hydrogen consumption compared to a straight hydrotreating process. A hybrid denitrogenation scheme for shale oil using an enhanced solvent system is illustrated in Figure 3.

## IV. CONCLUSIONS

In summary, the following conclusions are drawn from this experimental investigation. Polar solvents (e.g., furfural) removed nitrogenous compounds from mildly hydrotreated shale oil via continuous countercurrent extraction. Metal-enhanced solvent extraction for shale oil denitrogenation was found to be very effective. Regeneration of metals in the extract phase is complex and needs further development. Solvent extraction can remove both basic

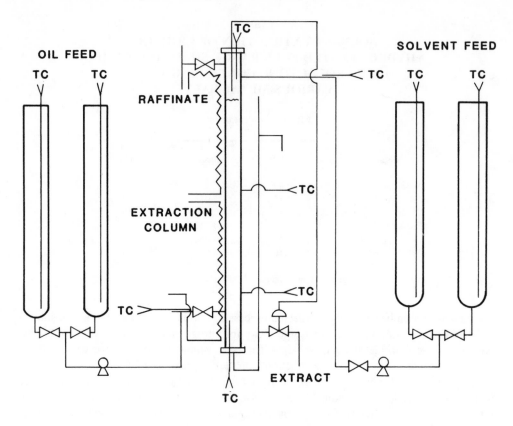

FIGURE 2.    Schematic flow diagram of multistage countercurrent, continuous extraction column.

## Table 9
### CONTINUOUS SOLVENT EXTRACTION OF PARTIALLY HYDROGENATED FULL RANGE SHALE OIL WITH FURFURAL

| Extraction time (min) | Raffinate | | |
| --- | --- | --- | --- |
| | Total N (ppm) | Basic N (ppm) | Yield (wt %) |
| 60 | 470 | 435 | 88 |
| 120 | 440 | 340 | 85 |

*Note*:  Total nitrogen content in feed = 3200 ppm; extract was analyzed by mass spectrometry (CEC 103) to reveal the various basic and nonbasic compounds discussed herein. Shale oil feed-to-solvent ratio = 1:1; extraction temperature = 40°C.

and nonbasic nitrogen compounds. Hence, it has the potential of greatly improving the stability of shale-derived oil products. Combined partial hydrogenation with solvent extraction merits consideration as an upgrading route for shale oil.

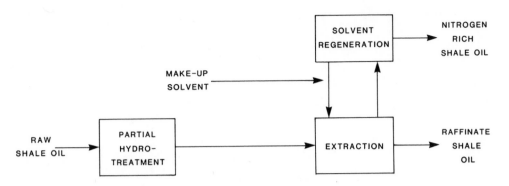

FIGURE 3.    Process block diagram for partial hydrogenation and solvent extraction for reduction of nitrogenous shale oil compounds.

# REFERENCES

1. **Frankenfeld, J. W. and Taylor, W. F.,** Fundamental Synthetic Fuel Stability Study, DOE/BC/10045-12, February 1980.
2. **Dellert, R. H.,** Two-Stage Catalytic Hydrogenation Process for Crude Shale Oil, U.S. Patent 3,481,867.
3. **Cullberson, S. F. and Rolinick, P. D.,** Shale oil likely prospect for refining, *Oil Gas J.,* 79, 43, 1981.
4. **Poulson, R. E., Frost, C. M., and Jensen, H. B.,** Characteristics of synthetic crude from crude shale oil produced by in situ combustion retorting, in *Advances in Chemistry Series,* No. 151, American Chemical Society, Washington, D.C., 1976.
5. **Holmes, S. A. and Latham, D. R.,** DOE/LETC/R.I.-80/12.
6. **Brown, D., Earnshaw, D. G., McDonald, F. R., and Jensen, H. B.,** *Anal. Chem.,* 42, 146, 1970.
7. **Stauffer, H. C. and Yanik, S. J.,** Shale oil: an acceptable refining syncrude, in ACS Natl. Meet., Miami, September 1978.
8. **Sullivan, R. F., Strangeland, B. E., Rudy, C. E., Green, D. C., and Framkin, H. A.,** Refining and Upgrading of Synfuels from Coal and Oil Shales by Advanced Catalytic Processes, DOE/HCP/T2315-25, July 1978.
9. **Gates, B. C., Katzer, J. R., and Schuit, G. C. A.,** *Chemistry of Catalytic Processes,* McGraw-Hill, New York, 1979.
10. **Tait, A. M. and Henseley, A. L.,** Jet fuels from shale oil by single stage hydrocracking, Tech. Rep. AFWAL-TR-81-2135, USAF, November 1981.
11. **Moor, H. J. and Tyler, A. L.,** The effect of an acidic catalyst on simultaneous HDS and HDN of heterocyclics sulfur and nitrogen, in AIChE Spring Meet., Houston, Tex., 1981.
12. **Lovell, P. F., Friback, M. G., Reif, H. E., and Schwedock, J. P.,** Maximize shale oil gasoline, *Hydrocarbon Process.,* 60, 125, 1981.
13. **Nowack, C. J., Delfosse, R. J., Speck, G., Solash, J., and Hazlett, R. N.,** Stability of oil shale derived jet fuel, in ACS Symp. Ser. No. 163, American Chemical Society, Washington, D.C., 1981.
14. **Reif, H. E., Schedock, J. P., and Schneider, A.,** Tech. Rep. AFWAL-TR-81-2135, USAF, November 1981.
15. **Stover, C. S.,** Methods for Reducing the Nitrogen Content of Shale Oil, U.S. Patent 4,271,009.
16. **Smith, R. H.,** Denitrogenation of Oils with Reduced Hydrogen Consumption, U.S. Patent 4,261,813.
17. **Jewell, D. M. and Snyder, R. E.,** Selective separation of non-basic nitrogen compounds from petroleum by anion exchange of ferric chloride complexes, *J. Chromatogr.,* 38, 351, 1968.
18. **Cronauer, D. C., Vogel, R. F., and Flinn, R. A.,** Denitrogenation of Shale Oil, U.S. Patent 4,238,320.
19. **Olzoeva, V. G.,** Removal of organonitrogen compounds from diesel distillates and fuels, *Khim. Tekhnol. Topl.,* 7, 8, 1978.

Chapter 4

# INTERFACIAL CHARGE AND MASS TRANSFER IN THE LIQUID-LIQUID EXTRACTION OF METALS

**K. L. Lin and K. Osseo-Asare**

## TABLE OF CONTENTS

# I. INTRODUCTION

The development of several new extraction reagents in the past 2 decades has made solvent extraction an increasingly active area for research. During hydrometallurgical solvent extraction, metal ions are extracted from an aqueous phase into an organic phase and vice versa. The liquid/liquid interface provides the environment within which the reacting species encounter one another. Therefore, a better understanding of interfacial properties should assist in elucidating the mechanisms of metal extraction processes.

The kinetics of metal extraction in liquid-liquid systems has been reviewed recently by Danesi and Chiarizia.[1] The extraction process was classified into three categories involving rate control by diffusion, chemical kinetics, and a combination of diffusion and chemical kinetics.[1] Rate models proposed for all of these types of rate processes often indicated a relationship between the metal flux and the concentrations of various interfacial species.

According to the Boltzmann equation, the interfacial concentration ($C_{int}$) of an aqueous species is proportional to the bulk concentration and is exponentially related to the electrical potential[2] as indicated by Equation 1:

$$C_{int} = C_o \exp(-ze\psi/kT) \tag{1}$$

where $C_o$ is the bulk concentration, z is the ionic charge including the sign, $\psi$ is the interfacial potential, k is the Boltzmann constant, and T is the absolute temperature.

The presence of the interfacial potential is a result of the unequal distribution of charged species at the two-phase boundary. Figure 1 presents a schematic illustration of the charge distribution at the liquid/liquid interface.[3] The surface potential, $\psi_o$, is the potential at the adsorbed layer of ionogenic surfactant species. The zeta potential, $\zeta$, is the potential at the plane of shear. In between the plane of shear and the adsorbed layer, there is another region, the Stern layer, in which reside specifically adsorbed counter-ions; the potential in this layer is called the Stern potential, $\psi_\delta$. The distance, $\delta$, of the Stern layer from the adsorbed layer is usually of the order of an ionic radius. The magnitude and the sign of $\psi_\delta$ and $\zeta$ depend on the degree of specific adsorption of the counter-ions, i.e., the number of counter-ions per unit area, $n_c$. The Stern and zeta potentials will decrease in magnitude and eventually change sign as $n_c$ increases.[3]

The interfacial charge of liquid-liquid dispersions has been studied for various systems, as will be discussed in the following sections. Nevertheless, the significance of interfacial electrostatic phenomena in relation to metal extraction is still not fully appreciated. De Ortiz[4] examined the work of Roddy et al.[5] and pointed out that the interfacial potential enhances the rate of iron and uranium extraction by HDEHP; a negatively charged interfacial film was suggested as a necessary condition for iron extraction. These same results have also been discussed by Cox and Flett.[6] Yagodin et al.[7] found that anionic surfactants (e.g., sodium lauryl sulfate) enhance the rate of metal extraction by LIX®64N (a proprietary benzophenone hydroxyoxime-based commercial extractant developed by Henkel Corp.), while cationic surfactants (e.g., cetyltrimethylammonium bromide) have the opposite effect. Similar observations have been reported by Miyake et al.[8] who found that copper extraction with 2-hydroxy-5-t-nonylacetophenone oxime (the active reagent in the Shell® reagent SME 529) was depressed by a cationic surfactant [$C_{12}H_{25}$-$N(CH_3)_3Br$], but was enhanced by an anionic surfactant ($C_{12}H_{25}OSO_3Na$). All of these authors attributed their findings to interfacial electrostatic effects.

It is well known that electrophoresis, which is the migration of charged particles in an electric field, provides a means of characterizing the interfacial electrification of colloidal systems.[9,10] For example, the electrophoretic mobility can be related to the zeta potential by the Smoluchowski equation,[9-14] although some corrections related to the relaxation effect

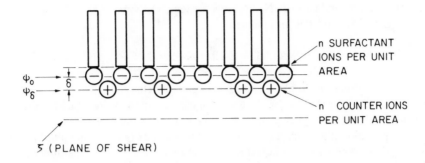

FIGURE 1.    The relative positions of surface potential ($\psi_o$), Stern potential ($\psi_\delta$), and zeta potential ($\zeta$). (After Davies, J. T. and Rideal, E. K., *Interfacial Phenomena*, 2nd ed., Academic Press, New York, 1963, chap. 2.)

— the deformation of the applied field, the liquid conductivity, and viscosity — may be necessary.[15,16] The surface pH, which might affect ion exchange at the interface, can also be described in terms of the zeta potential and the bulk pH.[17-20]

The present work provides a review of the electrophoresis studies which have been conducted in organic/aqueous systems. Additional results, based on the authors' electrokinetic and mass transfer studies with the solvent extraction reagents, LIX®63, HDNNS, and Aliquat® 336, are also presented. It is suggested that a preliminary step in the metal extraction process may be the electrostatic interaction between charged organic/aqueous interfaces and metal ions.

## II. MEASUREMENT OF ELECTROPHORETIC MOBILITY (EM)

Electrophoresis measures the migration of charged particles under the influence of an external electrical field.[9,10] The measured velocity per unit electric field strength is called the electrophoretic mobility (EM). The Smoluchowski equation relates this velocity with the zeta potential, $\zeta$:[9-14]

$$\frac{v}{E} = EM = \frac{\epsilon\zeta}{4\pi\eta} \tag{2}$$

where E is the electrical field, v is the particle velocity, and $\epsilon$ and $\eta$ are, respectively, the dielectric constant and the viscosity of the continuous phase.

Two common methods are available for measuring the EM of organic/aqueous systems: the macroscopic moving boundary and the microelectrophoresis techniques. The results obtained from these two methods have been declared to be in agreement with each other.[21] The macroscopic moving boundary method, initially described by Price and Lewis,[22] has been applied for measuring the EM of: paraffin wax and wax covered with aliphatic acids, amides, and alcohols;[23] various hydrocarbons and their hydroxyl, halide, and acidic derivatives; the effect of sugar additives and temperature on EM;[21,23-25] and polystyrene lattices in various ionic strength media.[26]

The microelectrophoresis method has been employed for the study of: the influence of alkalis and surface-active substances on the EM of Nujol drops;[27] the influence of octyl amine and SDS (sodium dodecyl sulfate) on the EM of Nujol drops;[17]*n*-decane in SDS solution;[28] Nujol in solutions of sulfosuccinic esters;[29] Decalin® droplets with and without ethyl alcohol and electrolytes;[30,31] Nujol in solutions of oleic acid, SDS and DDC (dodecylamine chloride) systems, and CTAB (cetyltrimethylammonium bromide);[32] the LIX®64N (a mixture of LIX®63 and LIX®65 N)-kerosene system;[33] LIX®63, LIX®64N, and LIX®65N in octane;[34] LIX®65N and P50 in toluene.[20]

## III. ORIGIN OF THE NEGATIVE CHARGE ON ORGANIC DROPLETS

It is well known that oil droplets tend to be negatively charged when dispersed in water. A common trend of the pH effect on the EM is an increase in the value of EM in the negative sign direction with increasing pH. In the case of nonpolar organic drops, this trend is believed to be due to the preferential adsorption of hydroxyl ions at the organic/water interface.[24,26,35,36] The hydroxyl ion adsorption is believed to occur in two ways:[26] one is the direct adsorption of the ions from the aqueous phase; the other is an initial adsorption of unimolecular water followed by a subsequent ionization with hydrogen ions migrating into the aqueous phase and leaving hydroxyl ions adsorbed at the interface. Parreira et al.[35] also proposed that the positive residual charge of C–H hydrogen adsorbs hydroxyl ions. In some cases, constant EM values attributable to surface adsorption saturation have also been observed.[18,21,24,26,32,35]

In the case of polar organic droplets, the negative charge on the oil drop can be attributed to the preferential distribution of the hydroxyl ion into the organic phase. The preferential distribution of the hydroxyl ion can, in turn, be attributed to its greater ionic radius compared with that of the hydrogen ion, and the generally smaller dielectric constant of the organic phase relative to that of the aqueous phase (many organic liquids have dielectric constants between 2 and 10, more polar liquids have values between 20 and 40, while water has a dielectric constant of ~80 at room temperature).[37] A general analysis of the distribution of charged species between two immiscible phases[37] can be used to show how the above factors give rise to the negative charge on polar organic droplets.[38]

The electrochemical potential, $\bar{\mu}_{ij}$, of a species i in a given phase j can be expressed by

$$\bar{\mu}_{ij} = \mu_{ij} + z_i e \phi_j \tag{3}$$

where $\bar{\mu}_{ij}$ is the chemical potential of species i in phase j, $z_i$ is the charge (including the sign) on species i, e is the electronic charge, and $\phi$ is the bulk phase potential. Assuming an ideal solution where the activity coefficient is unity, the chemical potential is given by

$$\mu_{ij} = \mu_{ij}^o + kT \ell n C_{ij} \tag{4}$$

where $\mu_{ij}$ and $C_{ij}$ are the standard chemical potential and the concentration of species i in bulk phase j, respectively.

When two phases (j = 1,2) are at equilibrium,

$$\bar{\mu}_{i1} = \bar{\mu}_{i2} \tag{5}$$

Hence, for the equilibrium distribution of a 1-1 electrolyte,

$$\mu_{+1} + e\phi_1 = \mu_{+2} + e\phi_2 \tag{6}$$

$$\mu_{-1} - e\phi_1 = \mu_{-2} - e\phi_2 \tag{7}$$

It follows from Equation 4 that

$$\mu_{+2} - \mu_{+1} = (\mu_{+2}^o - \mu_{+1}^o) - kT \ell n (C_{+1}/C_{+2}) \tag{8}$$

$$\mu_{-2} - \mu_{-1} = (\mu_{-2}^o - \mu_{-1}^o) - kT \ell n (C_{-1}/C_{-2}) \tag{9}$$

By combining Equations 6, 7, 8, and 9 with the Born equation

$$\mu_i = - \frac{e^2}{2r_i} \left( 1 - \frac{1}{\epsilon} \right)$$

(10)

where $r_i$ is the ionic radius and $\epsilon$ is the dielectric constant, the bulk phase potential difference can be obtained as[38]

$$\phi_1 - \phi_2 = \frac{e}{4} \left( \frac{r_- - r_+}{r_- r_+} \right) \left( \frac{\epsilon_1 - \epsilon_2}{\epsilon_1 \epsilon_2} \right)$$

(11)

where $r_-$ and $r_+$ are, respectively, the anion and cation radius ($OH^-$ and $H^+$ for the aqueous phase) and 1 and 2 stand for the organic phase and aqueous phase, respectively.

Equation 11 implies that the value of ($\phi_1 - \phi_2$) is negative if $r_- > r_+$ and $\epsilon_1 < \epsilon_2$. Accordingly, since the hydroxyl ion (radius = 1.40 Å)[39,40] is larger than that of the proton, and the dielectric constant of organic phases tends to be lower than that of the aqueous phase, it follows that a polar oil droplet surrounded by an aqueous phase would be negatively charged. This is in agreement with the general observation that increase in pH tends to enhance the magnitude of the negative electrophoretic mobility of oil droplets.

In the extreme case of nonpolar organic droplets, the analysis presented above for polar organic droplets is no longer valid, as there is negligible solubility of ionic species in nonpolar organic phases.[3] However, the dielectric constant of the interfacial region would be expected to have an intermediate value between that of the bulk organic and aqueous phases.[11,15] Accordingly, based on Equation 11, the hydroxyl ion would be preferentially soluble in the interfacial region. Thus, the negative charge on pure nonpolar organic droplets dispersed in water can be attributed to the selective adsorption of hydroxyl ions at the organic/water interface.

## IV. VARIATION OF EM DUE TO THE NATURE OF THE ORGANIC PHASE

### A. The Molecular Structure of the Organic Solvent: The Effects of C–H and C=C Bonds

The electrophoretic mobility is affected by the nature of the organic solvent as well as by the nature and concentration of dissolved organic phase species. The straight chain hydrocarbons, dodecane, octadecane, and paraffin wax, do not show much difference in EM in a 0.01-kmol m$^{-3}$ NaOH aqueous medium, as shown in Figure 2.[25] This indicates that the organic/water interfaces of these systems have the same affinity for hydroxyl ions. The similarity in the EM of straight chain hydrocarbons is probably due to the presence of the same end group ($-CH_3$) which is adjacent to the aqueous phase.

Douglas[25] reported the following order of increasing absolute EM at pH 9: straight chain paraffin ~ cycloparaffin (cyclohexane and Decalin®) < aromatic hydrocarbon (benzene) < olefin ($\Delta^{1,2}$ octadecene). The EM varied from 0.96 units ($\mu$m sec$^{-1}$/V cm$^{-1}$) for paraffin wax to 1.99 units for $\Delta^{1,2}$ octadecene. The higher EM of aromatic and olefin hydrocarbons has been attributed to the influence of the chain-end double bonds. According to Carruthers,[24] the presence of a double bond at the organic/water interface increases the primary adsorption of hydroxyl ions. However, the author did not give any theoretical explanation for this effect, aside from pointing out the possible importance of electrostatic forces.

The available data on bond moments provide some insight into the influence of double bonds and aromatic substances on the EM. The attachment of an aromatic ring on a $-CH_3$ group tends to induce a dipole moment with the negative direction away from the $-CH_3$ group.[42-44] The induced moment has a value of $+0.45$ compared to $+0.2$ of the H–C link.[44]

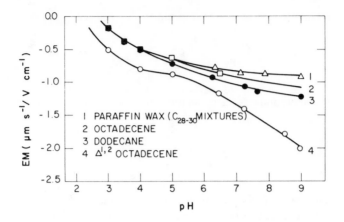

FIGURE 2.    EM-pH curves for hydrocarbons in the presence of 0.01 kmol m⁻³ sodium ions at 298 K. (From Douglas, H. W., *Trans. Faraday Soc.*, 39, 305, 1943. With permission.)

The effect of the aromatic ring on the bond moment has been described as an ''electromeric effect'' induced by the presence of the carbon-carbon double bond which tends to attract electrons through the orbitals.[42] This suggests that a hydrogen atom attached to a double bond would have a higher positive residual charge than an aliphatic hydrogen atom and, hence, the aromatic hydrogen would cause a greater adsorption of hydroxyl ions.

An EM study of various amounts of hexane, cyclohexane, benzene, and Decalin® in dodecane at pH 9 and 25°C indicated that variations in the concentrations of the various additives did not change the EM of dodecane solutions, except for benzene which gave a little increase in the EM of dodecane above 50 mol %.[30]

## B. The Molecular Structure of the Organic Solvent: The Effects of Inorganic Groups

The presence of certain inorganic groups in the molecular structure of hydrocarbon solvents is found to result in decreased EM. Thus, Douglas[21] reported lower absolute EM values for alcohols than for paraffin hydrocarbons, as shown in Figure 3. Similar results have been obtained by Carruthers.[24] Halide derivatives also give less negative EM values than the corresponding hydrocarbons.[23,24,26] In a study by Alty and Johnson,[23] it was found that the absolute value of the EM of constant chain length hydrocarbons decreased in the order (absolute value) $-CH_3 > -CONH_2 > -COOH > -CH_2OH$ (Table 1). It is likely that the hydroxyl and halide groups affect EM through the presence of negative residual charges which prevent aqueous hydroxyl ions from being adsorbed at the interface. The EM of the $-COOH$ series was found to become less negative with increasing hydrocarbon chain length, e.g., the absolute EM of carboxylic acids decreased in the following order:[23] lauric acid $(C_{12})$ > myristic acid $(C_{14})$ > palmitic acid $(C_{16})$. This decrease in EM with increasing chain length was ascribed to the decreasing interaction between the $-COOH$ and the aqueous solution, as the longer chain makes it more difficult for the $-COOH$ group to reach the aqueous phase.[23] The presence of a double bond near the middle of the chain length increased the absolute EM greatly, e.g., *cis*-euric acid $[C_8H_{15}CH=CH(CH_2)_{11}COOH]$ had a greater EM than lauric acid $(C_{11}H_{23}COOH)$.[23]

## C. The Effects of Organic-Phase Surface-Active Agents

The effect of surfactant concentration ratio on EM has been investigated by Despotović et al.[45] for the systems (1) [*n*-dodecylamine nitrate (LAN) + sodium-*n*-dodecyl sulfate (NaLS)], (2) [LAN + Triton® X 305 (TX305)], and (3) [NaLS + TX305]. The association

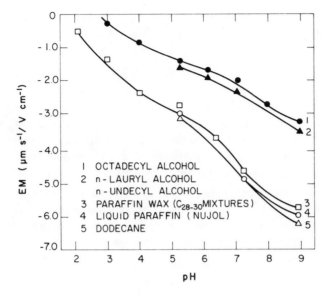

FIGURE 3. EM-pH curves for paraffins and *n*-alcohols at 298 K. (From Douglas, H. W., *Trans. Faraday Soc.*, 46, 1082, 1950. With permission.)

### Table 1
### EFFECT OF FUNCTIONAL GROUP ON EM OF HYDROCARBONS IN PURE WATER[23]

| Substance | EM ($\mu$m sec$^{-1}$/V cm$^{-1}$) $\times$ 10 |
| --- | --- |
| Paraffin wax (60—68°C) | $-19.60$ |
| Palmitamide ($C_{15}H_{31}CONH_2$) | $-16.87$ |
| Palmitic acid ($C_{15}H_{31}COOH$) | $-12.45$ |
| Cetyl alcohol ($C_{16}H_{33}OH$) | $-12.00$ |

of the ionic surfactants affects the magnitude and sign of the EM. Sodium dodecyl sulfate (NaLS) tends to enhance the negative value of the EM, while LAN increases the EM in the opposite direction. The nonionic surfactant TX305 affects the EM by excluding NaLS or LAN from the interface.

The recent commercial advances in the development of metal solvent extraction have given rise to some interest in examining the electrophoresis of solvent extraction systems, for the electrophoretic mobility of an extractant-containing organic drop should give some indication of the surface activity of the extractant. The EM of the oxime reagents LIX®64N in kerosene,[33] LIX®65N and P50 in toluene,[20] and LIX®65, LIX®64N, and LIX®63 in acetone[34] have been investigated. Interfacial tension data[34,46,47] indicate that these reagents adsorb at the organic/water interface. Thus, the observation[20,33,34] that the EM values of these systems do not deviate very much from the EM of their solvents suggests that the extractants are adsorbed primarily as their undissociated molecules.

## V. VARIATION OF EM DUE TO THE NATURE OF THE AQUEOUS PHASE

Aqueous phase factors which affect EM include the aqueous pH, the electrolyte concentration, the types of electrolyte, and the surface-active additives.

The EM is primarily controlled by the double layer thickness, $1/\kappa$, and the interfacial charge density (i.e., specific adsorption effects). The value of $\kappa$ is given by

$$\kappa^2 = \frac{4\pi e^2 \sum_i C_i z_i^2}{\epsilon kT} \tag{12}$$

where $C_i$ is the bulk phase concentration of aqueous species i, $z_i$ is the valence of species i, $\epsilon$ is the dielectric constant of the medium, k is the Boltzmann constant, and T is the absolute temperature. It is clear from Equation 12 that the double layer thickness, $1/\kappa$, decreases with increasing ionic strength, given by $1/2\sum_i C_i z_i^2$. The electrophoretic mobility is proportional to the zeta potential and, hence, to the surface potential. By assuming that the surface potential is simply due to an electrical field and that a charged ion in the aqueous medium can only approach a central charged particle to a closest distance a, one can obtain an expression for the potential distribution from Poisson's equation

$$\nabla^2 \psi = -\frac{4\pi\sigma}{\epsilon} \tag{13}$$

where $\sigma$ is the charge density of the central particle. Taking the charge density $\sigma'$ at any point as

$$\sigma' = -2C_i ze \, \sinh(ze\psi/kT) \tag{14}$$

the solution to Equation 13 is[48]

$$\psi(r) = \frac{ze}{r\epsilon} \exp(-\kappa r) \tag{15}$$

At the closest distance a, r = a. It is clear from Equation 15 that $\psi$ decreases with decreasing double-layer thickness. Subsequently, the absolute EM decreases with increasing species concentration $C_i$.

Thus, in surfactant-containing aqueous systems, the electrolyte concentration directly and inversely affects the interfacial charge density and the double-layer thickness, respectively. At low surfactant concentrations, the increase in charge density predominates, whereas the effect on double-layer thickness is predominant at high surfactant concentrations.[29]

The addition of a buffer to the Nujol/water system containing surface-active reagents, e.g., Aerosol O.T., Aerosol M.A., Aerosol A.Y., etc., was found[29] to enhance the charge density at low surfactant concentrations, giving rise to a sharp increase in EM. On the other hand, at high surfactant concentration, as the surface became saturated with long chain ions, the EM was affected through the compression of the double-layer thickness. For instance, surfactant solutions made up in 0.05 kmol m$^{-3}$ acetate buffer solution had a lower EM than unbuffered solutions at high surfactant concentrations. An EM concentration curve similar to that described above was observed for *n*-decane dispersions in sodium dodecyl sulfate (SDS) solution.[28] As shown in Figure 4, NaCl concentration was found to depress the EM; this effect may be attributed to the shrinkage of the double-layer thickness.

SDS and octadecylamine enhance the EM of Nujol drops at pH 1.65 in the negative and positive direction, respectively.[19] This was ascribed to the increase in surface charge density, as both of these two surfactants expose ionic polar head groups (sulfate and ammonium, respectively) to the aqueous phase. Another behavior of Nujol in SDS-water solution has

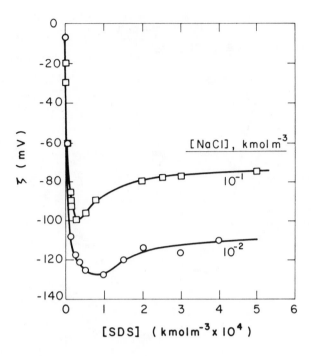

FIGURE 4. Effect of SDS on the zeta potential of *n*-decane dispersions. (From Anderson, P. J., *Trans. Faraday Soc.*, 55, 1421, 1959. With permission.)

been reported. In this case, a shallow trough occurs as SDS concentration increases.[27] The initial sharp increase in EM was said to be due to the adsorption of single long chain ions. A subsequent small decrease was suggested as caused by secondary adsorption of sodium ions. The rising portion of the shallow trough was explained[27] by proposing that an increasing proportion of sodium ions associates with the ionic micelle, as this EM behavior was found to occur around the critical micelle concentration (CMC) of the studied surfactants, including SDS, sodium laurate, and sodium oleate. How the secondary adsorption of sodium ions would occur was, however, not explained.[27] An alternative interpretation of the reported phenomenon might be in terms of the double-layer compression effect discussed above. The aggregation of SDS after its CMC decreases the total number of discrete ionic particles and, subsequently, decreases the ionic strength. This would result in double layer expansion and, hence, an increase in the EM of the oil drops.

The interfacial behavior of inorganic electrolytes has been reported for various systems. Positive adsorption of $I^-$ and $SCN^-$ anions and negative adsorption of $Cl^-$ ions at the cetyl acetate/water interface at pH 4 were observed through interfacial tension measurements.[17] As shown in Figure 5 the EM of the $I^-$-containing system is much greater than that of the $Cl^-$-containing system. A maximum in the absolute value of the EM of the $I^-$- and $Cl^-$-containing systems was observed and indicates that double-layer compression occurs as the concentration of these two species increases. Only a tiny hump, which is not visible in the curve because of its small value, was reported for the $SCN^-$-containing system when the $SCN^-$ concentration was increased in the low concentration range. The absolute EM values of this latter system keep increasing with increasing $SCN^-$ concentration. The difference in the adsorption behaviors of these ions is probably due to the difference in their ionic size. The adsorption of large ions at the oil/water interface tends to lower the free energy.[48] It has been reported that anions are adsorbed at the mercury/water interface in the order [49]

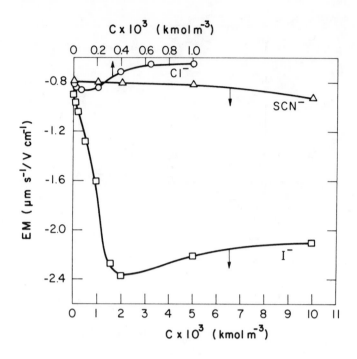

FIGURE 5.    EM of cetylacetate droplets in the presence of potassium salts at pH 4, 298 K. (From Dickinson, W., *Trans. Faraday Soc.*, 40, 48, 1944. With permission.)

$$S^{2-} > I^- > SCN^- > Br^- > Cl^- > OH^- > F^-$$

The above sequence is in very good agreement with the EM results as shown in Figure 5.

The presence of a maximum in the EM-electrolyte concentration curve has been reported for most of the electrolytes that have been studied. Powney and Wood[27] proposed that the decrease of EM after the sharp rise at low concentration is a result of the neutralization effect of sodium ion. These authors did not apply the concept of double-layer compression to their observations for NaOH, $Na_2SiO_3$, $Na_2CO_3$, $Na_4P_2O_7$, $NaHCO_3$, and NaCl. However, Taylor et al.,[31] in explaining their observations for the electrolytes NaOH, $Na_2CO_3$, $Na_4[Fe(CN)_6]$, NaF, NaCl, and HCl (Figure 6), employed both the concepts of surface charge density and double-layer compression. They suggested that the competition between these two factors controls the features of the EM concentration curves. The intrusion of positive ions at high ionic strength when the double layer becomes thinner was also suggested as a possible reason for the development of the maximum.

The presence of sugars (sucrose, lactose, and dextrose) in aqueous solution lowers the absolute EM of organic/aqueous systems.[21] It is likely that the adsorption of the nonionic sugar molecules at the interface displaces adsorbed anions and, thus, leads to a less negative EM. Metal ions dissolved in the aqueous solution can also modify the EM of droplets. The gradual formation of metal-hydroxo complexes with increasing pH can even bring the EM to the positive side in certain pH regions.[34]

## VI. INTERFACIAL CHARGE AND THE RATE OF METAL EXTRACTION

As mentioned above, the presence of a negative interfacial potential has been reported to enhance the rate of iron and uranium extraction with HDEHP.[4] The significance of interfacial

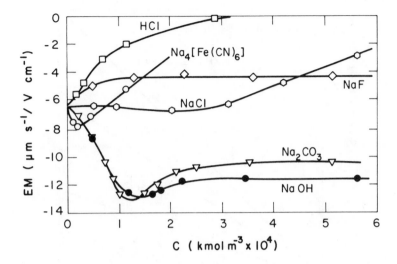

FIGURE 6. EM of Decalin® droplets in various electrolyte solutions. (From Taylor, A. J. and Wood, F. W., *Trans. Faraday Soc.*, 53, 523, 1957. With permission.)

concentration in reaction kinetics has also been mentioned by others.[6,50] A positively charged aqueous species would be attracted by a negatively charged organic/aqueous interface, whereas the opposite behavior would be observed for a positively charged interface. Thus, one can expect an enhancement or a decrease in extraction rate with increasing interfacial charge, depending on the signs of the interfacial charge and the charge on the extractable species.[50]

Assuming an interfacial reaction is first order with respect to the reacting species,

$$\frac{dC}{dt} = -k'C_{int} \tag{16}$$

where $k'$ is the rate constant and $C_{int}$ is given by Equation 1, the relationship between the reaction rate at the charged and the neutral interface can be given as

$$\frac{rate(\psi)}{rate(o)} = \exp(-ze\psi/kT) \tag{17}$$

A rate vs. potential curve for different valence species, Figure 7, determined according to Equation 17, clearly indicates that the presence of interfacial charge enhances the reaction rate of an oppositely charged aqueous species.

The presence of an interfacial potential can also influence reaction rates by its effect on the migration flux of aqueous species to and from the organic droplets. A general expression for the flux of a charged species is composed of three terms involving molecular diffusion, ionic migration, and convection:[51-53]

$$J_i = -D_i\nabla C_i - z_iu_iC_iF\nabla\psi + C_iv \tag{18}$$

where the subscript i stands for species i, D is the diffusion coefficient, C is the concentration, u is the ionic mobility, F is the Faraday constant, v is fluid velocity, and $\psi$ is the electrical potential. The electrical potential may arise from a variety of sources. For example, it may

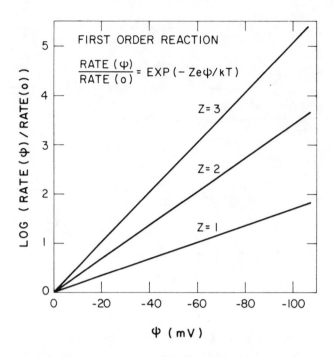

FIGURE 7.    Effect of the double-layer potential on the rate of a first-order interfacial reaction.

be the result of charge separation due to differences in the ionic mobilities; this is the diffusion potential. The electrical potential may also be the result of preferential adsorption (see Figure 1) or an externally applied electrical field.

Through an investigation of the diffusion potential, Tunison and Chapman[53] have suggested that fundamental investigations of hydrometallurgical solvent extraction rate processes may need to consider this potential term, although in practice it is unlikely to affect the overall flux by more than 10%. In the case of adsorption systems, the potential term is mainly contributed by the charged interface. The field strength right at the liquid/liquid interface is given by[51,54,55]

$$\nabla\psi = -\left(\frac{32\pi CRT}{\epsilon}\right)^{1/2} \sinh\frac{zF\psi}{2RT} \qquad (19)$$

Approximate calculations indicate that right at the liquid/liquid interface (in the absence of significant convection) the second term of Equation 18 contributes the major part to the total flux in the charged interface. For example, one obtains a value of $8.7 \times 10^2$ mol cm$^{-2}$ sec$^{-1}$ for the migration flux compared with a value of $2.5 \times 10^{-10}$ mol cm$^{-2}$ sec$^{-1}$ for the diffusion flux when the diffusivity is taken as $D = 5 \times 10^{-6}$ cm$^2$ sec$^{-1}$ for a charged species in the aqueous phase,[56] the diffusion film thickness is taken as 0.1 cm,[56] the concentration gradient is taken as 5 kmol m$^{-4}$, a charged interface with a surface potential of $-10$ mV is assumed, and the mobility of divalent ions in the aqueous phase is assumed to equal $6 \times 10^{-8}$ m$^2$ sec$^{-1}$ V$^{-1}$.[57]

From Equation 18 the ratio between the migration flux $J_m$ at two different potentials can be obtained as

$$\frac{J_m(\psi_1)}{J_m(\psi_2)} = \frac{\nabla\psi_1}{\nabla\psi_2} \qquad (20)$$

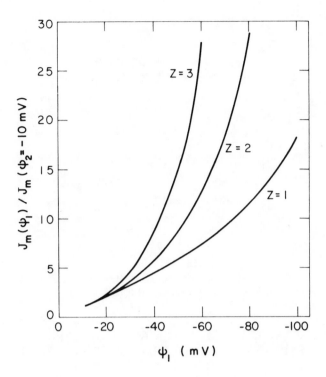

FIGURE 8.    Effect of a charged interface on the migration flux, taking $\psi_2 = -10$ mV.

assuming the concentration $C_i$ in Equation 18 is constant. Since the field strength is given by Equation 19, the ratio given in Equation 20 becomes

$$\frac{J_m(\psi_1)}{J_m(\psi_2)} = \frac{\sinh \dfrac{zF\psi_1}{2RT}}{\sinh \dfrac{zF\psi_2}{2RT}} \tag{21}$$

By taking $\psi_2 = -10$ mV and T = 298 K, the curves of this ratio for varying $\psi_1$ and z are obtained as shown in Figure 8. This plot indicates that the migration flux $J_m$ ($\psi_1$) increases with increase in the absolute value of the potential $\psi_1$ and the charge on the migrating ionic species.

Another role of interfacial potential on mass transfer in liquid-liquid extraction is in its effect on drop size. Bailes[58] indicated in a study of nickel extraction by HDEHP and of copper extraction by LIX®64N that the application of an electrostatic field enhances metal extraction significantly. This enhancement in extraction rate was ascribed to the production of daughter droplets through an electrostatic effect. This generation of smaller droplets provides more interfacial area for reactant contact. In traditional solvent extraction operations, mechanical stirring is used to produce the necessary large interfacial area. From the results of Bailes and the DLVO theory,[70] it can be seen that the presence of interfacial charge at the interface will assist in the formation of smaller droplets during dispersion and, hence, will increase interfacial area, thereby enhancing extraction rate.

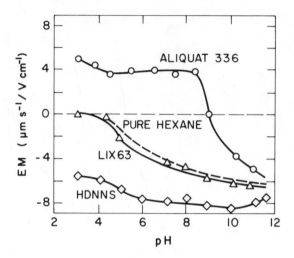

FIGURE 9.    EM-pH curves for hexane droplets in the presence and absence of extractants ([LIX®63] = [ALIQUAT®] = 1.5 × 10⁻⁴ kmol m⁻³; [HDNNS] = 1.4 × 10⁻⁴ kmol m⁻³).[66]

## VII. INTERFACIAL CHARGE AND MASS TRANSFER IN THE LIX®63-HDNNS MIXED EXTRACTANT SYSTEM

LIX®-63-HDNNS mixtures provide a synergistic system for nickel, cobalt, and copper extraction at low pH values.[46,59-62] The extremely high interfacial activity of HDNNS has been demonstrated by interfacial tension studies.[46] This pronounced interfacial activity had led to the suggestion of phase transfer catalysis (PTC) as an important mechanism for metal extraction in the mixed LIX®63-HDNNS system.[62] In this particular system, the PTC process is possible only if a positively charged metal ion can be captured by an appropriate surface-active extractant. Under these circumstances, the existence of a negative interfacial charge will certainly be an advantage. It was, therefore, decided to examine the characteristics of the interfacial electrification of the LIX®63-HDNNS system.

In this work the EM was measured with the Riddick Zeta Meter (Zeter Meter Inc., New York, N.Y.) by using a molybdenum anode and a platinum cathode. A 0.35-cm³ hexane solution of the extractants was dispersed with the aid of a mechanical wrist-action shaker in 500 cm³ aqueous solution of the desired pH. The ionic strength of the aqueous phase was 4.0 × 10⁻³ kmol m⁻³ controlled with KNO₃; doubly distilled water was used. Commercial LIX®63, supplied by Henkel Corp., was purified by the copper oximate method.[63] HDNNS from King Industries was purified by an ion exchange method.[64] Aliquat® 336 was supplied by Henkel Corp., and its concentration was obtained by the Volhard titration method.[65] All the electrophoretic mobility measurements were conducted at room temperature (~25°C).

Within the pH range investigated, the hexane/water interface was found to be negatively charged; the EM-pH curve presented in Figure 9 shows the same trend as that observed by other workers in the different systems discussed above. LIX®63 does not seem to effect significant deviations in the EM relative to the pure hexane droplet. A similar EM behavior was reported for this hydroxyoxime in toluene,[20] kerosene,[33] and octane[34] systems.

It can be seen from Figure 9 that HDNNS has a marked effect on the EM, enhancing the negative interfacial charge, even at pH 3. Extrapolation of this EM-pH curve suggests that the organic drops maintain their negative charge at even lower pH values. The high surface activity[46] and strong acidity of HDNNS give rise to the observed results. The large negative EM is caused by the deprotonation of HDNNS, followed by the adsorption of the anionic

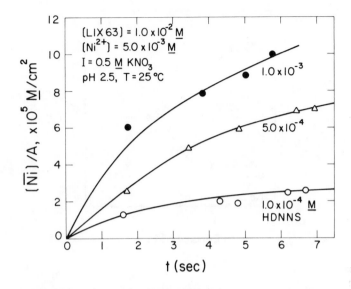

FIGURE 10. The effect of HDNNS on the rate of nickel extraction by LIX®63-HDNNS mixtures. A = droplet surface area; [Ni] = organic phase nickel concentration.[66]

sulfonate groups. LIX®63 is known to interact with HDNNS through hydrogen bonding.[46] This interaction, which decreases the adsorption of HDNNS, also results in a decrease in the absolute value of the EM of the LIX®63-HDNNS-hexane/water system.[66]

A study of nickel extraction in the LIX®63-HDNNS system with the rising-drop technique[66] shows that an enhancement in extraction rate occurs with increasing HDNNS concentration, as illustrated in Figure 10. In view of the electrophoresis results presented in Figure 9 and the rate vs. potential curve, Figure 7, it is reasonable to suggest[66] that the electrostatic interaction between metal ions and the negative interfacial charge is the first step in the metal extraction process.

The importance of the negatively charged interface in the extraction of cations is further demonstrated by the effects of basic extractants on the extraction behavior of anionic extractants. Aliquat® 336 (tricaprylylmethylammonium chloride) is an anion exchanger used for metal extraction.[67,68] Osseo-Asare and Keeney[62] observed a poor extraction of nickel in a mixed extractant system consisting of HDNNS and Aliquat® 336, Figure 11. As can be seen from Figure 9, the electrophoresis results indicate a positive interfacial charge for the Aliquat® 336-containing hexane drops. This positive charge presented to positively charged nickel ions would be expected to inhibit metal extraction. The concentration of $R_4N^+$ groups at the organic/aqueous interface has also been reported for the *n*-decane/water system, where a free energy change of $-695$ cal/mol was found for this adsorption.[69]

## VIII. SUMMARY AND CONCLUSIONS

The electrophoretic mobility behavior of organic/aqueous systems has been reviewed. The factors affecting EM arise from both the organic phase and the aqueous phase properties.

Aqueous dispersions of organic drops are generally negatively charged. An increase in the aqueous pH value tends to increase the EM in the negative direction through hydroxyl ion adsorption, although surface saturation of hydroxyl ion adsorption appears to occur for some cases. The negative charge on polar organic droplets has been explained thermodynamically with the aid of the Born equation. As a result of the lower dielectric constant of

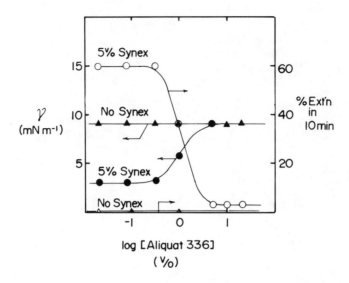

FIGURE 11.   Comparison of interfacial tension and nickel extraction at 298 K for HDNNS-Aliquat® 336 systems. Synex 1040 (King Industries): 40% HDNNS in heptane.[66]

the organic phase compared with the aqueous phase, large ions have a greater affinity for the organic phase ($r_{OH^-} > r_{H^+}$). In the case of nonpolar organic droplets, the negative surface charge is explained in terms of the desorption of proton away from the interface.

The absolute EM of hydrocarbons decreases in the following order: end chain double bond hydrocarbon > aromatic > straight chain paraffin ~ cycloparaffin. Chain length does not significantly affect the EM of these hydrocarbons. Paraffin derivatives which have a negative residual charge tend to have a smaller EM. The end group of constant chain length hydrocarbons decreases the absolute EM in the following order: $-CH_3 > -CONH_2 > -COOH > -CH_2OH$, due to the decreasing magnitude of the negative residual charge in this order: $-CH_2OH > -COOH > -CONH_2 > -CH_3$. The absolute EM of $-COOH$ molecules increases as the hydrocarbon chain length decreases; apparently, the longer chain reduces the ability of the $-COOH$ group to reach the aqueous side of the interface.

Addition of electrolytes affects the EM through the surface adsorption of anions and double-layer thickness contraction. The presence of surfactants in the organic or the aqueous phase affects the EM of oil drops through interfacial adsorption of the surfactant molecules at the interface with their hydrophilic groups extending into the aqueous phase. The dissociation of these groups gives rise to charged interfaces. The hydroxyoxime metal extractants LIX®63, LIX®64N, and LIX®65N do not significantly affect the EM of octane, toluene, hexane, and kerosene droplets, indicating that the compounds are adsorbed primarily as neutral species. On the other hand, the anionic extractant, HDNNS, and the cationic extractant, Aliquat® 336, impart strong negative and positive charges, respectively, to the organic drops.

The presence of a charged organic/aqueous interface alters the interfacial concentration of extractable aqueous species. Where there is an enhanced interfacial concentration (i.e., aqueous species and interface oppositely charged, e.g., the $Ni^{2+}$/HDNNS-LIX®63 system) an increase in extraction rate is to be expected. On the other hand, in the presence of an interface with the same sign as the extractable species (e.g., $Ni^{2+}$/HDNNS-Aliquat® 336), reaction inhibition is to be expected.

## ACKNOWLEDGMENT

Support of the National Science Foundation under Grant No. CPE8110756 is gratefully acknowledged.

## NOMENCLATURE

| | |
|---|---|
| $a$ | Drop radius |
| $C_{int}$ | Interfacial concentration |
| $C_o$ | Bulk phase concentration |
| $C_{ij}$ | Concentration of species i in phase j |
| $D$ | Diffusion coefficient |
| $e$ | Electronic charge |
| $E$ | Electric field |
| $EM$ | Electrophoretic mobility |
| $F$ | Faraday constant |
| $J$ | Flux |
| $J_m$ | Migration flux |
| $k'$ | Rate constant |
| $k$ | Boltzmann constant |
| $n$ | Number of adsorbed surfactant ions per unit area |
| $n_c$ | Number of counter-ions per unit area |
| $r$ | Radial distance from drop center |
| $r_i$ | Ionic radius |
| $R$ | Universal gas constant |
| $t$ | Time |
| $T$ | Absolute temperature |
| $u$ | Ionic mobility |
| $v$ | Velocity |
| $z$ | Ionic charge including sign |
| $\epsilon$ | Dielectric constant |
| $\epsilon'$ | Effective dielectric constant of the Stern layer |
| $\delta$ | Thickness of the Stern layer |
| $\sigma$ | Charge density |
| $\eta$ | Viscosity |
| $\psi$ | Interfacial potential |
| $\psi_o$ | Surface potential |
| $\psi_\delta$ | Stern layer potential |
| $\zeta$ | Zeta potential |
| $\bar{\mu}_{ij}$ | Electrochemical potential of species i in phase j |
| $\mu_{ij}$ | Chemical potential of species i in phase j |
| $\mu^o_{ij}$ | Standard chemical potential of species i in phase j |
| $\phi_j$ | Bulk phase potential of phase j |
| $\kappa$ | Reciprocal of the double-layer thickness |
| $\gamma$ | Interfacial tension |

# RECOMMENDED READING

**Adamson, A. W.,** *Physical Chemistry of Surfaces,* 3rd ed., John Wiley & Sons, New York, 1976.
**Davies, J. T. and Rideal, E. K.,** *Interfacial Phenomena,* 2nd ed., Academic Press, New York, 1963.
**Shaw, D. J.,** *Electrophoresis,* Academic Press, New York, 1963.
**Hunter, R. J.,** *Zeta Potential in Colloid Science, Principles and Applications,* Academic Press, New York, 1981.
**Bier, M.,** *Electrophoresis — Theory, Methods and Applications,* Vol. 2, Academic Press, New York, 1967.
**Dean, J. A.,** *Chemical Separation Methods,* Van Nostrand Reinhold, New York, 1969.
**Huheey, J. E.,** *Inorganic Chemistry,* 2nd ed., Harper & Row, New York, 1978.
**Bockris, J. O'M. and Reddy, A. K. N.,** *Modern Electrochemistry,* Vol. 1, Plenum Press, New York, 1970.
**Sidgwick, N. V.,** *The Covalent Link in Chemistry,* Cornell University Press, Ithaca, N.Y., 1933.
**Brown, M. G.,** *Carbon Chemistry — Some Aspects of Covalent Chemistry,* English University Press, London, 1965.
**Millich, F. and Carraher, C. E., Jr.,** *Interfacial Synthesis,* Vol. 1, *Fundamentals,* Marcel Dekker, New York, 1977.
**Skoog, D. A. and West, D. M.,** *Fundamentals of Analytical Chemistry,* 2nd ed., Holt, Rinehart & Winston, New York, 1969.
**Newman, J. S.,** *Electrochemical Systems,* Prentice-Hall, Englewood Cliffs, N.J., 1973.
**Verwey, E. J. W. and Overbeek, J. Th. G.,** *Theory of the Stability of Liophobic Colloids,* Elsevier, Amsterdam, 1948.
**Moore, W. J.,** *Physical Chemistry,* 4th ed., Prentice-Hall, Englewood Cliffs, N.J., 1972.

# REFERENCES

1. **Danesi, P. R. and Chiarizia, R.,** The kinetics of metal solvent extraction, *CRC Crit. Rev. Anal. Chem.,* 10, 1, 1980.
2. **Adamson, A. W.,** *Physical Chemistry of Surfaces,* 3rd ed., John Wiley & Sons, New York, 1976, 197.
3. **Davies, J. T. and Rideal, E. K.,** *Interfacial Phenomena,* 2nd ed., Academic Press, New York, 1963, chap. 2.
4. **de Ortiz, E. S. P.,** Interfacial potential effects on the solvent extraction of metals, *J. Appl. Chem. Biotechnol.,* 28, 149, 1978.
5. **Roddy, J. W., Coleman, C. F., and Sumio, A.,** Mechanism of the slow extraction of iron (III) from acid perchlorate solutions by di(2-ethylhexyl) phosphoric acid in n-octane, *J. Inorg. Nucl. Chem.,* 33, 1099, 1973.
6. **Cox, M. and Flett, D. S.,** The significance of surface activity in solvent extraction reagents, in *Proc. Int. Solvent Extraction* Conf., Lucas, B. H., Ritcey, G. M., and Smith, H. W., Eds., Canadian Institute of Mining and Metallurgy, 1977, 63.
7. **Yagodin, G. A., Ivakhno, S. Y., and Tarasov, V. V.,** Adsorption phenomena and their effects on the mechanism of copper extraction by hydroxyoximes, in *Proc. Int. Solvent Extraction* Conf., Paper No. 80-140, 1980.
8. **Miyake, Y., Takenoshita, Y., and Teramoto, M.,** Extraction rates of copper with SME529. Mechanism and effects of surfactants, in *Proc. Int. Solvent Extraction Conf.,* 1983, 301.
9. **Shaw, D. J.,** *Electrophoresis,* Academic Press, New York, 1963, 1.
10. **Hunter, R. J.,** *Zeta Potential in Colloid Science, Principles and Applications,* Academic Press, New York, 1981, chap. 1.
11. **Haydon, D. A.,** The electrical double layer and electrokinetic phenomena, in *Recent Progress in Surface Science,* Vol. 1, Danielli, J. F., Pankhurst, K. G. A., and Riddiford, A. C., Eds., Academic Press, New York, 1964, chap. 3.
12. **Adamson, A. W.,** *Physical Chemistry of Surfaces,* 3rd ed., John Wiley & Sons, New York, 1976, 210.
13. **Davies, J. T. and Rideal, E. K.,** *Interfacial Phenomena,* 2nd ed., Academic Press, New York, 1963, 129.
14. **Shaw, D. J.,** *Electrophoresis,* Academic Press, New York, 1963, 16.
15. **Booth, F.,** The cataphoresis of spherical fluid droplets in electrolytes, *J. Chem. Phys.,* 19, 1331, 1951.
16. **Overbeek, J. Th. G. and Wiersema, P. H.,** The interpretation of electrophoretic mobilities, in *Electrophoresis Theory, Methods and Applications,* Vol. 2, Bier, M., Ed., Academic Press, New York, 1967, chap. 1.
17. **Dickinson, W.,** The determination of ionic adsorption in the Helmholtz-Gouy (electrical) layers by the combination of electro-kinetic and interfacial tension measurements, *Trans. Faraday Soc.,* 40, 48, 1944.
18. **Ottewill, R. H. and Shaw, J. N.,** Studies on the preparation and characterization of monodisperse polystyrene latices, *Kolloid-Z. Z. Polym.,* 218, 34, 1967.

19. **Chattoraj, D. K. and Bull, H. B.,** Electrophoresis and surface charge, *J. Phys. Chem.,* 63, 1809, 1959.
20. **Hughes, M. A. and Middlebrook, P. D.,** Variation of zeta-potential with pH for purified hydroxyoximes dissolved in toluene, *Trans. Inst. Min. Metall.,* 90, C126, 1981.
21. **Douglas, H. W.,** The influence of sugars on the electrokinetic potential and interfacial tension between aqueous solutions and certain organic compounds. I. The electrophoretic behavior of organic dispersions, *Trans. Faraday Soc.,* 46, 1082, 1950.
22. **Price, C. W. and Lewis, W. C. M.,** The electrophoretic behavior of lecithin and certain fats, *Trans. Faraday Soc.,* 29, 775, 1933.
23. **Alty, J. and Johnson, O.,** The cataphoresis of particles of the fatty acids and related compounds, *Phil. Mag.,* 20, 129, 1935.
24. **Carruthers, J. C.,** The electrophoresis of certain hydrocarbons and their simple derivatives as a function of pH, *Trans. Faraday Soc.,* 34, 300, 1938.
25. **Douglas, H. W.,** The electrophoretic behaviour of certain hydrocarbons and the influence of temperature thereon, *Trans. Faraday Soc.,* 39, 305, 1943.
26. **Dickinson, W.,** The effect of pH upon the electrophoretic mobility of emulsions of certain hydrocarbons and aliphatic halides, *Trans. Faraday Soc.,* 37, 140, 1941.
27. **Powney, J. and Wood, L. J.,** The properties of detergent solutions. IX. The electrophoretic mobility of oil drops in detergent solutions, *Trans. Faraday Soc.,* 36, 57, 1940.
28. **Anderson, P. J.,** The relation of electrokinetic potential to adsorption at the oil/water interface, *Trans. Faraday Soc.,* 55, 1421, 1959.
29. **Powell, B. D. and Alexander, A. E.,** The mobility of oil drops, interfacial tension measurements, and gegen ion adsorption in soap solutions, *Can. J. Chem.,* 30, 1044, 1952.
30. **Jordan, D. O. and Taylor, A. J.,** The electrophoretic mobilities of hydrocarbon droplets in water and dilute solutions of ethyl alcohol, *Trans. Faraday Soc.,* 48, 346, 1952.
31. **Taylor, A. L. and Wood, F. W.,** The electrophoresis of hydrocarbon droplets in dilute solutions of electrolytes, *Trans. Faraday Soc.,* 53, 523, 1957.
32. **Mackenzie, J. M. W.,** Electrokinetic properties of Nujol-flotation collector emulsion drops, *Trans. Soc. Mining Eng. AIME,* 244, 393, 1969.
33. **Wie, J. M.,** Ph.D. thesis, University of California, Berkeley, 1975.
34. **Fleming, C. A.,** NIM Report No. 1793, National Institute for Metallurgy, South Africa, 1976.
35. **Parreira, H. C. and Shulman, J. H.,** Stream potential measurements on paraffin wax, in *Adv. Chem. Series,* No. 33, American Chemical Society, Washington, D.C., 1961, 160.
36. **Davies, J. T. and Rideal, E. K.,** *Interfacial Phenomena,* 2nd ed., Academic Press, New York, 1963, 144.
37. **Dean, J. A.,** *Chemical Separation Methods,* Van Nostrand Reinhold, New York, 1969, 26.
38. **Verwey, E. J. W.,** Electrical double layer and stability of emulsions, *Trans. Faraday Soc.,* 36, 192, 1940.
39. **Huheey, J. E.,** *Inorganic Chemistry,* 2nd ed., Harper & Row, New York, 1978, 74.
40. **Waddington, T. C.,** Lattice energies and their significance in inorganic chemistry, *Adv. Inorg. Chem. Radiochem.,* Vol. 1, Emeleus, H. J. and Sharpe, A. G., Eds., Academic Press, New York, 1959, 180.
41. **Bockris, J. O'M. and Reddy, A. K. N.,** *Modern Electrochemistry,* Vol. 1, Plenum Press, New York, 1970, 461.
42. **Sidgwick, N. V.,** *The Covalent Link in Chemistry,* Cornell University Press, Ithaca, N.Y., 1933, 179.
43. **Brown, M. G.,** *Carbon Chemistry — Some Aspects of Covalent Chemistry,* English University Press, London, 1965, 117.
44. **Sidgwick, N. V.,** *The Covalent Link in Chemistry,* Cornell University Press, Ithaca, New York, 1933, 155.
45. **Despotović, R., Despotović, Lj. A., Filipović-Vinceković, N., Horvat, V., and Mayer, D. Z.,** On "surfactant/surfactant" aqueous systems, *Tenside Detergents,* 12, 323, 1975.
46. **Osseo-Asare, K. and Keeney, M. E.,** Aspects of the interfacial chemistry of nickel extraction with LIX63-HDNNS mixtures, *Metall. Trans. B,* 11B, 63, 1980.
47. **Keeney, M. E. and Osseo-Asare, K.,** Molecular interaction in a mixed α-hydroxyoxime-sulfonic acid solvent extraction system, *Polyhedron,* 1, 453, 1982.
48. **Adamson, A. W.,** *Physical Chemistry of Surfaces,* 3rd ed., John Wiley & Sons, New York, 1976, 197.
49. **Davies, J. T. and Rideal, E. K.,** *Interfacial Phenomena,* 2nd ed., Academic Press, New York, 1963, 98.
50. **MacRitchie, F.,** Interface effect on chemical reaction rate, in *Interfacial Synthesis,* Vol. 1, Millich, F. and Carraher, C. E., Jr., Eds., Marcel Dekker, New York, 1977, 103.
51. **Scibona, G., Danesi, P. R., and Fabiani, C.,** *Ion Exch. Solvent Extr.,* 8, 95, 1981.
52. **Newman, J. S.,** *Electrochemical Systems,* Prentice-Hall, Englewood Cliffs, N.J., 1973, 217.
53. **Tunison, M. E. and Chapman, T. W.,** The effect of a diffusion potential on the rate of liquid-liquid ion exchange, *Ind. Eng. Chem. Fundam.,* 15, 196, 1976.

54. **Verwey, E. J. W. and Overbeek, J. Th. G.,** *Theory of the Stability of Liophobic Colloids,* Elsevier, Amsterdam, 1948, 31.
55. **Lyklema, J. and Overbeek, J. Th. G.,** On the interpretation of electrokinetic potentials, *J. Colloid Sci.,* 16, 501, 1961.
56. **Danesi, P. R., Chiarizia, R., and Sanad, W. A. A.,** Transfer rate of some tervalent cations in the biphasic system $HClO_4$, water-dinonylnaphthalenesulfonic acid, tolune. II, *J. Inorg. Nucl. Chem.,* 39, 519, 1977.
57. **Moore, W. J.,** *Physical Chemistry,* 4th ed., Prentice-Hall, Englewood Cliffs, N.J., 1972, 435.
58. **Bailes, P. J.,** Electrostatic extraction for metals and nonmetals, in *Proc. Int. Solv. Extr. Conf.,* Lucas, B. H., Ritcey, G. M., and Smith, H. W., Eds., Canadian Institute of Mining and Metallurgy, 1977, 233.
59. **Fekete, S. O., Meyer, G. A., and Wicker, G. R.,** The Selective Extraction of Nickel and Cobalt from Acid Leach Solutions Using a Mixed Solvent System, *Trans. Metall. Soc. AIME,* preprint A77-95, 1977.
60. **Osseo-Asare, K., Leaver, H. S., and Laferty, J. M.,** Extraction of nickel and cobalt from acidic solutions using LIX63-HDNNS mixtures, in *Process and Fundamental Considerations of Selected Hydrometallurgical Systems,* Kuhn, M. C., Ed., Society of Mining Engineers, AIME, New York, 1981, 195.
61. **Gallacher, L. V.,** U.S. Patent 4,018,865, April 19, 1977.
62. **Osseo-Asare, K. and Keeney, M. E.,** Phase transfer and micellar catalysis in hydrometallurgical liquid-liquid extraction systems, in *Proc. Int. Solvent Extraction Conf.,* Paper No. 80-121, 1980.
63. **Tammi, T. T.,** Separation of the isomers of the commercial α-hydroxyoxime LIX63, *Hydrometallurgy,* 2, 371, 1976/1977.
64. **Danesi, P. R., Chiarizia, R., and Scibona, G.,** A simple purification method for the liquid cation exchanger dinonylnaphthalene sulfonic acid (DNNSA), *J. Inorg. Nucl. Chem.,* 35, 3926, 1973.
65. **Skoog, D. A. and West, D. M.,** *Fundamentals of Analytical Chemistry,* 2nd ed., Holt, Rinehart & Winston, New York, 1969, chap. 4.
66. **Lin, K. L., and Osseo-Asare, K.,** to be published.
67. **Vieux, A. S., Bibombe, K., and Nsele, M.,** Organic phase species in the extraction of molybdenum (VI) by Aliquat 336 from chloride media, *Hydrometallurgy,* 6, 35, 1980.
68. **Sato, T.,** The extraction of zirconium (IV) from hydrochloric acid solutions by tricaprylylmethylammonium chloride, *Anal. Chim. Acta,* 49, 463, 1970.
69. **Haydon, D. A. and Taylor, F. H.,** Adsorption of sodium octyl and decyl sulphates and octyl and decyl trimethylammonium bromides at the decane-water interface, *Trans. Faraday Soc.,* 58, 1233, 1962.
70. **Adamson, A. W.,** *Physical Chemistry of Surfaces,* 3rd ed., John Wiley & Sons, New York, 1976, 323.

Chapter 5

# EXTRACTION WITH SUPERCRITICAL FLUIDS

**Mark A. McHugh**

## TABLE OF CONTENTS

# I. INTRODUCTION

Separation processes play a vital role in the chemical process industries. Recently, a new separations technique, supercritical fluid solvent extraction, has emerged.[1] This technique combines features of both distillation (i.e., separation due to differences in component volatilities) and liquid extraction (i.e., separation of components which exhibit little difference in their relative volatilities or which are thermally labile). Supercritical fluid solvent extraction is based on the experimental observation that a gas, when compressed isothermally at a temperature greater than its critical temperature to pressures greater than its critical pressure, exhibits enhanced solvating power.[2] Although this experimental fact has been known for more than 100 years,[3-5] it is only in the past decade or so that supercritical fluid (SCF) solvents have been the focus of active research and development programs. Numerous symposia[6-8] as well as an increasing number of review papers[1,9-14] on SCF solvent extraction attest to the magnitude of interest in this technique. The factors that have supplied the impetus for the development of SCF solvent extraction as a viable separations technique include:

1. Traditional separation techniques have become more expensive within the last 10 years due to the sharp increase in energy costs.
2. Increased government scrutiny and regulation of common industrial solvents, such as chlorinated hydrocarbons, has made nontoxic environmentally acceptable supercritical fluid solvents such as $CO_2$ very attractive as alternative industrial solvents.
3. Tougher pollution control legislation has forced industry to consider alternative means of waste removal.[15]

The various review papers concerning SCF solvent extraction offer a spectrum of information about the technique. The review paper of Paul and Wise[9] offers a very straightforward introduction to the basic concepts of this technique. A number of areas in which supercritical solvent extraction may be applied are suggested in this review. Irani and Funk[10] present a review of the available data for a number of supercritical fluid solvents. They also compare the energy requirements for distillation vs. SCF solvent extraction. They conclude that energy savings are realized with SCF solvent extraction if the process is operated at low gas compression ratios.

Another recent review[1] is actually a compilation of the papers presented at the sessions devoted to extraction with supercritical gases held as part of a conference in Germany in 1978.[8] This review contains an extensive amount of experimental information on the extraction of natural products such as hops, spices, caffeine, tobacco, and flavors, extracted with a number of different SCF solvents such as ethane, ethylene, $CO_2$, and $N_2O$. Although numerous candidate SCF solvents are presented, carbon dioxide is by far the most extensively used SCF solvent. This is because carbon dioxide is nontoxic, nonflammable, environmentally acceptable, and inexpensive. Hence, carbon dioxide is an ideal solvent for processing natural products.

The papers presented at the symposium entitled "Chemical Engineering at Supercritical Fluid Conditions", held in 1981 at the annual AIChE meeting in New Orleans, La.,[6] recently appeared as a book with the same title as the symposium.[16] The material in this book covers three major areas: experimental data on phase equilibrium behavior, thermodynamic theories and equations of state, and practical applications of supercritical fluid solvent extraction. More will be said about specific papers in this book in a later section of this review paper.

Paulaitis et al.[14] present one of the most comprehensive review papers in the area of SCF solvent extraction. An extensive tabulation of experimental studies can be found in this paper. Also, the area of transport properties of supercritical fluid solvents is addressed and a number of practical applications are described.

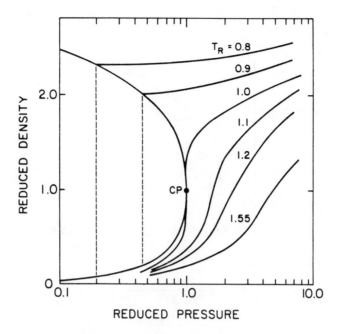

FIGURE 1.   Schematic representation of the reduced density as a function of reduced pressure and reduced temperature in the vicinity of the pure component critical point. (From Paulaitis, M. E., Krukonis, V. J., Kurnik, R. T., and Reid, R. C., *Rev. Chem. Eng.*, 1, 179, 1982. With permission.)

Another current review paper in this area is by Randall.[17] Although the focus in this review is the application and development of supercritical fluid chromatography, an extensive review of the available experimental data on phase equilibria studies involving SCF solvents and the available experimental data on supercritical fluid chromatography extends and complements the tabulation of experimental studies found in the previously mentioned review paper by Paulaitis et al.[14] Also, recently, Johnston[18] has written a chapter for the *Encyclopedia of Chemical Technology* on "Supercritical Fluids" in which he reviews the basic principles of SCF solvent extraction.

A SCF solvent extraction process operates, by definition, at a temperature which is greater than or equal to the pure SCF solvent's critical temperature and at a pressure which is greater than or equal to the pure SCF solvent's critical pressure. However, as noted by Paulaitis[14] and others,[12] for practical application of SCF solvents, the SCF region of interest is bounded by a $T_R$ between 0.9 and 1.2 and a $P_R$ greater than 1.0. The reasons for this range of reduced temperatures and pressures can be understood by considering the phase behavior of a pure SCF solvent shown in Figure 1.

Notice in this figure that in the region $0.9 < T_R < 1.2$ and $P_R > 1$ the SCF solvent is highly compressible. In fact, the density of the SCF solvent approaches the density of a liquid and, as a first approximation, the solvent power of the SCF solvent approaches the solvent power of a liquid. Note, however, that as the reduced temperature is increased to a value of 1.55, the reduced pressure must now be increased to values greater than about 10.0 to reach liquid-like densities. Hence, for all practical purposes the SCF region of interest is that region very near the pure substance critical conditions where small changes in either pressure and/or temperature cause large changes in the SCF solvent density and, therefore, in the SCF solvent power.

By operating near the critical point of the pure solvent, enhanced solvent properties can

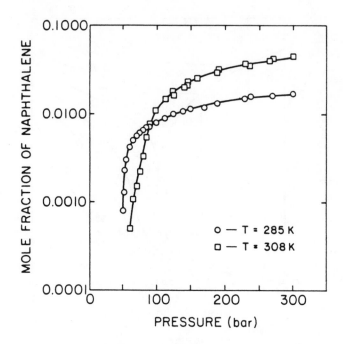

FIGURE 2.    The solubility of naphthalene in supercritical ethylene at temperatures of 285 and 308 K.[21,22] (Copyright© 1983 by Freund Publishing House Ltd. and D. Reidel Publishing Company.)

be obtained. As an example of the features of a SCF solvent extraction process, consider the solubility of solid naphthalene (Tm = 353.4 K) in supercritical ethylene (Tc = 282.4 K, Pc = 50.4 bar) near the critical point of ethylene as shown in Figure 2. Following the 285 K ($T_R$ = 1.01) isotherm at pressures slightly greater than 50 bar ($P_R$ = 1.0), conditions where ethylene is highly compressible, the solubility increases dramatically for small changes in pressure. As a consequence of the increased density of ethylene at 100 bar relative to that at 1 bar, a solubility enhancement of more than four orders of magnitude occurs at 100 bar ($P_R$ = 2.0) and 285 K ($T_R$ = 1.01).

The 308 K ($T_R$ = 1.09) isotherm is not as pressure sensitive near the critical pressure of ethylene as compared to the 285 K isotherm. However, at pressures greater than 100 bar at liquid-like densities, the 308 K isotherm reaches a higher limiting solubility value as a result of the increased vapor pressure of naphthalene relative to the slight decrease in ethylene density. Near the critical point of pure ethylene (i.e., near $T_R$ = 1.0 and $P_R$ = 1.0) where small changes in temperature cause large changes in ethylene density, the solubility of naphthalene in supercritical ethylene decreases by as much as an order of magnitude when the temperature is increased from 285 to 308 K.

Novel separation schemes are possible with SCF solvents as a result of the solubility behavior of the solvent in the supercritical fluid region. For example, thermally labile, nonvolatile substances, such as coal, can be extracted using a SCF solvent with a critical temperature below the temperature at which extensive thermal degradation of the substance occurs. The heavy solute can then be easily separated from the supercritical solvent phase by adjusting the density and, hence, solvent power of the supercritical fluid. If the pressure is lowered isothermally, the density decreases and the solute will precipitate from solution. Alternatively, isobarically increasing the temperature near the critical pressure of the solvent decreases the solvent density, and, hence, precipitation of the solute can occur.

In addition to the unique solubility behavior of a SCF solvent, there are certain desirable

**Table 1**
**ORDER OF MAGNITUDE COMPARISON OF THE PROPERTIES OF**
**SUPERCRITICAL CARBON DIOXIDE ($T_R = 1.0$, $P_R = 1.9$) TO THOSE**
**OF A GAS AND A LIQUID**

| | Phase | | |
| Property | Liquid | $CO_2$ | Gas |
|---|---|---|---|
| Density (kg/m³) | 1000 | 700 | 1 |
| Viscosity (N·s/m²) | $10^{-3}$ | $10^{-4}$ | $10^{-5}$ |
| Diffusion coefficient (cm²/sec) | $10^{-5}$ | $10^{-4}$ | $10^{-1}$ |

physicochemical properties which make it a good solvent. As shown in Table 1, while a SCF solvent has a liquid-like density[19] and, hence, solvent loading comparable to a liquid, the diffusivity[20] and viscosity are intermediate to that of a liquid and a gas. Thus, a SCF solvent has better mass transfer characteristics and, therefore, better separation efficiencies.

The information concerning SCF solvent extraction that is presented in this review paper covers three main areas. First, the classification of the phase diagrams for SCF solvent systems is described. The phase diagrams for polymer-SCF solvent systems are also considered. Next, the thermodynamic analysis and data correlation of SCF solvent systems are discussed. Finally, a number of recent applications of SCF solvent extraction are described.

## II. THEORY

An attractive feature of SCF solvent extraction is that, in some cases, certain relatively nonvolatile substances can be selectively extracted from a multicomponent mixture. New applications for SCF solvent extraction are dependent on our ability to understand the phase behavior which occurs for mixtures in the critical region. This can be quite a formidable task since a variety of phase behaviors can occur, even for simple binary mixtures in which the components are chemically similar but differ only in molecular size, such as methane-*n*-hexane.[23] For a given practical application of SCF solvent extraction, it is highly likely that not only will the mixture components differ in molecular size, but they will also differ in molecular shape, structure, and/or polarity. In fact, the components of the mixture may even be ill defined (e.g., petroleum residuum). When the mixture constituents exhibit this much asymmetry, and/or when the mixture has numerous components, the resultant phase behavior can be quite complex. It is important, then, to first understand the limiting phase behavior of a binary mixture which consists of a single SCF solvent and a single solute.[24] An understanding of the phase behavior of this limiting case provides a basis for understanding and generalizing the important phase equilibrium principles which dictate the SCF solvent extraction process of mixtures.

Complex phase behavior can be interpreted in a relatively straightforward manner with the phase rule. As described by Streett,[24] the phase rule imposes certain geometrical constraints on the construction of phase diagrams for mixtures. The phase rule is given by the simple relation:

$$f = c + 2 - p \tag{1}$$

where f is the number of independent variables, c is the number of components, and p is the number of phases. Shown in Table 2 are the geometrical constraints imposed by the phase rule for the phase diagram representation of multiphase-multicomponent equilibria.

**Table 2**

**SUMMARY OF THE GEOMETRICAL FEATURES OF PHASE
DIAGRAMS FOR ONE AND TWO COMPONENT SYSTEMS**

| Number of equilibrium phases | | Degrees of freedom | Geometrical features |
|---|---|---|---|
| One-component system | Two-component system | | |
| 3 | 4 | 0 | Points |
| 2 | 3 | 1 | Lines |
| 1 | 2 | 2 | Surfaces |
| — | 1 | 3 | Volumes |

For a SCF solvent extraction process, the most important regions in PTx space are those of two-phase liquid-vapor (LV) or liquid-liquid (LL) equilibria, three-phase liquid-liquid-vapor (LLV), solid-liquid-vapor (SLV), or solid-solid-vapor (SSV) equilibria, or sometimes four-phase liquid-liquid-solid-vapor (LLSV) or liquid-solid-solid-vapor equilibria (LSSV). When these regions of phase equilibria are projected onto a two-dimensional PT diagram, their geometrical representation simplifies since pressure and temperature are field varia-bles,[24] that is, variables which are the same in each of the equilibrium phases. Hence, two surfaces representing equilibrium between two phases in PTx space project as a single surface in PT space; three lines representing three equilibrium phases in PTx space project as a single line in PT space, and four points which represent four equilibrium phases in PTx space project as a single point in PT space. A detailed discussion of three-dimensional PTx phase diagrams can be found elsewhere.[24-26]

Scott and van Konynenburg have shown that the van der Waals equation of state quali-tatively describes most of the experimentally observed binary fluid phase diagrams.[27,28] They also have shown that all known experimentally observed binary phase diagrams can be classified into five basic types as described by the van der Waals equation. (Streett[24] describes a sixth classification not predicted by the van der Waals equation; however, that type of phase diagram is not described here.) They simplify their classification scheme by using the two-dimensional projection of critical mixture curves and three phase lines from three-dimensional PTx diagrams. The five classes of possible fluid phase diagrams are shown in Figure 3.

**A. Phase Diagrams for Binary Mixtures**
**Type I** — Shown in Figure 3a is the simplest possible phase diagram for this type of system. The liquids are miscible in all proportions and the critical mixture curve runs continuously from the critical point of the heavier component to the critical point of the lighter component. Examples of mixtures which exhibit this type of phase behavior include $CO_2$-$C_3H_8$ and $C_2H_6$-$n$-$C_7H_{16}$.[23] In this class of phase behavior, it is also possible to have critical lines with azeotropes.[29]

**Type II** — The phase diagram for this type of system exhibits certain similarities to that of the type I system. In this case, the critical mixture curve is again a continuous curve which runs between the critical points of the two pure components. Now, however, the liquids are no longer always miscible. A liquid-liquid-gas line ending at an upper critical end point (i.e., a point at which one of the liquid phases becomes critically identical with the gas phase in the presence of the other liquid phase) is now evident at temperatures lower than the critical point of either component. A three-dimensional PTx representation of the

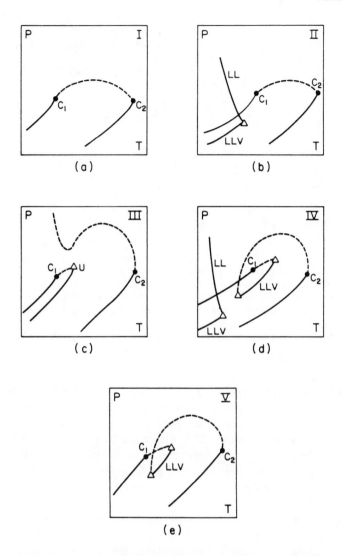

FIGURE 3.   The five classes of binary phase diagrams predicted by van der Waals equation.[24]

three-phase liquid-liquid-gas region is shown in Figure 4. Notice that the three-phase liquid-liquid-gas region projects as a single line on a PT diagram. Again, this is a consequence of using two field variables, pressure and temperature, to represent the phase behavior.[24] The $CO_2$-$n$-octane system is an example of a binary mixture which exhibits this type of phase behavior.

**Type III** — In this instance, the critical mixture curve no longer runs continuously between the critical points of the two components. The branch of the critical mixture curve which starts at the critical point of the component with the higher critical temperature can take on a number of different characteristic shapes,[24] although only one shape is shown in Figure 3c. The branch of the critical mixture curve, which starts at the critical point of the other component, intersects a region of three-phase equilibria at an upper critical end point (UCEP). Examples of binary mixtures which exhibit this type of phase behavior include $CO_2$-squalane and $CO_2$-$n$-$C_{16}H_{34}$.

**Type IV** — The branch of the critical mixture curve, which starts at the critical point of

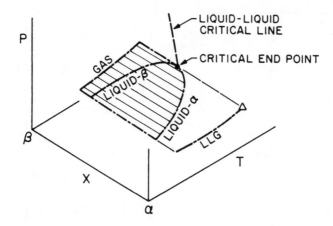

FIGURE 4. The three-dimensional PTX representation of a three-phase liquid-liquid-gas region. The triangle on the two-dimensional PT face of the diagram represents an upper critical end point (UCEP) (i.e., the two liquid phases become critically identical in the presence of the gas phase). (From Streett, W. B., *Chemical Engineering at Supercritical Fluid Conditions,* Paulaitis, M. E., Penninger, J. M. L., Gray, R. D., and Davidson, P., Eds., Ann Arbor Science, Ann Arbor, Mich., 1983. With permission.)

the component with the higher critical temperature, intersects a region of liquid immiscibility at the lower critical solution temperature (LCST) (i.e., at the LCST the two liquid phases of the LLV line become critically identical in the presence of the gas phase). As with type III behavior, the other branch of the critical mixture curve, which starts at the critical point of the other component, intersects the LLV line at the UCEP. At temperatures below the LCST, a region of liquid immiscibility again appears similar to that in type II phase behavior. This type of phase behavior has been observed for binary mixtures of $CO_2$-nitrobenzene. It is also interesting to note that this type of phase behaivor has been found for polymer-solvent binary mixtures.[30-35] The fact that binary polymer-solvent systems exhibit type IV phase behavior has important practical consequences which will be discussed in a later section of this paper.

**Type V** — This phase behavior is very similar to type IV. Now, however, there is no region of liquid immiscibility at temperatures below the LCST. Examples of systems which exhibit this type of behavior include $C_2H_6$-$C_5H_5OH$.

As has been previously noted,[12,14] SCF extraction of solids is of tremendous practical importance. The phase behavior for mixtures which consist of a single SCF solvent and a heavy nonvolatile solid whose melting temperature is greater than the critical temperature of the SCF solvent can be considered as a subset of type III phase behavior.[24] Shown in Figure 5 is a PT diagram for such a solid-SCF solvent system.

CD and MH are the pure component vapor pressure curves, MN the heavy component melting curve, and EM the pure heavy component sublimation curve. Points D and H represent pure component critical points. For this type of system, the critical mixture curve, which represents the critical conditions for mixtures of different composition, has two branches. One branch, which starts at the critical point of the heavy component, H, intersects the three-phase solid-liquid-gas (SLG) freezing point depression curve at the UCEP. The other branch of the critical mixture curve starts at the critical point of the light component, D, and intersects the SLG line at the lower critical end point (LCEP). The freezing point depression of the heavy solid is a consequence of the solubility of the light component in the heavy liquid. If this solubility were large, the SLG curve would start at the melting

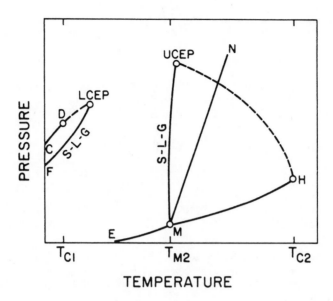

FIGURE 5. Pressure-temperature diagram for a mixture consisting of a heavy nonvolatile solid and a light supercritical fluid solvent whose critical temperature is less than the melting temperature of the solid.[25]

point of the heavy solid and run continuously to lower temperatures. Also, the critical mixture curve would run continuously between the critical points of the two pure components.[25]

In the present case, the light component is only slightly soluble in the heavy liquid phase and, hence, the freezing point depression is also only slight. However, the solubility is still sufficient to offset the increase in freezing temperature with pressure for the pure solid. As shown in Figure 5, the SLG line rises steeply and intersects the critical mixture curve at the LCEP and the UCEP. At these critical end points the liquid and gas phase of the SLG line merge into a single fluid phase in the presence of excess solid.[36] Supercritical solvent extraction of solids would occur in the gas-solid region which exists between the two branches of the SLG line.

As noted by McHugh and Paulaitis,[37-40] a variety of high pressure phase behavior can occur for binary mixtures near the mixture UCEP. Consider the phase behavior for the naphthalene-ethylene system[21] shown in Figure 6. The 323 K isotherm exhibits a large solubility enhancement which is sensitive to small changes in pressure around 175 bar. Also, at 200 bar, the increase in the solubility of naphthalene in ethylene from 318 to 323 K is substantially greater than that from 298 to 318 K. The sensitivity of the solubility behavior for the 323 K isotherm near 175 bar is a consequence of operating extremely close to the naphthalene-ethylene UCEP (i.e., 325.3 K and 176.3 bar).[21] Note, also, that the loading of solid naphthalene in supercritical ethylene can be quite substantial near the UCEP (i.e., 15 mol % naphthalene corresponds to 45 wt % naphthalene in supercritical ethylene). Hence, operating near the mixture UCEP has important practical significance.

Now compare the previously described naphthalene-ethylene solubility behavior with that observed for the biphenyl-$CO_2$ system as shown in Figure 7. Biphenyl is a heavy solid which melts at 342.7 K. The UCEP for the biphenyl-$CO_2$ system is 328.3 K and 475.2 bar.[37,41,42] At pressures below the UCEP pressure, the behavior of the 328 K solubility isotherm is quite similar to that exhibited by the 323 K naphthalene-ethylene isotherm. However, at pressures greater than the UCEP pressure the solubility behavior of the 328 K isotherm is radically different from that observed for the 323 K naphthalene-ethylene isotherm. The

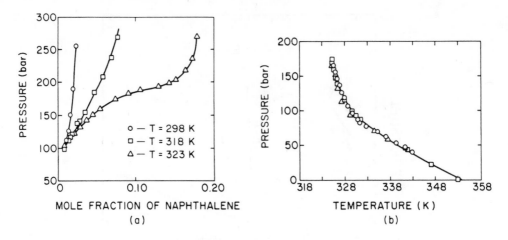

FIGURE 6.    Phase behavior of the naphthalene-ethylene system. (a) The solubility of naphthalene in supercritical ethylene at temperatures of 298, 318, and 323 K;[21] (b) the PT projection of the naphthalene-ethylene three-phase SLG line ending at the mixture UCEP.[37,103] (Copyright© 1983 by Freund Publishing House Ltd. and D. Reidel Publishing Company.)

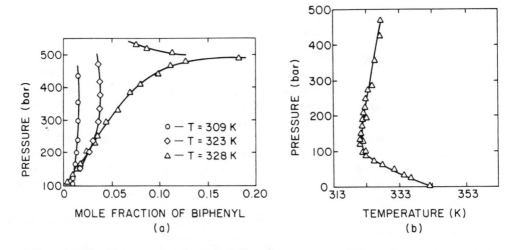

FIGURE 7.    Phase behavior of the biphenyl-carbon dioxide system. (a) The solubility of biphenyl in supercritical carbon dioxide at temperatures of 309, 323, and 328 K;[37] (b) the PT projection of the biphenyl-carbon dioxide three-phase SLG line ending at the mixture UCEP.[41]

reasons for the differences in solubility behavior near the UCEP for these two systems can be understood by considering the shapes of their SLG lines shown in the PT diagrams in Figures 6b and 7b.

As shown in Figure 7b, the 328 K biphenyl-$CO_2$ solubility isotherm actually intersects the SLG line at two different pressures. At pressures below 475 bar the 328 K solubility isotherm shown in Figure 7a actually represents the fluid phase of a fluid-liquid equilibria state.[37] As the pressure increases to the mixture UCEP pressure, the solubility of liquid biphenyl in supercritical $CO_2$ increases dramatically. In fact, at the UCEP pressure, this fluid phase isotherm must exhibit a horizontal inflection on a pressure-composition diagram due to the presence of the liquid-fluid criticallity.[26] At pressures greater than the UCEP pressure the 328 K solubility isotherm now represents the fluid phase of a solid-fluid equilibria state. At these extreme pressures the solubility of biphenyl decreases dramatically with increasing pressure due to the large size disparity between $CO_2$ and biphenyl.[42] Also, as the pressure is increased further, the influence of the liquid-fluid criticality is greatly diminished.

For the naphthalene-ethylene 323 K isotherm the SLG (see Figure 6b) is never intersected at high pressures. Hence, there is always solid-fluid equilibria in this instance. The fact that these two systems exhibit such different phase behavior near their mixture UCEPs emphasizes the need for companion PT studies when doing solubility studies near phase boundaries.

It is interesting to note that the phase behavior of certain polymer-supercritical solvent systems are very similar to that described in Figure 5,[43-47] although there are some important differences. For pure polymer, the solid melting curve shown in Figure 5 is now replaced by a crystallization surface. Also, unlike crystalline solids such as naphthalene, solid polymers can sorb a considerable amount of solvent and as a result undergo morphology changes (i.e., swelling as well as internal structure rearrangement).[48,49] Finally, a disperse polymer molecular weight distribution can often mask certain phase behavior effects.[44]

de Loos et al.[47] have recently published a study of the system polyethylene-ethylene at high pressures which investigates the effects of molecular weight and molecular weight distribution of polyethylene on the observed phase behavior. Both of these variables have dramatic effects on the polymer-solvent phase behavior. A polydisperse molecular weight distribution shifts the critical point for various polyethylene-ethylene mixtures from the maximum of the pressure-composition curve to a point on the curve at much higher concentrations of polymer. Also, an increase in the average molecular weight of the polymer sample causes the pressure-composition curve to shift to higher pressures.

## B. Phase Diagrams for Ternary Mixtures

The most comprehensive source of ternary phase diagrams is the work of Francis.[50] However, ternary phase diagrams in which one of the components is a supercritical fluid are much more scarce. A compilation of supercritical ethylene-water-organic solvent ternary phase diagrams can be found in the work of Elgin and co-workers.[51] In their work they classified ternary phase diagrams into three basic types. Elgin and co-workers have also studied a large number of binary SCF solvent-organic solvent mixtures with the objective of generalizing the phase behavior of these binary systems at conditions very near the critical point of the SCF solvent.[52-54] They discuss how to use the binary phase behavior information to develop a SCF solvent-liquid extraction process of a single liquid from a binary liquid mixture.[55] It is instructive to consider their phase behavior classification scheme for ternary systems which consist of organic solvent(s), water ($H_2O$), and ethylene ($C_2H_4$).

*Type I* ternary phase behavior is shown in Figure 8. The three ternary phase diagrams in this figure are all at a fixed temperature which is slightly higher than the critical temperature of ethylene. The phase behavior for this system at atmospheric pressure is shown in Figure 8a. At this condition water is miscible in all proportions with the organic solvent, while ethylene is virtually insoluble in water and slightly soluble in the organic solvent. The gas solubility curve has a slight curvature to it reflecting the solubility of ethylene in the solvent. The tie line shown in this diagram indicates that the gas phase is virtually pure ethylene at this temperature and pressure.

In Figure 8b the pressure has been increased to a point slightly below the critical pressure of ethylene. At this pressure ethylene still remains virtually insoluble in water although it now is more soluble in the organic solvent. Hence, the gas solubility curve begins to bend more toward the ethylene apex.

In Figure 8c the pressure has increased to a value which is greater than the critical pressure of the ethylene-solvent binary. Therefore, ethylene is now miscible in all proportions with the organic solvent. However, even at this elevated pressure ethylene still remains virtually insoluble in the water. This is not a surprising result for water-hydrocarbon mixtures.[56] As

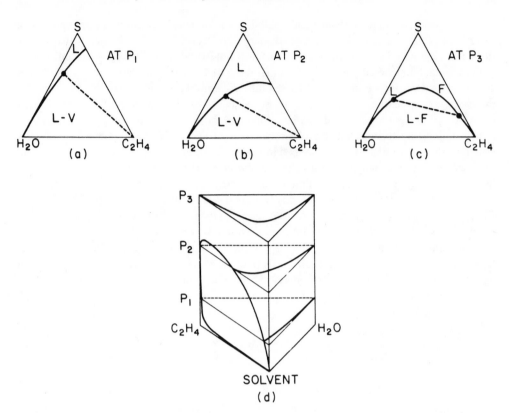

FIGURE 8.    Ternary ethylene (C₂H₄)-water (H₂O)-organic solvent(s) phase diagrams following the type I classification scheme of Elgin and Weinstock.[51,55]

shown in Figure 8c, the gas solubility curve now intersects the ethylene-water binary axis in two locations. Hence, the tie lines for this system now indicate that there is a liquid phase which is mostly water and organic solvent in equilibrium with a fluid phase which is mainly ethylene and organic solvent. Therefore, the organic solvent which is miscible with water at low pressures is now selectively extracted from the water-organic solvent mixture with supercritical ethylene. An example of this type of extraction process is suggested by the phase behavior shown in Figure 9 for the ethanol-water-supercritical ethane system.[57] This phase behavior has also been observed for the ethanol-water-CO₂ system,[58] the ethanol-water-ethylene system,[58] and the isopropanol-water-CO₂ system.[59] Elgin and Weinstock also present a number of organic solvent-water-ethylene systems which exhibit this type of phase behavior.[51] Finally, to construct a pressure-composition phase diagram, a solid prism is used as shown in Figure 8d.[51,60]

*Type II* phase behavior is depicted in the ternary phase diagrams shown in Figure 10. In this instance, there is a liquid phase miscibility gap which appears within the pressure-composition prism but which does not extend to the ethylene-organic solvent face of the prism. The phase behavior at atmospheric pressure shown in Figure 10a is identical to that previously described for type I phase behavior shown in Figure 8a. At $P_2$, a pressure below the critical pressure of ethylene, a miscibility gap appears at intermediate water-organic solvent compositions, thus, creating both liquid-liquid (LL) and liquid-liquid-vapor (LLV) regions in the phase diagram (see Figure 10b). Notice that the tie lines in the liquid-liquid region are somewhat parallel to the water-organic solvent axis. Hence, a very good separation of organic solvent from water is realized at this condition.

As the pressure is increased further (see Figure 10c), the LL and LLV regions expand

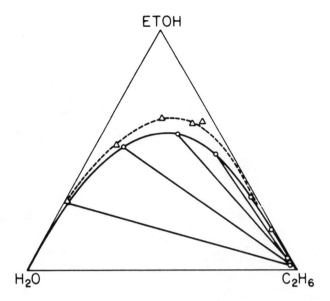

FIGURE 9. Phase behavior for the ternary ethanol (EtoH)-water ($H_2O$)-ethane ($C_2H_6$) system at a temperature of 323 K and pressures of 80 bar (○) and 50 bar (△). (From McHugh, M. A., Mallett, M. W., and Kohn, J. P., *Chemical Engineering at Supercritical Conditions,* Paulaitis, M. E., Penninger, J. M. L., Gray, R. D., and Davidson, P., Eds., Ann Arbor Science, Ann Arbor, Mich., 1983, 113. With permission.)

considerably. The separation of organic solvent from water also increases. However, the loading of organic solvent in the organic solvent-rich phase decreases slightly.

When the pressure is finally increased above the ethylene-organic solvent critical pressure (see Figure 10d), the miscibility gap disappears and the phase behavior becomes identical to that described for a type I system at the same pressure. An example of type II phase behavior is shown in Figure 11 for the *n*-propanol-water-ethylene system.[55]

In *type III*, the miscibility gap of the three-component system intersects the water-organic solvent face of the pressure-composition diagram. This large miscibility gap is consequence of either the chemical nature of the solvent or the low temperature of the system.[55]

At atmospheric pressure the water-organic solvent binary system already shows immiscibility (see Figure 12). As the pressure increases (see Figure 12b), the LL region grows until finally at pressures above the organic solvent-ethylene critical pressure a single liquid-fluid solubility curve exists. In Figure 12c the liquid-fluid solubility curve intersects the water-organic solvent axis indicating that a miscibility region still exists in the water-organic solvent binary system.

For Type III systems the LL region increases with increasing ethylene content (see Figure 12b). Although the solvent is only slightly miscible with water in the absence of ethylene, it is now possible to use supercritical ethylene to separate even more water from the solvent.

The methyl ethyl ketone (MEK)-water-ethylene system exhibits type III phase behavior. Elgin and Weinstock use this phase behavior as a basis for a process for dehydrating MEK-water mixtures using supercritical ethylene.[51]

Recently, Kohn, Luks, and co-workers have completed a study of the phase behavior occurring near the critical point of methane for a number of ternary mixtures. These ternary systems are shown in Table 3. While their work has application in the liquefied natural gas industry, the phase behavior observed for these systems is entirely analogous to that for systems at much higher temperatures near the critical point of one of the components.

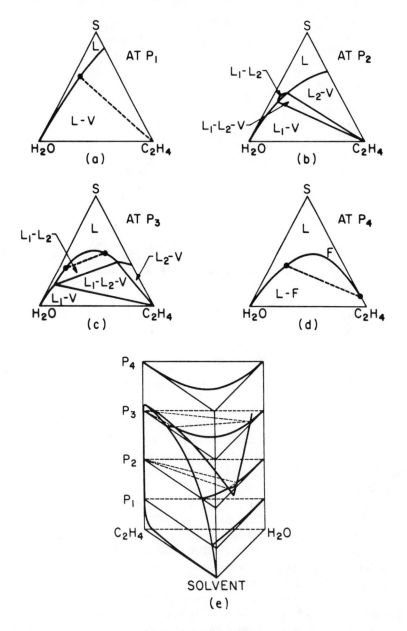

FIGURE 10.   Ternary ethylene ($C_2H_4$)-water ($H_2O$)-organic solvent(s) phase diagrams following the type II classification scheme of Elgin and Weinstock.[51,55]

Kohn and co-workers have also studied the phase behavior of the ternary systems consisting of ethane (C2) with *n*-hexadecane (C16) + *n*-eicosane (C20)[65] and with *n*-nonadecane (C19) + *n*-eicosane (C20).[66] Both of these systems exhibit liquid-liquid-vapor (LLV) behavior in the vicinity of the critical point of ethane. They found that supercritical ethane will separate C16 from C20 more effectively than C19 from C20 along the LLV curve (i.e., the distribution coefficient for the two liquid phases along the LLV curve on an ethane-free basis is approximately 1.3 for the C2-C16-C20 system as compared to approximately 1.1 for the C2-C19-C20 system).

The amount of experimental information available for ternary mixtures consisting of a

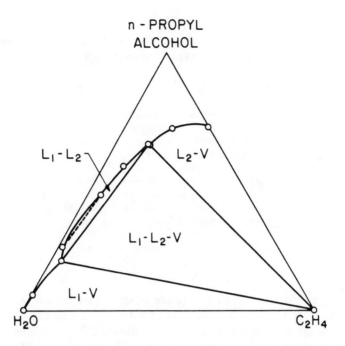

FIGURE 11.   Phase behavior for the ternary *n*-propanol-water-ethylene system at a temperature of 288 K and a pressure of 49.3 bar.[55]

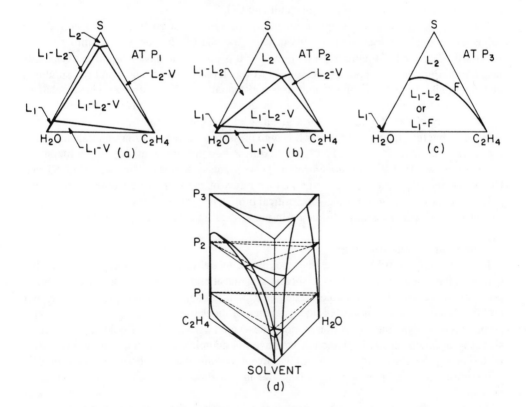

FIGURE 12.   Ternary ethylene ($C_2H_4$)-water($H_2O$)-organic solvent(s) phase diagrams following the type III classification scheme of Elgin and Weinstock.[51,55]

## Table 3
### TERNARY METHANE MIXTURES STUDIED BY KOHN, LUKS, AND CO-WORKERS

| System | Ref. |
|---|---|
| Methane + ethane + n-octane | 64 |
| Methane + propane + n-octane | 63 |
| Methane + n-butane + n-octane | 63 |
| Methane + n-butane + nitrogen | 61 |
| Methane + n-pentane + n-octane | 62 |
| Methane + n-pentane + nitrogen | 61 |
| Methane + n-hexane + n-octane | 62 |
| Methane + n-hexane + nitrogen | 61 |
| Methane + n-hexane + carbon dioxide | 62 |
| Methane + n-octane + carbon dioxide | 63 |

single supercritical solvent and two nonvolatile solids is very limited. Kurnik and Reid[67] investigated the phase behavior for a number of binary solid mixtures consisting of combinations of phenanthrene, naphthalene, benzoic acid, and 2,3 (and 2,6) dimethylnaphthalene. The supercritical fluids which they used were $CO_2$ and ethylene. They found that for a binary solid mixture, the solubility of one of the solids increases significantly if the other solid is also soluble in the supercritical fluid solvent phase. Holder[68] also found similar results for the system naphthalene-phenanthrene-$CO_2$.

Diepen et al.[69] describe the phase diagrams for systems consisting of two nonvolatile solids and a single SCF solvent. In this instance the phase behavior is very similar to that previously described in Figure 5. Now, however, it is possible to have a liquid-gas criticality occur in the presence of both of the solids. At this ternary critical end point the solubility of both solids in the SCF solvent increases substantially. Due to complex phase behavior which occurs for this type of ternary system, no general guidelines are given as to determining the selectivity of the SCF solvent for a given solid in a binary solid mixture.

It is also interesting to note that very few high pressure ternary phase diagrams are available for polymer-organic solvent-SCF solvent mixtures. In one study available in the literature, the phase behavior of the butonone-acetone-polystyrene system is investigated.[70] LCST and UCST phenomena are determined for various ratios of butonone to acetone. Data on the poly-(ethylene-propylene)-hexane-supercritical propylene system is also available in the patent literature.[71] This system is discussed in more detail in a later section of this paper.

## C. Thermodynamic Modeling

It is extremely useful to have a thermodynamic model or correlation scheme to be able to predict the phase behavior for SCF solvent systems at high pressures. There are two major problems in developing a reliable model of a SCF solvent extraction process. The first problem has to do with our lack of understanding of the dense fluid state. The second major problem is describing the interaction of molecules which significantly differ in their molecular size, shape, and/or polarity. Although we have not completely overcome these difficulties in describing the phase behavior for SCF solvent systems, certain thermodynamic models will describe the high pressure phase behavior in a qualitative,[27-28] if not semiquantitative, manner.[39]

Any thermodynamic model which is used to calculate the phase equilibria of SCF solvent systems must satisfy the following relationships for the two equilibrium phases, ' and " :

$$f_i' = f_i'' \qquad i = 1, 2, 3, \ldots m \tag{2}$$

where f denotes the fugacity of component i in a multicomponent mixture. For most SCF solvent applications the phases at equilibrium are either liquid-SCF solvent or solid-SCF solvent.

Various modeling procedures are proposed in the literature to predict the phase behavior of liquid-SCF solvent systems.[72-74] The most computationally straightforward method is to model the equilibrium liquid and SCF solvent phases using the following relationship:

$$y_i \, \phi_i^{SCF} = x_i \, \phi_i^L \qquad i = 1, 2, 3, \ldots m \tag{3}$$

where $\phi_i^{SCF}$ is the fugacity coefficient of component in in the SCF solvent phase, and $\phi_i^L$ is the fugacity coefficient of component i in the liquid phase. These fugacity coefficients are defined by the thermodynamic relationship[75]

$$RT \, Ln \, \phi_i = \int_{\infty}^{ZRT/P} \left[ \frac{RT}{v} - N \left( \frac{\partial P}{\partial Ni} \right)_{T,v,N_{j \neq i}} \right] dv - LnZ \tag{4}$$

A single equation of state can be used in Equation 4 to calculate both fugacity coefficients. The most commonly used equations of state are the Peng-Robinson[76] and the Soave-Redlich-Kwong equations.[77]

Usually a single adjustable mixture parameter which accounts for specific binary molecular interactions is incorporated into the equation of state. In some cases, it has been necessary to make this parameter a function of temperature when specific binary interactions between mixture constituents are especially strong.[57,67] It is also possible to incorporate a second adjustable binary interaction parameter to account for the large size disparity between the mixture constituents.[78,79]

Recently, Prausnitz[80] and others[81,82] have combined local composition models, such as the NRTL equation,[83] with a simple cubic equation of state to account for the nonrandom mixing of the mixture constituents as well as specific binary interactions between mixture constituents. This approach results in a three-parameter equation of state which appears to work well with both nonpolar asymmetric mixtures (i.e., mixtures with components which differ significantly in size) and with polar mixtures.

For solid-SCF solvent equilibria, the fugacity of component i in the SCF solvent phase can be calculated in the previously described manner, that is,

$$f_i^{SCF} = y_i \, \phi_i^{SCF} \, P \qquad i = 1, 2, 3, \ldots m \tag{5}$$

In this instance the solid phase is normally considered a pure solid so that the fugacity of component i as a pure solid phase is

$$f_i^{solid} = p_i^{sub} \, \phi_i^{sub} \, exp \left\{ \frac{1}{RT} \int_{P_i^{sub}}^{P} v_i^{solid} \, dP \right\} \tag{6}$$

where $P_i^{sub}$ is the sublimation pressure of the pure solid at the system temperature, $\phi_i^{sub}$ is the fugacity coefficient at T and $p_i^{sub}$, and the exponential term is the Poynting correction for the fugacity of the pure solid. Therefore, the solubility of a heavy nonvolatile solid in the supercritical fluid solvent phase now becomes

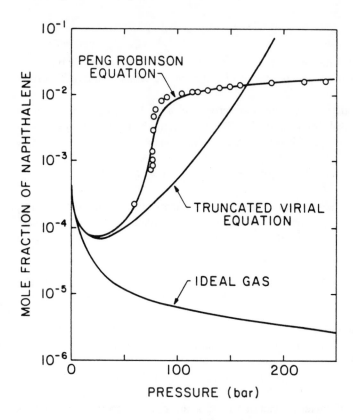

FIGURE 13.    Comparison of experimental and calculated solubilities for solid naphthalene in compressed carbon dioxide (experimental data of Tsekhanskaya et al.[22]). (From Paulaitis, M. E., Krukonis, V. J., Kurnik, R. T., and Reid, R. C., *Rev. Chem. Eng.*, 1, 179, 1982. With permission.)

$$y_2 = \frac{P_2^{sub}}{P} \left\{ \frac{\phi_2^{sub} \exp\left[ \dfrac{1}{RT} \displaystyle\int_{P_2^{sub}}^{P} v_2^{solid} \, dP \right]}{\phi_2^{SCF}} \right\} \tag{7}$$

As noted by a number of authors,[12,14,84] the term in brackets, which is called the enhancement factor, accounts for the large increase in solubility of the heavy component in the SCF solvent compared to the solubility of the heavy component in an ideal gas. The enhancement factor is a strong function of $\phi_i^{SCF}$ which, as previously mentioned, can be calculated from a simple cubic equation of state. Although a number of authors[10,25] use the virial equation to calculate $\phi_i^{SCF}$ as shown in Figure 13, the virial equation does not adequately predict the fugacity coefficient at the conditions of liquid-like density exhibited by a SCF solvent. Kurnik et al.[67] show that the Peng-Robinson equation with a single adjustable mixture parameter, which accounts for specific molecular interactions of the mixture components, can adequately represent solid solubility behavior. However, they need to make the interaction parameter a weak function of temperature to fit their data. To account for the large size disparity between a heavy nonvolatile solid and a light SCF solvent, Chai[79] and others[78] incorporate a second adjustable parameter into the equation of state. Johnston and Eckert[85,86] use an augmented van der Waals equation of state to predict solid solubilities in SCF solvents. They combine the Carnahan-Starling equation for the repulsive contribution to the pressure with the attractive term of the van der Waals equation. Their approach is reasonably successful, especially for extremely nonvolatile solids for which no critical property information exists.

Kurnik and Reid[87] also use the Peng-Robinson equation of state to correlate their data on the solubility of two pure solids in a SCF solvent. Their results are very sensitive to the value of the binary solid$_1$-solid$_2$ interaction parameter which they use. While Holder and Gopal[68] find similar results, they also show that the calculated solid solubilities are dependent on whether they have a mixture of two pure solids or whether the solids have melted and formed a single mixed solid phase.

At the high pressures and liquid-like densities encountered with a SCF solvent extraction process, the distinction between gas and liquid phases becomes less clear. It is, therefore, also possible to model the SCF solvent phase as an expanded liquid rather than a compressed gas.[14,72,88] MacKay and Paulaitis[88] use this approach to calculate solid solubilities. They state that although an activity coefficient parameter and an equation of state parameter are needed, it may be possible to estimate the solubility of a solid in a SCF solvent phase from the solid solubility in the corresponding liquid solvent.

Although Scott and van Konynenburg have shown that the van der Waals equation of state will qualitatively predict virtually all of the known binary phase behavior,[27,28] simple cubic equations of state fail to quantitatively predict the phase behavior of SCF solvent systems. However, a simple cubic equation of state can be used to correlate data for SCF solvent systems quite well if empirical parameters are incorporated into the equation. Further advances in modeling SCF solvent systems will eventually depend on our ability to adequately describe the dense fluid state and the interactions of highly asymmetric mixture components.

## III. RECENT TRENDS IN RESEARCH AND DEVELOPMENT OF SUPERCRITICAL FLUID (SCF) SOLVENT EXTRACTION

As previously mentioned, the reviews by Paulaitis et al.[14] and Randall[17] cover virtually all of the experimental activity which has occurred with SCF solvent extraction from about the turn of this century, and especially within the last decade, to about early 1982. The focus of this section will be to review the experimental studies completed since early 1982 until early 1983. Also, a brief description of some of the more interesting SCF solvent patents will be presented.

The recent review by Williams[12] gives an indication of the breadth of problems to which SCF solvent extraction has been applied. No attempt is made here to repeat his comprehensive review; however, it is instructive to note the wide range of separation problems for which SCF solvent extraction has been applied. These problems include the decaffeination of coffee, extraction of hops, spices, and tobacco, fractionation of high boiling mixtures, deasphalting of heavy petroleum fractions, extraction of mineral deposits, and tertiary oil recovery with SCF solvents, especially with supercritical carbon dioxide.

It is also instructive to note the major areas in which Paulaitis et al.[14] consider that SCF solvent processes have had a recent impact. These six categories are

1. Separation of chemicals from water streams
2. Oil extraction from seeds and foods
3. Coffee and tea decaffeination processes
4. Processing low-vapor pressure oils
5. Extraction of solvents, monomers, oligomers, fractionation of polymers
6. Regeneration of activated carbon and redistribution of particle size via SCF nucleation

Their review of SCF solvent process development is also quite comprehensive, and it focuses on current applications of SCF solvent technology.

**Table 4**
## COMPANIES INVOLVED IN SUPERCRITICAL FLUID SOLVENT PROCESS DEVELOPMENT

| Company | Activity |
|---|---|
| General Foods | Extraction of caffeine from coffee |
| Nestlé | Extraction of caffeine from solution |
| du Pont | Extraction of organic acids from water; separation of polymer from solvents[96-98] |
| Proctor and Gamble | — |
| Archer-Daniels-Midland | — |
| A. E. Staley | — |
| Dow Chemical | — |
| Exxon Chemical Company | Separation of polymer from solvent[71] |
| General Electric Co. | Polymer fractionation |
| Champion International | Extraction of tall oil from wood |
| Modar, Inc. | Waste detoxification using supercritical water |
| Critical Fluid Systems | Alcohol-water separation; activated carbon regeneration; potato chip degreasing; organic solvent-water separation[89] |
| Air Products and Chemicals | Air Products has been granted U.S. licensing rights for the Max Planck Institute's patented supercritical fluid extraction technology; they are developing a crank case oil rerefining process and in the future they may offer processes for extraction and deodorization of seed oils, extraction of tobacco, spices, hops, coffee, and tea, and purification of aluminum alkyls |
| Phasex Corporation | Fractionation of low vapor pressure oils; extraction of monomers and oligomers from polymers; production of solid materials of controlled particle size; extraction of refractory or heat-labile organics from water; separation of high melting point aromatic isomers; deposition of materials in microporous substrates |
| Suprex Corporation | Purification of biochemicals and food additives[101] |

Adapted from Paulaitis, M. E., Krukonis, V. J., Kurnik, R. T., and Reid, R. C., *Rev. Chem. Eng.*, 1, 179, 1982. With permission.

An indication of the current industrial commitment to SCF solvent process development is indicated by the list of companies shown in Table 4. This list is by no means exhaustive, since in many instances companies either have not publicized their activities in SCF solvent process development or, if it is known that they are involved in SCF solvent process development, they have not authorized disclosure of such information. The fact that SCF solvent technology is in a state of rapid development is indicated by the fact that no less than five of the companies listed in Table 3 (i.e., Phasex Corporation, Air Products and Chemicals, Inc., Critical Fluid Systems, Modar, Inc., and Suprex Corporation) are developing SCF solvent technology for the marketplace.

Since late 1981, the bulk of SCF solvent process development has occurred in the areas of separation of industrial solvents, separation and purification of pharmaceuticals and natural

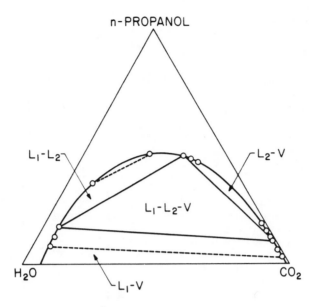

FIGURE 14.    Phase behavior for the ternary *n*-propanol-water-carbon dioxide system at a temperature of 313 K and a pressure of 103 bar.[90]

products, and polymer and/or high molecular weight oil processing. Each of these three areas are discussed separately.

## A. Industrial Solvents Separation

As noted earlier, the government is currently scrutinizing the dumping of industrial wastes much more closely. Also, the utility costs associated with conventional solvent separation techniques, such as distillation, have increased quite rapidly within the past few years. These developments have supplied the impetus for the development of SCF solvent separation processes.

Recently, Critical Fluid Systems, Inc. announced a new SCF separations process for recovering industrial chemicals such as 2-propanol, methyl ethyl ketone, 2-butanol, vinyl acetate, methyl isobutyl ketone, 1-butanol, and acetone from aqueous solutions.[89] They contend that using supercritical $CO_2$ to recover these solvents from water uses substantially less energy than distillation. Although Critical Fluid Systems does not give any technical information concerning their process, the basic principles of an industrial solvent separation with a SCF solvent can be understood by considering the *n*-propanol-water-$CO_2$ system shown in Figure 14.[90]

Two possibilities exist for splitting *n*-propanol-water mixtures which are in the range of approximately 20 to 75% (w/w) *n*-propanol. As shown in Figure 14, the separation process can operate either in the two-phase liquid-liquid (LL) region or in the three-phase liquid-liquid-vapor (LLV) region. If it is assumed that the process operates in the LLV region, then there are no degrees of freedom for this system (see Table 2) since the temperature is fixed at 40°C and the pressure is fixed at 1500 psia. Although the relative amounts of the three equilibrium phases will depend on the overall mixture composition within this region, their compositions (based on weight) always remain fixed at approximately 18% *n*-propanol-72% water-10% $CO_2$, 45% *n*-propanol-20% water-35% $CO_2$, and 12% *n*-propanol-1% water-87% $CO_2$. Thus, to recover *n*-propanol, either the $CO_2$-rich phase can be removed (i.e., the ratio of *n*-propanol to water in this phase is 12 to 1) or the middle *n*-propanol-rich phase can be removed (i.e., here the ratio of *n*-propanol to water is 2.25:1). The particular process

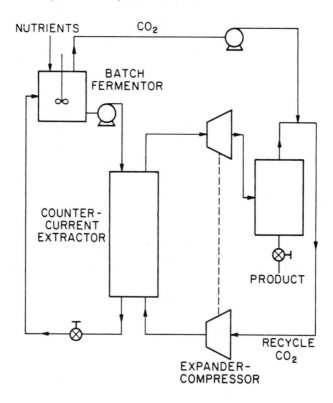

FIGURE 15.    Process schematic for the SCF solvent extraction of fermentation products.[40,102]

economics will dictate whether it is more economical to recover *n*-propanol from the middle liquid phase, which has a high loading of *n*-propanol but a low selectivity of *n*-propanol to water, or the $CO_2$-rich phase, which has less alcohol but a much higher alcohol-to-water ratio. A large number of organic solvent-water-SCF solvent systems can be found in the work of Elgin and co-workers.[51,55]

Another area which is receiving current attention is the formation of various chemical feedstocks, especially ethanol, from fermentation processes. In this instance, it is possible to use SCF solvents to extract the fermentation products from dilute aqueous solutions. Again, it appears that carbon dioxide is the preferred SCF solvent, since, in addition to $CO_2$'s other favorable attributes, it is actually produced in the fermentation process and it only exhibits limited mutual solubility with water at high pressures.[40] A proposed process schematic for such an extraction process is shown in Figure 15.

The fermentation broth from a batch fermentor is continuously fed to a countercurrent SCF solvent extraction unit in which supercritical $CO_2$, at mild temperatures, extracts the fermentation products. The loaded supercritical $CO_2$ phase is then expanded to low pressure through a turbine to recover the fermentation products in a flash vessel. The energy produced in the turbine can be used to recompress the $CO_2$ which is recycled back to the process. The carbon dioxide produced in the fermentation process can also be used to supply make-up solvent replacing the $CO_2$ which is lost with the separated fermentation products.

## B. SCF Solvent Processing of Pharmaceuticals and Natural Products

The Germans have developed a large body of experimental data on the solubility of natural products such as oils from castor seeds, steroids, and caffeine from coffee beans in various

SCF solvents such as $CO_2$, ethane, ethylene, and $N_2O$. Carbon dioxide is by far the most widely used SCF solvent, since its critical temperature (Tc = 304.2 K) makes it an ideal solvent for extracting natural products which are thermally labile. Also, $CO_2$ is nontoxic, nonflammable, environmentally acceptable, and inexpensive.

Stahl and co-workers[91,92] have developed a microextraction apparatus which they directly couple with thin layer chromatography for quickly screening the solubilities of a wide range of natural products in candidate SCF solvents. In a recent study[91] they established a number of variables which control the solubility of natural products in supercritical $CO_2$. They found that:

1.    Fractionation of condensed phases is possible if the mixture constituents exhibit large differences in vapor pressure, mass, or polarity.
2.    Low molar mass hydrocarbons and lipophilic organic compounds such as esters, ethers, and lactones are easily extractable.
3.    Hydroxyl and carboxyl substituent groups on the mixture constituents makes the extraction extremely difficult.
4.    Sugars and amino acids are not extracted.

Using supercritical $CO_2$, Stahl and Quirin[91] measured the solubilities of tetracyclic steroids which differed in molecular structure but which all had virtually the same vapor pressure at the system operating temperature. They found that carboxyl groups on the steroids, such as bile acids, rendered the steroids virtually insoluble in supercritical $CO_2$, while carbonyl groups had very little effect on steroid solubilities. The differences in the masses and melting points of the steroids had no direct influence on the solubility behavior. Hence, they argue that high pressure extraction constitutes a new separation and fractionation technique which does not depend on differences in the mixture components' vapor pressure and which exhibits selectivity for certain classes of compounds.

It is also interesting to note that while supercritical $CO_2$ will selectively extract caffeine from raw coffee beans,[93] this selectivity is lost if roasted coffee beans are used.[94] Entrainers can be added to the SCF solvent phase to make it more selective for caffeine,[95] although, in general, more work is needed with entrainers to fully understand their role in the extraction process.[96]

## C. Polymer and High Molecular Weight Oil Processing

As mentioned earlier, the fact that polymer-solvent systems exhibit lower critical solution (LCST) phenomena — that is, miscible polymer-solvent mixtures will separate into a polymer-rich phase and a solvent-rich phase at a temperature close to the solvent critical temperature (Tc) — has obvious practical industrial significance. LCST phenomena are a consequence of the chemical nature of the mixture components, the molecular sizes of the components, especially the polymer, and the Tc and Pc of the solvent.[30] If the polymer molecular weight is on the order of 10,000, the LCST will occur within approximately 20 to 30 K of the solvent's Tc. However, if the polymer molecular weight is closer to 1 million, then the phase separation can occur as low as 120 K below the Tc of the solvent. Since polymer solvents tend to be very dense liquids, it is often necessary to raise the system temperature to well over 473 K to obtain a phase separation. Separating polymer from solvent at very high temperatures may cause thermal degradation of the polymer.

Recently, Irani et al.[71] have patented a process for lowering the LCST of a polymer-solvent mixture by adding a SCF solvent to the system. An example of the effect of the SCF solvent additive is shown in Figure 16. The phase behavior depicted in this figure is similar to that described for the type IV system shown in Figure 3d. The major difference between the phase diagrams in Figures 16 and 3d is that the branch of the LLV line near

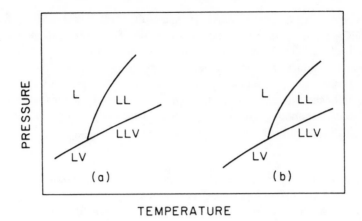

FIGURE 16.   Schematic representation of the effect of a SCF solvent on the phase behavior of polymer-organic solvent mixtures.[71] (a) PT projection of various phase boundary curves for a polymer-organic solvent mixture with a SCF solvent added to the system; (b) same as (a) but without the SCF solvent.

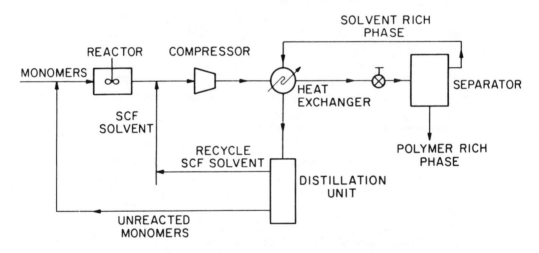

FIGURE 17.   Process schematic for the recovery of polymer from solution using a SCF solvent.[71]

C1, shown in Figure 3d, is now superimposed on the pure solvent vapor pressure curve in Figure 16. Also, the LL line in Figure 3d has been left off of Figure 16. By adding approximately 5 to 15% (w/w) supercritical propylene to a poly-(ethylene propylene)-hexane mixture, the phase border curves for this system are shifted to lower temperatures. For example, Irani et al.[71] state that adding supercritical propylene reduces the phase separation temperature by approximately 30°. The polymer-rich phase that is recovered now contains about 40% polymer. Conventional techniques, such as steam stripping, are used to recover the polymer from this phase. In addition, the added SCF solvent reduces the amount of residual polymer remaining in the solvent-rich phase by about a factor of three (e.g., in some instances from about 0.22 [w/w]% polymer to about 0.08 [w/w]%).

A process schematic for such a polymer-solvent-SCF solvent additive process is shown in Figure 17. Subcritical propylene is added to the reactor product stream prior to the compressor. The entire mixture is first compressed to the desired operating pressure and then heated to much higher temperatures. The mixture is now a single liquid phase. Also,

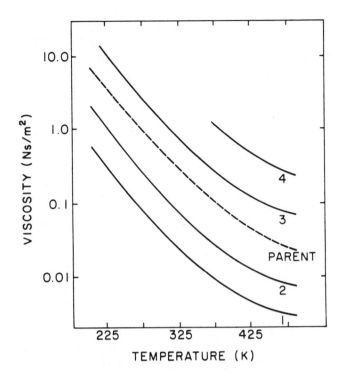

FIGURE 18.   Viscosity of a parent perfluoralkylpolyether oil and of the various fractions of the oil fractionated with supercritical carbon dioxide.[100]

the propylene is now supercritical (Tc = 365 K, Pc = 46.2 bar). The polymer mixture is subsequently throttled to lower pressures. As a consequence of reducing the pressure, the system now forms three phases (LLV) in the separator. At this point the process could be operated in the two-phase LL region of the phase diagram. However, Irani et al.[71] state that a better solvent-polymer split is obtained in the LLV region. Also, lower operating pressures are realized, and the small amount of vapor which is formed can be vented to maintain a constant system operating pressure. The lower polymer-rich phase is recovered as a bottoms stream from the separator for further processing. Conventional techniques can be used to further remove solvent from this stream. The lighter solvent-rich phase is recovered as an overhead stream, integrated with the heat exchanger to recover its heat content, and purified and recycled back to the front end of the process. When compared to recovering polymer and solvent by steam stripping the entire reactor product stream, this SCF solvent additive process should prove to be less energy intensive. Also, less thermal degradation of the polymer should occur at the more moderate process operating temperature. Therefore, product quality control (i.e., polymer molecular weight distribution) should be easier, and less polymer degradation inhibitors would be needed.[97-99]

Krukonis[100] has recently published data on a number of polymers which he separates and fractionates using SCF solvents. In one study, perfluoroalkyl polyethers, which are used as vacuum oil and for specialty lubrication applications, are fractionated with supercritical $CO_2$. In this study, viscosity is used to characterize the polymer fractions (i.e., the viscosity is approximately proportional to the square root of the molecular weight of the polymer). Shown in Figure 18 are the viscosity data (i.e., the degree of separation) of a parent 10,000-molecular-weight perfluoroalkyl polymer which is fractionated with supercritical $CO_2$ at 353 K and pressures ranging from 80 to 275 bar. Krukonis notes that although these polymers do not dissolve in aliphatic or aromatic liquid solvents, they are readily soluble in supercritical

## Table 5
### MOLECULAR WEIGHT ANALYSIS OF MALEIC ANHYDRIDE-TRIAMINE COPOLYMER FRACTIONS SEPARATED WITH SUPERCRITICAL ETHYLENE

| Fraction | Molecular weight | % of Parent |
|---|---|---|
| Parent | 4000 | 100 |
| 1 | 480 | 15.9 |
| 2 | 480 | 9.7 |
| 3 | 530 | 2.9 |
| 4 | 1500 | 7.6 |
| 5 | 2000 | 7.6 |
| 6 | 3000 | 11.3 |
| 7 | 5500 | 26.4 |
| 8 | 7100 | 5.3 |
| 9 | 8000 | 5.1 |
| 10 | 9000 | 5.1 |
| 11 | 9000 | 3.3 |

$CO_2$. This perfluoroether oil is used as a lubricant for advanced aircraft engines. Although it exhibits a very low vapor pressure, the vapor pressure is directly related to the amount of low molecular components in the oil. Hence, removing the low-molecular-weight components with supercritical $CO_2$ greatly improves the performance of the oil.

Polymers of silicon and carbon (i.e., polycarbosilane polymers) have also been fractionated with SCF solvents.[100] These polymers are being considered in the production of silicon carbide shapes and filaments for high temperature application.[100]

In another application, supercritical ethylene at a temperature of 393 K and a pressure range of 60 to 320 bar is found to fractionate a copolymer of polyalkene maleic anhydidetriamine.[100] This particular copolymer is an excellent surfactant for stabilizing suspensions of 200-Å particles in a hydrocarbon solution. As a surfactant it is necessary to have a narrow polymer molecular weight range to avoid adverse suspension viscosities. Shown in Table 5 are the results of a gel permeation chromatographic analysis of the parent polymer and the polymer fractions after processing with supercritical ethylene. It is evident that SCF solvent processing can be applied in this instance to obtain narrow molecular weight fractions of this copolymer and thus a better surfactant.

## IV. CONCLUSION

As noted in this and other review papers,[1,9-14,16] extraction with SCF solvents has emerged as a viable alternative to conventional separation techniques such as distillation and liquid extraction. Future applications of this technology are dependent on the progress which we make in understanding the high pressure fluid phase behavior of mixtures, since many SCF solvent extraction processes operate in the critical or near-critical region of one or more of the mixture components.

Although the mass transfer properties of SCF solvents are not considered in this paper, a review of this area can be found elsewhere.[14] While it is often mentioned that the diffusion coefficient of a SCF solvent is higher than that of a liquid, and that the viscosity of a SCF solvent is closer to that of a gas than a liquid, much more work is needed to fully capitalize on the unique transport properties of SCF solvents.

Finally, considering the unique solvent and transport properties of SCF solvents, it is surprising that very little experimental information is available on the use of SCF solvents as a reaction medium.[104-106] With a SCF solvent reaction medium, it may be possible to run reactions in a homogeneous fluid and to adjust the pressure and temperature to selectively precipitate reactants, products, or, perhaps, catalysts. However, much more experimental information is needed to develop a better understanding of this technology.

## ADDENDUM

As with any new area of research in its early development, supercritical fluid extraction has experienced an explosion of information in the period between the time that this review paper was accepted for publication (June 1983) to the time that it was published. While it is not possible to review all of the new information for this paper three major publications should be mentioned. Two of these publications are the direct result of two symposia dealing exclusively with supercritical fluid technology which were held in this time period. The papers from one of these symposia were published as a special volume of a journal.[107] The papers from the other symposium were published as a compendium[108] in book form. The third publication, a monograph dealing with the principles and practice of supercritical fluid extraction,[109] is scheduled for publication at about the same time that this article appears. These sources of information complement this review paper and offer a large amount of new information of interest to the practitioner as well as to the technical manager who may be interested in pursuing this new technology.

## REFERENCES

1. **Schneider, G. M., Stahl, E., and Wilke, G., Eds.,** *Extraction with Supercritical Gases,* Verlag Chemie, Deerfield Beach, Fla., 1980.
2. **Diepen, G. A. M. and Scheffer, F. E. C.,** The solubility of naphthalene in supercritical ethylene, *J. Am. Chem. Soc.,* 70, 4085, 1948.
3. **Hannay, J. B. and Hogarth, J.,** On the solubility of solids in gases, *Proc. R. Soc. London,* 29, 324, 1879.
4. **Hannay, J. B. and Hogarth, J.,** On the solubility of solids in gases, *Proc. R. Soc. London,* 30, 178, 1880.
5. **Hannay, J. B.,** On the solubility of solids in gases, *Proc. R. Soc. London,* 30, 484, 1880.
6. **Paulaitis, M. E., Penninger, J. M. L., Davidson, P., and Gray, R. D., Jr.,** Parts I to V, AIChE Natl. Meet., New Orleans, November 1981.
7. Supercritical Fluids: Their Chemistry and Applications, Meeting of the Royal Society of Chemistry, Faraday Div., Cambridge, England, September 13 to 15, 1982.
8. Extraction with Supercritical Gases, Essen, Germany, June 5 to 6, 1978.
9. **Paul, P. M. F. and Wise, W. S.,** *The Principles of Gas Extraction,* Mills and Boon, London, 1971.
10. **Irani, C. A. and Funk, E. W.,** Separations using supercritical gases, in *Recent Developments in Separation Science,* Vol. 3, Part A, CRC Press, Boca Raton, Fla., 1977, 171.
11. **Gangoli, N. and Thodos, G.,** Liquid fuels and chemical feedstocks from coal by supercritical gas extraction, *Ind. Chem. Prod. Res. Dev.,* 16, 208, 1977.
12. **Williams, D. F.,** Extraction with supercritical gases, *Chem. Eng. Sci.,* 36, 1769, 1981.
13. **Brunner, G. and Peter, S.,** Zum Stand der Extraktion mit Komprimierten Gasen, *Chem. Ing. Tech.,* 53, 529, 1981.
14. **Paulaitis, M. E., Krukonis, V. J., Kurnik, R. T., and Reid, R. C.,** Supercritical fluid extraction, *Rev. Chem. Eng.,* 1, 179, 1982.
15. **Modell, M.,** Process Using a Supercritical Fluid for Regenerating Synthetic Organic Polymeric Adsorbents and Wastewater Treatment Embodying the Same, U.S. Patent 4,016,506, 1977.
16. **Paulaitis, M. E., Penninger, J. M. L., Gray, R. D., and Davidson, P., Eds.,** *Chemical Engineering at Supercritical Fluid Conditions,* Ann Arbor Science, Ann Arbor, Mich., 1983.
17. **Randall, L. B.,** The present status of dense (supercritical) gas extraction and dense gas chromatography: impetus for DGC/Ms development, *Sep. Sci. Technol.,* 17, 1, 1982.
18. **Johnston, K.,** Supercritical fluids, in *Encyclopedia of Chemical Technology,* 3rd ed., suppl. vol., John Wiley & Sons, New York, 1984, 872.

19. **Schneider, G. M.,** Physicochemical principles of extraction with supercritical gases, *Angew. Chem. Int. Ed. (Eng.),* 17, 716, 1978.

20. **Swaid, I. and Schneider, G. M.,** Determination of binary diffusion coefficients of benzene and some alkylbenzenes in supercritical $CO_2$ between 308 and 328 K in the pressure range 80 to 160 bar with supercritical fluid chromatography (SFC), *Ber. Bunsenges. Phys. Chem.,* 83, 969, 1979.

21. **Diepen, G. A. M. and Scheffer, F. E. C.,** The solubility of naphthalene in supercritical ethylene. II, *J. Phys. Chem.,* 57, 575, 1953.

22. **Tsekhanskaya, Yu. Y., Iomtev, M. B., and Mushkina, E. V.,** Solubility of naphthalene in ethylene and carbon dioxide under pressure, *Russ. J. Phys. Chem.,* 38, 1173, 1964.

23. **Hicks, C. P. and Young, C. L.,** The gas-liquid critical properties of binary mixtures, *Chem. Rev.,* 75, 119, 1975.

24. **Streett, W. B.,** Phase equilibria in fluid and solid mixtures at high pressure, in *Chemical Engineering at Supercritical Fluid Conditions,* Paulaitis, M. E., Penninger, J. M. L., Gray, R. D., and Davidson, P., Eds., Ann Arbor Science, Ann Arbor, Mich., 1983.

25. **Rowlinson, J. S. and Richardson, M. J.,** The solubility of solids in compressed gases, *Adv. Chem. Phys.,* 2, 85, 1958.

26. **Rowlinson, J. S. and Swinton, F. L.,** *Liquids and Liquid Mixtures,* 3rd ed., Butterworths, Boston, 1982, chap. 6.

27. **Scott, R. L.,** Thermodynamics of critical phenomena in fluid mixtures, *Ber. Bunsenges. Phys. Chem.,* 76, 296, 1972.

28. **Scott, R. L. and van Konynenburg, P. H.,** II. Static properties of solutions — Van der Waals and related models for hydrocarbon mixtures, *Discuss. Faraday Soc.,* 49, 87, 1970.

29. **Schneider, G. M.,** Phase equilibria in fluid mixtures at high pressures, *Adv. Chem. Phys.,* 17, 1, 1970.

30. **Allen, G. and Baker, C. H.,** Lower critical solution phenomena in polymer-solvent systems, *Polymer (London),* 6, 181, 1965.

31. **Freeman, P. I. and Rowlinson, J. S.,** Lower critical points in polymer solutions, *Polymer (London),* 1, 20, 1960.

32. **Baker, C. H., Clemson, C. S., and Allen, G.,** Polymer fractionation at a lower critical solution temperature phase boundary, *Polymer (London),* 1, 525, 1966.

33. **Zeman, L., Biros, J., Delmas, G., and Patterson, D.,** Pressure effects in polymer solution phase equilibria. I. The lower critical solution temperature of polyisobutylene and polydimethylsiloxane, *J. Phys. Chem.,* 76, 1206, 1972.

34. **Zeman, L. and Patterson, D.,** Pressure effects in polymer solution phase equilibria. II. Systems showing upper and lower critical solution temperatures, *J. Phys. Chem.,* 76, 1214, 1972.

35. **Siow, K. S., Delmas, G., and Patterson, D.,** Cloud-point curves in polymer solutions with adjacent upper and lower critical solution temperatures, *Macromolecules,* 5, 29, 1972.

36. **Diepen, G. A. M. and Scheffer, F. E. C.,** On critical phenomena of saturated solutions in binary systems, *J. Am. Chem. Soc.,* 70, 4081, 1948.

37. **McHugh, M. A.,** An Experimental Investigation of the High Pressure Fluid Phase Equilibrium of Highly Asymmetric Binary Mixtures, Ph.D. thesis, University of Delaware, Newark, 1981.

38. **McHugh, M. A., Ayarza, J., and Yogan, T. J.,** An Experimental Investigation of the High Pressure Phase Behavior of Octacosane and Carbon Dioxide, presented at the Natl. AIChE Meet., Houston, March 1983.

39. **Paulaitis, M. E., McHugh, M. A., and Chai, C. P.,** Solid solubilities in supercritical fluids at elevated pressures, in *Chemical Engineering at Supercritical Fluid Conditions,* Paulaitis, M. E., Penninger, J. M. L., Gray, R. D., and Davidson, P., Eds., Ann Arbor Science, Ann Arbor, Mich., 1983, 139.

40. **Paulaitis, M. E. and McHugh, M. A.,** Practical Application of High Pressure Phase Equilibria Near the Critical Region, presented at the 17th State-of-the-Art Symp., High Pressure as a Reagent and an Environment, American Chemical Society, Washington, D.C., June 10, 1981.

41. **McHugh, M. A. and Yogan, T. J.,** A study of three phase solid-liquid-gas equilibria for three $CO_2$-solid, two ethane-solid, and two ethylene-solid systems, *J. Chem. Eng. Data,* 29, 112, 1984.

42. **McHugh, M. A. and Paulaitis, M. E.,** Solid solubilities of naphthalene and biphenyl in supercritical carbon dioxide, *J. Chem. Eng. Data,* 25, 326, 1980.

43. **Ehrlich, P. and Takahashi, T.,** Absolute Rate Constants for the Free Radical Polymerization of Ethylene in Supercritical Monomer/Polymer Solutions, presented at the Annual AIChE Meet., New Orleans, November 1981.

44. **Ehrlich, P.,** Phase equilibria of polymer-solvent systems at high pressures near their critical loci. II. Polyethylene-ethylene, *J. Polym. Sci. A,* 3, 131, 1965.

45. **Ehrlich, P. and Graham, E. B.,** Solubility of polymers in compressed gases, *J. Polym. Sci.,* 45, 246, 1960.

46. **Ehrlich, P. and Kurpen, J. J.,** Phase equilibria of polymer-solvent systems at high pressures near their critical loci: polyethylene with n-alkanes, *J. Polym. Sci. A,* 1, 3217, 1963.

47. **de Loos, T. W., Poot, W., and Diepen, G. A. M.,** Fluid phase equilibria in the system polyethylene + ethylene. I. Systems of linear polyethylene + ethylene at high pressure, *Macromolecules,* 16, 111, 1983.
48. **Ham, J. S., Bolen, M. C., and Hughes, J. K.,** The use of high pressure to study polymer-solvent interactions, *J. Polym. Sci.,* 57, 25, 1962.
49. **Enscore, D. J., Hopfenberg, H. B., and Stannett, V. T.,** Diffusion, swelling, and consolidation in glossy polystyrene microspheres, *Polym. Eng. Sci.,* 20, 102, 1980.
50. **Francis, A. W.,** Ternary systems of liquid carbon dioxide, *J. Phys. Chem.,* 58, 1099, 1954.
51. **Elgin, J. C. and Weinstock, J. J.,** Phase equilibrium at elevated pressures in ternary systems of ethylene and water with organic liquids, *J. Chem. Eng. Data,* 4, 3, 1959.
52. **Chappelear, D. C.,** Phase Equilibria in the Critical Region — Binary Systems with Chlorotrifluoromethane, Ph.D. thesis, Princeton University, 1960.
53. **Close, R. E.,** Vapor-Liquid Equilibrium at the Critical Region — Systems of Aliphatic Alcohols with Propane and Propylene, Ph.D. thesis, Princeton University, 1951.
54. **Todd, D. B.,** Phase Equilibria in Systems with Supercritical Ethylene, Ph.D. thesis, Princeton University, 1952.
55. **Weinstock, J. J.,** Phase Equilibrium at Elevated Pressure in Ternary Systems of Ethylene and Water and Organic Liquids, Ph.D. thesis, Princeton University, 1954.
56. **Culberson, O. L. and McKetta, J. J.,** Phase equilibria in hydrocarbon-water systems. III. The solubility of methane in water at pressures of 10,000 psia, *Petrol. Trans. AIME,* 192, 223, 1951.
57. **McHugh, M. A., Mallett, M. W., and Kohn, J. P.,** High pressure fluid phase equilibria of alcohol-water-supercritical solvent mixtures, in *Chemical Engineering at Supercritical Conditions,* Paulaitis, M. E., Penninger, J. M. L., Gray, R. D., and Davidson, P., Eds., Ann Arbor Science, Ann Arbor, Mich., 1983, 113.
58. **Paulaitis, M. E., Gilbert, M. L., and Nash, C. A.,** Separation of Ethanol-Water Mixtures with Super-critical Fluids, presented at the 2nd World Congr. of Chemical Engineers, Montreal, October 4 to 9, 1981.
59. **Kuk, M. S. and Montagna, J. C.,** Solubility of oxygenated hydrocarbons in supercritical carbon dioxide, in *Chemical Engineering at Supercritical Fluid Conditions,* Paulaitis, M. E., Penninger, J. M. L., Gray, R. D., and Davidson, P., Eds., Ann Arbor Science, Ann Arbor, Mich., 1983, 101.
60. **Treybal, R. E.,** *Mass-Transfer Operations,* 2nd ed., McGraw-Hill, New York, 1968, chap. 10.
61. **Merrill, R. C.,** Liquid-Liquid-Vapor Phenomena in Cryogenic Liquefied Natural Gas Systems, Ph.D. thesis, University of Notre Dame, Notre Dame, Ind., 1983.
62. **Merrill, R. C., Luks, K. D., and Kohn, J. P.,** Three phase liquid-liquid-vapor equilibria in the methane + n-pentane + n-octane, methane + n-hexane + n-octane, and methane + n-hexane + carbon dioxide systems, *J. Chem. Eng. Data,* 28, 210, 1983.
63. **Hottovy, J. D., Kohn, J. P., and Luks, K. D.,** Partial miscibility behavior of the ternary systems methane + propane + n-octane, methane + n-butane + n-octane, and methane + carbon dioxide + n-octane, *J. Chem. Eng. Data,* 27, 298, 1982.
64. **Hottovy, J. D., Kohn, J. P., and Luks, K. D.,** Partial miscibility behavior of the methane + ethane + n-octane system, *J. Chem. Eng. Data,* 26, 135, 1981.
65. **Wagner, J. R., McCaffrey, D. S., and Kohn, J. P.,** Partial miscibility phenomena in the ternary system ethane-n-hexadecane-n-eicosane, *J. Chem. Eng. Data,* 13, 22, 1968.
66. **Kim, Y. J., Carfagno, J. A., McCaffrey, D. S., and Kohn, J. P.,** Partial miscibility phenomena in the ternary system ethane + n-nonadecane + n-eicosane, *J. Chem. Eng. Data,* 12, 289, 1967.
67. **Kurnik, R. T., Holla, S. J., and Reid, R. C.,** Solubility of solids in supercritical carbon dioxide and ethylene, *J. Chem. Eng. Data,* 26, 47, 1981.
68. **Gopal, J. S., Holder, J. D., Wender, I., and Bishop, A. A.,** Supercritical Behavior in Multicomponent Systems, presented at AIChE Meet., Houston, March 1983.
69. **Koningsveld, R. and Diepen, G. A. M.,** Supercritical phase behavior involving solids, *Fluid Phase Equilibria,* 10, 159, 1983.
70. **Wolf, B. A., Breitenbach, J. W., and Senftl, H.,** Upper and lower solubility gaps in the system butanone-acetone-polystyrene, *J. Polym. Sci. C,* 31, 345, 1970.
71. **Irani, C. A., Cosewith, C., and Kasegrande, S.,** New Method for High Temperature Phase Separation of Solutions Containing Copolymer Elastomers, U.S. Patent 4,319,021, 1982.
72. **Balder, J. R. and Prausnitz, J. M.,** Thermodynamics of ternary, liquid-supercritical gas systems with applications for high pressure vapor extraction, *Ind. Eng. Chem. Fund.,* 5, 449, 1966.
73. **Wenzel, H. and Rupp, W.,** Calculation of phase equilibria in systems containing water and supercritical components, *Chem. Eng. Sci.,* 33, 683, 1978.
74. **Peng, D.-Y. and Robinson, D. B.,** Two and three phase equilibrium calculations for systems containing water, *Can. J. Chem. Eng.,* 54, 595, 1976.
75. **Sandler, S. I.,** *Chemical and Engineering Thermodynamics,* 1st ed., John Wiley & Sons, New York, 1977, chap. 7.

76. **Peng, D.-Y. and Robinson, D. B.,** A new two constant equation of state, *Ind. Eng. Chem. Fund.,* 15, 59, 1976.

77. **Soave, G.,** Equilibrium constants from a modified Redlich-Kwong equation of state, *Chem. Eng. Sci.,* 27, 1197, 1972.

78. **Deiters, U. and Schneider, G. M.,** Fluid mixtures at high pressures. Computer calculations of the phase equilibria and the critical phenomena in fluid binary mixtures from the Redlich-Kwong equation of state, *Ber. Bun. Ges.,* 12, 1316, 1976.

79. **Chai, C. P.,** Phase Equilibrium Behavior for Carbon Dioxide and Heavy Hydrocarbons, Ph.D. thesis, University of Delaware, Newark, 1981.

80. **Prausnitz, J. M.,** Phase equilibria for complex mixtures, presented at the 3rd Int. Conf. Fluid Properties and Phase Equilibria for Chemical Process Design, Callaway Gardens, Ga., April 10 to 15, 1983.

81. **Mathias, P. M. and Copeman, T. W.,** Extension of the Peng-Robinson equation of state to complex mixtures: evaluation of the various forms of the local composition models, presented at the 3rd Int. Conf. Fluid Properties and Phase Equilibria for Chemical Process Design, Callaway Gardens, Ga., April 10 to 15, 1983.

82. **Wong, J. M. and Johnston, K. P.,** Thermodynamic models for nonrandom and strongly-nonideal liquid mixtures, *Ind. Eng. Chem. Fund.,* submitted.

83. **Renon, H. and Prausnitz, J. M.,** Local compositions in thermodynamic excess functions for liquid mixtures, *AIChE J.,* 14, 135, 1968.

84. **Prausnitz, J. M.,** *Molecular Thermodynamics of Fluid-Phase Equilibria,* 1st ed., Prentice-Hall, Englewood Cliffs, N.J., 1969, chap. 5.

85. **Johnston, K. P. and Eckert, C. A.,** An analytical Carnahan-Starling van der Waals model for solubility of hydrocarbon solids in supercritical fluids, *AIChE J.,* 27, 773, 1981.

86. **Johnston, K. P., Zieger, D. H., and Eckert, C. A.,** Solubilities of hydrocarbon solids in supercritical fluids: the augmented van der Waals treatment, *Ind. Eng. Chem. Fund.,* 21, 191, 1982.

87. **Kurnik, R. T. and Reid, R. C.,** Solubility of solid mixtures in supercritical fluids, *Fluid Phase Equilibria,* 8, 93, 1982.

88. **MacKay, M. E. and Paulaitis, M. E.,** Solid solubilities of heavy hydrocarbons in supercritical solvents, *Ind. Eng. Chem. Fund.,* 18, 149, 1979.

89. Science/technology concentrates, critical fluid process separates solvents, *Chem. & Eng. News,* p. 44, May 9, 1983.

90. **Kuk, M. S. and Montagna, J. C.,** Bench scale experiments with supercritical fluids — heavy oil upgrading and other applications, presented at AIChE Meet., Houston, March 1983.

91. **Stahl, E. and Quirin, K. W.,** Dense gas extraction on a laboratory scale: a survey of some recent results, *Fluid Phase Equilibria,* 10, 269, 1983.

92. **Stahl, E., Schily, W., Schutz, E., and Willing, E.,** A quick method for the microanalytical evaluation of the dissolving power of supercritical gases, in *Extraction with Supercritical Gases,* Schneider, G. M., Stahl, E., and Wilke, G., Eds., Verlag Chemie, Deerfield Beach, Fla., 1980, 93.

93. **Zosel, K.,** Decaffeination of Coffee, U.S. Patent 4,260,639, 1981.

94. **Roselius, W., Vitzthum, O., and Hubert, P.,** Method for the Production of Caffeine-Free Coffee Extract, U.S. Patent 3,843,824, 1974.

95. **Roselius, W.,** Method for Selective Extraction of Caffeine from Vegetable Materials, U.S. Patent 4,255,458.

96. **Brunner, G.,** Selectivity of supercritical compounds and entrainers with respect to model substances, *Fluid Phase Equilibria,* 10, 289, 1983.

97. **Anolick, C. and Slocum, E. W.,** Process for Isolating EPDM Elastomers from their Solvent Solutions, U.S. Patent 3,726,843, 1973.

98. **Caywood, S. W.,** Crude EPDM Copolymer Stabilized with a Lewis Base, U.S. Patent 3,496,135, 1970.

99. **Anolick, C. and Goffinet, E. P.,** Separation of Ethylene Copolymer Elastomers from their Solvent Solutions, U.S. Patent 3,553,156, 1971.

100. **Krukonis, V.,** Supercritical fluid fractionation — an alternative to molecular distillation, presented at National AIChE Meet., Houston, March 1983.

101. **Houck, R. K.,** private communication, Suprex Corporation, Pittsburgh, Pa., 1983.

102. **Shimshick, E. J.,** Removal of Organic Acids from Dilute Aqueous Solutions of Salts of Organic Acids by Supercritical Fluids, U.S. Patent 4,250,331, 1981.

103. **van Gunst, C. A., Scheffer, F. E. C., and Diepen, G. A. M.,** On critical phenomena of saturated solutions in binary systems. II, *J. Phys. Chem.,* 57, 578, 1953.

104. **Kramer, G. M. and Leder, F.,** Paraffin Isomerization in Supercritical Fluids, U.S. Patent 3,880,945, 1975.

105. **Squires, T. G., Venier, C. G., and Aida, T.,** Supercritical fluid solvents in organic chemistry, *Fluid Phase Equilibria,* 10, 261, 1983.

106. **Metzger, J. O., Hartmans, J., Malwitz, D., and Koll, P.,** Thermal organic reactions in supercritical fluids, in *Chemical Engineering at Supercritical Fluid Conditions,* Paulaitis, M. E., Penninger, J. M. L., Gray, R. D., and Davidson, P., Eds., Ann Arbor Science, Ann Arbor, Mich., 1983, 515.
107. *Ber. Busenges. Phys. Chem.,* 88, 1984.
108. **Penninger, J. M. L., Radosz, M., Mc Hugh, M. A., and Krukonis, V. J., Eds.,** *Supercritical Fluid Technology,* Elsevier, Amsterdam, 1986.
109. **McHugh, M. A. and Krukonis, V. J.,** *Supercritical Fluid Extraction: Principles and Practice,* Butterwort, Stoneham, Mass., 1986.

Chapter 6

# SEPARATION OF METAL CHELATES BY CHARGED COMPOSITE MEMBRANES

## D. Bhattacharyya* and C. S. Cheng

### TABLE OF CONTENTS

---

* Author to whom correspondence should be addressed.

# I. INTRODUCTION

Membrane processes represent a broadly applicable technique for the separation and concentration of various inorganic and organic compounds from aqueous systems.[1-3] Since the practical cellulose acetate reverse osmosis membranes came into existence 20 years ago, a significant attempt has been made to improve membrane capability and performance. The development of low pressure composite membranes has provided membranes with high solute separation characteristics and high flux at moderate operating pressure. Charged, noncellulosic, low pressure (ultrafiltration) membranes provide the concomitant advantages of moderate water flux ($8 \times 10^{-4}$ to $20 \times 10^{-4}$ cm/sec) at low transmembrane pressures of $3.0 \times 10^5$ to $8.0 \times 10^5$ Pa and the selective (based on species charge) separation of specific inorganic ions from multisalt systems.

For charged low pressure membranes, the Donnan exclusion mechanism is primarily responsible for the rejection of ionic solutes, and rejection is particularly high at low concentrations. Bhattacharyya et al. have studied the rejection characteristics of negatively charged membranes for single and multisalt systems[4-6] and actual wastewaters.[7-10] These studies showed that charged membranes provide good water flux at low pressures and 80 to 97% rejections of solutes. The rejection of co-ions was dependent on charge and types of species. For example, the rejection trend for some common anions were $PO_4^{3-}$ (98% rejection) $> HPO_4^{2-} > SO_4^{2-} > H_2AsO_4^-$ (88% rejection). The rejections of divalent metal salts were found to be $ZnCl_2 > CaCl_2 > PbCl_2$. Gregor and co-workers[11,12] have reported the use of various polystyrene sulfonic acid-polyvinyledene fluoride-charged ultrafiltration membranes with sewage effluents and paper pulp wastewater.

Nomura et al.[13] have investigated the properties of charged membranes prepared by the sulfonation and amination of resins. The sulfonated membranes showed relatively low salt rejection and high water flux, whereas aminated membranes showed high rejection and high water flux. Kimura and Jitsuhara[14] recently reported the transport characteristics of simple inorganic solutes (KBr and $Na_2SO_4$) by charged ultrafiltration membranes (18 to 24 Å pore diameter). These membranes were made from sulfonated polysulfone with a charge density of 0.8 to 1.6 meq/g.

The separation of heavy metals by charged ultrafiltration membranes represents a promising application. In metal finishing effluents, heavy metals (such as Cu, Cd, Zn, Ni, etc.) are often present as complexes (e.g., cyanide, ethylenediamine tetraacetic acid, etc.). Although the rejection behavior of metal-EDTA complexes with reverse osmosis membranes[15,16] has been reported, the rejection behavior of metal cyanide and oxalate complexes has never been studied with low pressure reverse osmosis or with conventional ultrafiltration membranes. In addition, none of these studies included the separation behavior of the associated free complexing agent. This chapter will primarily deal with the results of extensive experimental investigations of negatively charged ultrafiltration membranes to separate heavy metals ($Cu^{2+}$, $Cu^{1+}$, $Zn^{2+}$, and $Cd^{2+}$) in the presence of complexing agents (EDTA, cyanide, and oxalate). The extent of metal complexation and species charge are calculated in order to explain the membrane separation results of total metal and free complexing agents.

# II. CHARGED MEMBRANE SEPARATION MECHANISMS

Solute rejection by membrane processes can be defined in terms of a rejection parameter, R:

$$R = 1 - C_f/C_i = 1 - J_s/(J_w C_i) \qquad (1)$$

where $C_i$ = feed solute concentration, $C_f$ = permeate solute concentration, $J_s$ = solute flux, and $J_w$ = water flux.

If rejection is defined in terms of anion concentration, then Equation 1 could be rewritten in the following form:

$$R = 1 - J_y/(J_w C_y) \tag{2}$$

where $C_y$ = anion concentration in the feed and $J_y$ = anion flux.

## A. Donnan Equilibrium Model

When a charged membrane is immersed in a salt solution, as in the case of charged membrane ultrafiltration, a dynamic equilibrium condition is maintained.[17,18] The counter-ion concentration is higher while the co-ion concentration is lower in the membrane phase than in the bulk feed solution. The equilibrium Donnan potential can be expressed as the result of equilibrium (i.e., equal chemical potentials) of the components in the solution and the membrane phases; viz.,

$$E = [R'T/(Z_j F)][\ln(C_j \gamma_j/[C_{j(m)} \gamma_{j(m)}])] \tag{3}$$

For a salt, $M_{z_y} Y_{z_m}$, which ionizes to $M^{z_m}$ and $Y^{-z_y}$, the electroneutrality condition can be described as follows:

$$\text{in the membrane phase:} \quad \Sigma Z_j C_{j(m)} - C_m^* = 0 \tag{4}$$

$$\text{in the bulk feed solution:} \quad \Sigma Z_j C_j = 0 \tag{5}$$

where $C_m^*$ is the charge capacity of the membrane.

The charge capacity, $C_m^*$, for negatively charged ultrafiltration membranes typically varies between 300 and 2000 m$M$; thus, the salt distribution between an aqueous solution and membrane is expressed as:

$$(C_{y(m)}/C_y)^{Z_m}(\gamma_m/\gamma)^{Z_y + Z_m} = [Z_y C_y/(C_m^* + Z_y C_{y(m)})]^{Z_y} \tag{6}$$

The salt distribution coefficient is defined as:

$$K^* = C_{y(m)}/C_y \tag{7}$$

Since $Z_y C_{y(m)} \ll C_m^*$, the following expression can be derived:

$$K^* = C_{y(m)}/C_y = [Z_y^{Z_y}(C_y/C_m^*)^{Z_y}(\gamma/\gamma_m)^{Z_y + Z_m}]^{1/Z_m} \tag{8}$$

Thus, as an approximation (at high water flux), rejection could be expressed as:

$$R = 1 - K^* \tag{9}$$

This model predicts that the ultrafiltrate concentration is a function of the membrane charge capacity, the feed concentration, and the charge of the co-ion. Experiments showed that ultrafiltrate concentration also depends on the type of counter-ions as well as co-ions in the feed solution. The model predicts poorer rejection of the salt with increasing feed concentration, which has also been proven by experiments.

Although the model does not take into account the diffusive and convective fluxes which influence the salt rejection in charged membrane ultrafiltration processes, it does provide a simple qualitative picture of the solute rejection in charged membrane ultrafiltration processes.

## B. Nernst-Planck Model

Dresner[19] used the extended Nernst-Planck equation to describe salt rejection by charged membranes. According to this model, the flux of ions through the membrane is expressed as:

$$J_j = \beta_j J_w C_{j(m)} + Z_j C_{j(m)} D_{j(m)}[FE/(R'T)]$$
$$- D_{j(m)}(dC_{j(m)}/dx) - C_{j(m)} D_{j(m)}(d1n\gamma_{j(m)}/dx) \tag{10}$$

In Equation 10, the first term describes the convective solute flux, the second term accounts for the flux due to the Donnan potential, while the last two terms describe the diffusional salt flux.

In Equation 10, $\beta_j$ is a correction factor for the possibility that in the pure convection mode the ions may not be swept along with the velocity of the permeating water. Dresner has integrated Equation 10 involving permeation of multicomponent solutions through charged membranes for the case of good co-ion exclusion.

At high water flux, $(dC_{j(m)}/dx) = 0$, and the above solute transport equation can be written in the following simpler form:

$$J_j = \beta_j J_w C_{j(m)} + Z_j C_{j(m)} D_{j(m)}[FE/(R'T)] \tag{11}$$

With good co-ion exclusion (i.e., $C_m^* \gg C_{y(m)}$), the co-ion flux for single salt systems can be written as:

$$J_y = J_w C_{y(m)}[1 - \beta(Z_y D_y/(Z_m D_m)] \tag{12}$$

According to this equation, the flux of solute is a function of feed concentration, membrane charge density, charge of the ion, and the diffusion coefficients of the ions in the membrane. The effectiveness of selective metal recovery simply depends on the magnitude of the fluxes of the various solutes in the solution.

The Nernst-Planck model differs from the Donnan equilibrium model in that it accounts for salt-solvent coupling. This model also makes allowance for the effect of convective flow and for variation in the properties of different ions in terms of diffusion coefficients. Both the Nernst-Planck and Donnan equilibrium models show a dependence of salt rejection on the feed concentration.

In general, the relationship between solute ultrafiltrate and feed concentration can be expressed empirically in the following form:

$$C_f = kC_i^n \quad \text{or} \quad R = 1 - KC_i^{n-1} \tag{13}$$

in which k and n are two parameters related to the nature of solute and the membrane. For simple metal salts, "n" varies between 1.0 and 1.5.

## III. METAL COMPLEXATION REACTION MODELS

Metal complexes are compounds that contain a central metal ion surrounded by a cluster of ions or molecules. Depending on the complexing agents, it may form cationic, anionic, or nonionic species. Compounds such as EDTA, cyanide, and oxalic acid are commonly used as complexing agents.

### A. Dissociation of Complexing Agents

In the absence of metal ions, the complexing agent forms ionized species in the solution.

For a polyprotic acid, the equilibria involved and the stepwise dissociation constants can be expressed as follows:

$$H_nL \rightleftarrows H + H_{n-1}L; \qquad K_1' = [H][H_{n-1}L]/[H_nL] \qquad (14)$$

$$H_{n-1}L \rightleftarrows H + H_{n-2}L; \qquad K_2' = [H][H_{n-2}L]/[H_{n-1}L] \qquad (15)$$

$$\bullet \qquad \bullet \qquad \bullet \qquad\qquad \bullet$$

$$\bullet \qquad \bullet \qquad \bullet \qquad\qquad \bullet$$

$$\bullet \qquad \bullet \qquad \bullet \qquad\qquad \bullet$$

$$HL \rightleftarrows H + L; \qquad K_n' = [H][L]/[HL] \qquad (16)$$

The stepwise dissociation constants for EDTA, HCN, and $H_2C_2O_4$ have been reported in the literature.[20] Using stepwise dissociation constants, the material balance equation and charge balance relationship, a set of simultaneous equations can be solved to obtain the distribution of species in the solution under various pH conditions. The fraction of each species can be calculated from:

$$\alpha_0 = [H_2L]/C_L = [H^+]^2/([H^+]^2 + K_1[H^+] + K_1K_2) \qquad (17)$$

$$\alpha_1 = [HL^-]/C_L = K_1[H^+]/([H^+]^2 + K_1[H^+] + K_1K_2) \qquad (18)$$

$$\alpha_2 = [L^{2-}]/C_L = K_1K_2/([H^+]^2 + K_1[H^+] + K_1K_2) \qquad (19)$$

The species distributions of typical complexing agents (EDTA and oxalate) are computed and shown in Figures 1 and 2, as a function of pH.

## B. Metal Complexation Calculations

In the presence of metal ions, additional equilibria have to be taken into account for complex formation. As an example, the following additional equilibria occur in aqueous solutions:

$$M + L \rightleftarrows ML; \qquad K_1 = [ML]/([M][L]) \qquad (20)$$

$$ML + L \rightleftarrows ML_2; \qquad K_2 = [ML_2]/([M][L]) \qquad (21)$$

$$ML_{n-1} + L \rightleftarrows ML_n; \qquad K_n = [ML_n]/([ML_{n-1}][L]) \qquad (22)$$

where $K_i$'s are the successive stability constants characterizing the formation of the different species in the metal complex solution.

Denoting the products of the stepwise stability constants by $\beta_i$ (overall stability constant) and the corresponding subscript, we have:

$$\beta_1 = [ML]/([M][L]) = K_1 \qquad (23)$$

$$\beta_2 = [ML_2]/([M][L]^2) = K_1K_2 \qquad (24)$$

$$\beta_n = [ML_n]/([M][L]^n) = K_1K_2.....K_n \qquad (25)$$

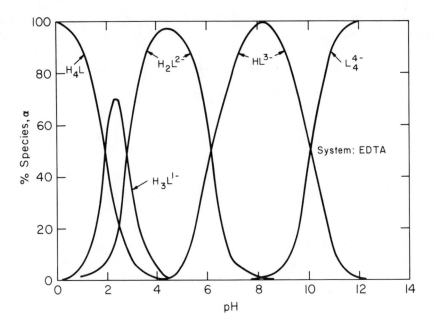

FIGURE 1.   Distribution of various species of EDTA in aqueous solutions as a function of pH.

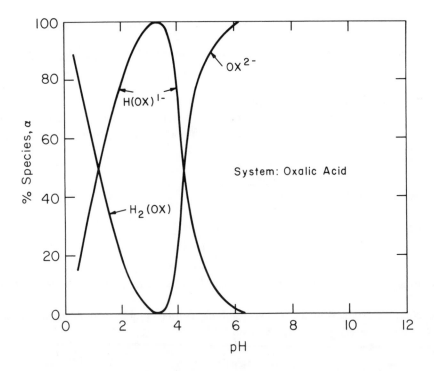

FIGURE 2.   Distribution of various species of oxalic acid in aqueous solutions as a function of pH.

The $\beta_i$ values for the cases in this study are obtained from the literature.[21] The mole fraction of the $i^{th}$ complex species in the solution, which gives the fraction of the total metal ion in the complex of composition $ML_i$, is given by:

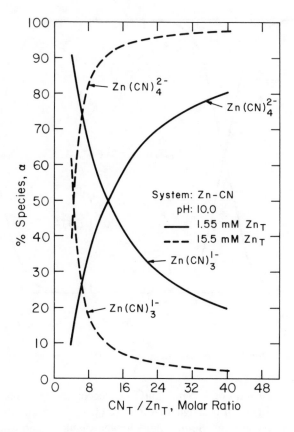

FIGURE 3.   Distribution of various species in Zn-CN system as a function of $CN_T/Zn_T$ molar ratio.

$$\alpha_i = [ML_i]/C_M$$

$$= \beta_i[M][L]^i/([M] + \beta_1[M][L] + \beta_2[M][L]^2 + \ ....)$$

$$= \beta_i[L]^i/(1 + \sum_1^N \beta_i[L]^i) \tag{26}$$

The mole fraction of the free metal ion is given by:

$$\alpha_0 = [M]/C_M = 1/(1 + \beta_1[L] + \beta_2[L]^2 + \ ....) \tag{27}$$

In addition to metal-ligand reactions, metal hydrolysis reactions (such as $M[OH]$, $M[OH]_2$, etc.) are also included in all the calculations. In the case of the copper-EDTA complexation study, species such as $CuHL^{1-}$ and $Cu(OH)L^{3-}$ exist in solution in the presence of $CuL^{2-}$.

The equilibrium concentration of all species in multimetal, multiligand systems can be calculated by solving a series of simultaneous equations (Equations 14 to 27 and metal hydrolysis equilibria), the stability constant expressions, pH of the solution, total metal concentration, total ligand concentration, and the relevant equilibrium constants. A computer program was developed for these calculations.

## 1. Calculated Species Distributions

Species distributions in solutions of zinc and cyanide mixtures at various $CN_T/Zn_T$ molar ratios are shown in Figure 3. At 1.55 m$M$ total zinc concentration, the predominant species

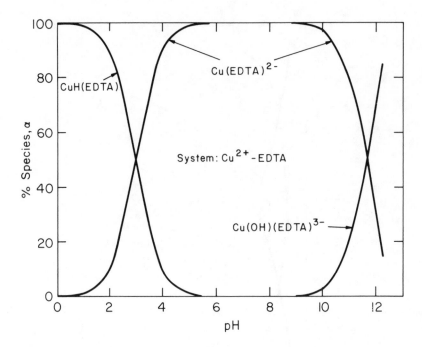

FIGURE 4.    Distribution of various species in $Cu^{2+}$-EDTA system as a function of pH.

is $Zn(CN)_3^{1-}$ for a molar ratio of 4, while $Zn(CN)_4^{2-}$ becomes predominant at a ratio of 14 or above. In the same figure, it is shown that for fixed $CN_T/Zn_T$, more $Zn(CN)_4^{2-}$ is formed at high total metal concentration. Similar plots for the cadmium and cyanide system were also calculated. At a ratio of 4 or above, $Cd(CN)_4^{2-}$ always predominates in the solution. For $Cu^{1+}$-CN system at a concentration of 1.55 m$M$ total copper, $Cu(CN)_3^{2-}$ predominates at a ratio of 4, while $Cu(CN)_4^{3-}$ predominates at a ratio of 18 or above. At the higher metal concentration, $Cu(CN)_4^{3-}$ predominates above a $CN_T/Cu_T$ ratio of 10.

In the presence of EDTA, heavy metal could form various species (depending on pH) in solutions. Distributions of species for copper complexes at various pH values are shown in Figure 4. Since the only type of complex is $Cu(EDTA)^{2-}$ over the range of interest (pH 4 to 10), the species distribution is not influenced by the concentration of ligand present in the solution. It should also be noted that although complexes of different forms may exist, EDTA always forms 1:1 complex with heavy metals.

In the case of metal-oxalate complexation reactions, a precipitation problem was encountered for cadmium and zinc even at very high total ligand-to-total metal ratio in the pH range of 4 to 10. Copper oxalate precipitates above pH 7. Experimental results confirmed this behavior. The species distribution of copper-oxalate complexes is shown in Figure 5. $Cu(OX)_2^{2-}$ always predominates for $OX_T/Cu_T \geq 8$ at pH 4. At pH 6, practically all the species present in the solution are $Cu(OX)_2^{2-}$. At high total metal concentration (15.5 m$M$), the calculation shows that $Cu(OX)_2^{2-}$ predominates (at both pH 4 and 6) at all $OX_T/Cu_T$ ratios greater than 2.

## 2. Average Charge Calculations

The behavior of the solution which contains metal complexes could also be characterized in terms of species charge by defining a new parameter — average charge :

$$\bar{n} = Z_i \alpha_i \tag{28}$$

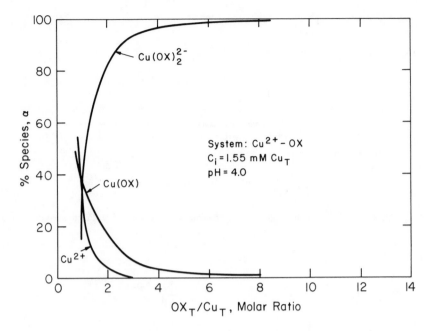

FIGURE 5. Distribution of various species in $Cu^{2+}$-OX system as a function of $OX_T/Cu_T$ molar ratio.

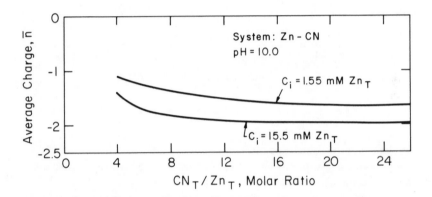

FIGURE 6. Average charge ñ in Zn-CN system as a function of $CN_T/Zn_T$ molar ratio.

where $Z_i$ = charge of metal-containing species i, and $\alpha_i$ = fraction of metal-containing species in the solution.

Plots of ñ vs. ratio of total complexing agent to total metal for typical metal-cyanide, and metal-oxalate systems are shown in Figures 6 to 8. Average charges are higher for higher total metal concentration at a fixed ratio in all cases. At pH 4, the average charge for copper (II)-oxalate system is $-2$ at $OX_T/Cu_T \geq 6$. At pH 6 (not shown in the figure), the average charge is $-2$ for $OX_T/Cu_T \geq 2$.

## IV. MEMBRANE RESULTS WITH VARIOUS METAL-LIGAND SYSTEMS

The membrane experiments were conducted in a batch cell (10 $cm^2$ membrane area; 2.0 $\ell$ solution volume) pressurized by nitrogen. A mixing speed of about 600 rpm was used to minimize concentration polarization effects. Membrane runs were made with free complexing

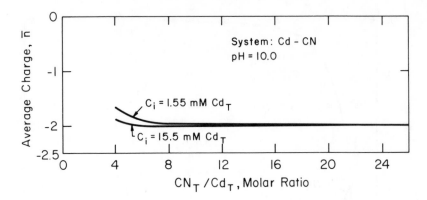

FIGURE 7.   Average charge ñ in Cd-CN system as a function of $CN_T/Cd_T$ molar ratio.

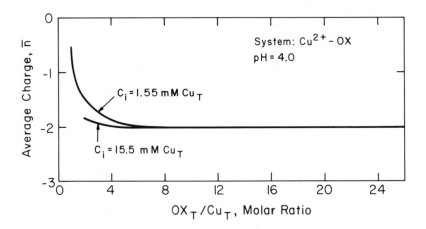

FIGURE 8.   Average charge ñ in $Cu^{2+}$-OX system as a function of $OX_T/Cu_T$ molar ratio.

agents (ethylenediamine tetracetate, EDTA; cyanide, CN; and oxalate, OX) and metal-complexing agents (metals: $Cu^{2+}$, $Cu^{1+}$, $Zn^{2+}$, $Cd^{2+}$) systems over a broad concentration and pH range. For metal cyanide systems cyanide salts were used, whereas for EDTA and oxalate systems metal chloride salts were used. Negatively charged composite (Millipore® PTAL) membranes used in this study had the following characteristics:

- Membrane: noncellulosic skin with noncellulosic backing
- Pore diameter: 15 to 20 Å
- Fixed charge: negative sulfonate group
- Charge capacity: 300 to 400 m$M$
- pH limit: 2 to 12
- Typical water flux at $5.5 \times 10^5$ Pa: $1.8 \times 10^{-3}$ cm/sec

Membrane performance is characterized by two parameters: rejection and water flux. The influence of concentration polarization on these parameters was minimized by maintaining good mixing for the high pressure feed solutions. The comparison of distilled water flux values before and after each run showed no concentration polarization effects. In addition to water flux measurements, standard runs with $CuCl_2$ solution (at $\Delta P = 5.6 \times 10^5$ Pa) were made to determine the rejections of various membrane samples and the rejection stability

**Table 1**
## CHARACTERISTICS OF CHARGED MEMBRANES

| Membrane batch no. | $R_m{}^a$ | $J_w$ (cm/sec)$^b$ | $R_{CuCl2}{}^c$ |
|---|---|---|---|
| 1 | $2.35 \times 10^8$ | $23.8 \times 10^{-4}$ | 0.33 |
| 2 | $2.37 \times 10^8$ | $23.6 \times 10^{-4}$ | 0.34 |
| 3 | $3.0 \times 10^8$ | $18.6 \times 10^{-4}$ | 0.64 |
| 4 | $3.57 \times 10^8$ | $15.7 \times 10^{-4}$ | 0.57 |
| 5 | $3.33 \times 10^8$ | $16.8 \times 10^{-4}$ | 0.61 |

$^a$  $R_m$ = membrane resistance, $\dfrac{Pa}{cm/sec}$.

$^b$  At $\Delta P = 5.6 \times 10^5$ Pa.

$^c$  $CuCl_2$ concentration = 1.55 m$M$, pH = 4.

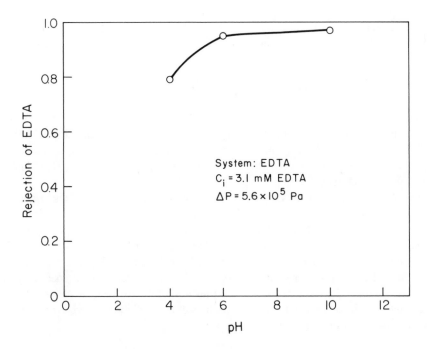

FIGURE 9.    Effect of pH on EDTA rejection.

before and after each run with metal-ligand systems. The standard rejections and flux behavior of different membrane batch products are shown in Table 1.

## A. Rejection of Individual Complexing Agents

Cyanide rejection at pH 10 was found to be very low at all concentration levels. The rejection dropped significantly as the feed concentration of $CN_T$ increased. For example, CN rejection dropped from 28 to 17% as the feed concentration was increased from 3.1 to 8.0 m$M$. The rejection of $EDTA_T$ at various pH values is shown in Figure 9. The plot shows that the rejection of $EDTA_T$ is relatively high above pH 4. The rejection increases as pH increases. These observations can be rationalized by looking back to Figure 1, where it

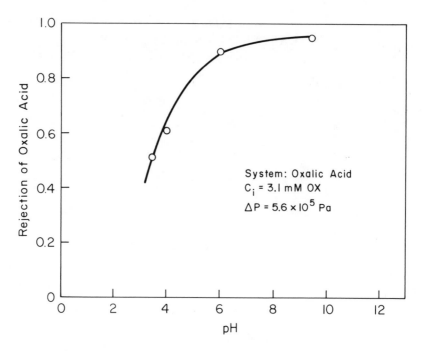

FIGURE 10.    Effect of pH on oxalic acid rejection.

shows that higher charge species such as $H(EDTA)^{3-}$ or $EDTA^{4-}$ become predominant as the pH increases. The relatively large size of the EDTA molecules contributes to high rejection behavior even at low pH, where the lower charge species such as $H_3(EDTA)^{1-}$ or $H_2(EDTA)^{2-}$ predominate.

The rejection of oxalic acid at various pH values is shown in Figure 10. This figure shows that the rejection of oxalic acid is greatly enhanced as the pH value increases from 4 to 7. This behavior can be explained on the basis of ionization (Figure 2). As pH increases, the higher rejection is due to the conversion of the low charge species $H(OX)^{1-}$ into the higher charge species $OX^{2-}$. The Nernst-Planck model can be used to explain the difference in the rejection behavior of EDTA or oxalate when the charge on predominant species are equal in both cases. It should also be noted that the rejection of monovalent species may be lower in the presence of di- or trivalent species, and hence the prediction of overall rejection may not conform to the additive rule.

## B. Metal-Cyanide Rejection Studies

Rejections of total zinc ($Zn_T$) and total cyanide ($CN_T$) at various total cyanide-to-total metal ratios ($CN_T/Zn_T$) were studied quite extensively; similar experiments were carried out for Cd and Cu for comparison. In order to explain the results of the above studies, the ionic strength effect was investigated for the Zn-CN system. The concentration effect was also investigated in order to provide useful information on water recovery.

A pressure of $5.6 \times 10^5$ Pa was used for most runs to obtain maximum (asymptotic) rejection. Above a $\Delta P$ of $5 \times 10^5$ Pa, the diffusional contribution was found to be negligible.

Rejections of $M_T$ (M = Zn, Cd, or Cu) and $CN_T$ at various $CN_T/M_T$ ratios were established at pH 10. Typical rejections of Zn-CN and Cu-CN systems are shown in Figures 11 and 12. The rejections of $Zn_T$, $Cd_T$, and $Cu_T$ were found to be increased considerably in the presence of complexing agents. This is due to the fact that the complexed metal cyanide anions are rejected better than the $M^{2+}$ ions. The rejections of $Cu_T$ and $CN_T$ were found to

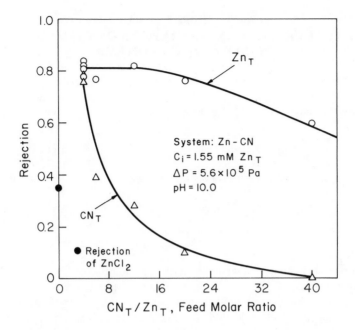

FIGURE 11.   Rejection behavior of $Zn_T$ and $CN_T$ as a function of feed molar ratio ($CN_T/Zn_T$) for Zn-CN system.

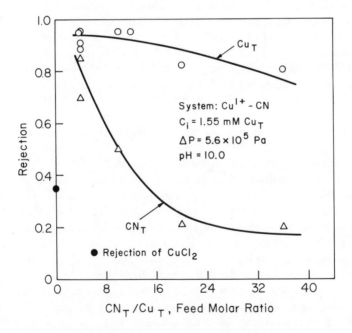

FIGURE 12.   Rejection behavior of $Cu_T$ and $CN_T$ as a function of feed molar ratio ($CN_T/Cu_T$) for $Cu^{1+}$-CN system.

be highest for the copper (1)-cyanide solution at pH 10 under various $CN_T/M_T$ ratios. The average charge "n" for the $Cu^{1+}$-CN system is always higher than the Zn-CN and Cd-CN systems at all $CN_T/M_T$ ratios.

If the metal-containing complex is treated as a co-ion $[M^{Z_m+}(CN)_Y]^{-(Y-Z_m)}$, the Nernst-

## Table 2
## CALCULATED CONCENTRATIONS OF VARIOUS
## SPECIES IN Zn-CN SYSTEMS
(pH = 10, $C_i$ = 1.55 m$M$ $Zn_T$)

| $CN_T/Zn_T$ ratio | Species | Feed concentration (m$M$) |
|---|---|---|
| 4 | $Zn(CN)_3^-$ | 1.41 |
| 4 | $Zn(CN)_4^{2-}$ | 0.14 |
| 4 | Free $CN^-$ | 1.41 |
| 4 | Complexed cyanide (as CN) | 4.80 |
| 40 | $Zn(CN)_3^-$ | 0.30 |
| 40 | $Zn(CN)_4^{2-}$ | 1.25 |
| 40 | Free $CN^-$ | 56.0 |
| 40 | Complexed cyanide (as CN) | 5.90 |

Planck equation predicts higher metal and cyanide rejection for the $Cu^{1+} - CN$ system. The rejection of total cyanide is shown to be approximately equal to the rejection of metal ion at a low ratio because of low free cyanide in water. As $CN_T/M_T$ increases, the metal rejection should increase; however, the rejection actually decreased. The rejection decrease was particularly substantial for Zn. This is due to the fact that rejection is a function of both (average charge) and ionic strength. Ionic strength increases as free cyanide increases. Total cyanide rejection drops drastically as $CN_T/M_T$ increases because of the large increase in poorly rejecting free cyanide ions. The free cyanide concentrations in the case of $CN_T/Zn_T$ = 4 and 40 can be calculated by using Figure 5, and the results are shown in Table 2.

In order to prove the effect of excess free $CN^-$, studies of the effect of ionic strength (using $NaNO_3$) on total zinc rejection in the $Zn - CN$ system were carried out. $NaNO_3$ was selected for the adjustment of ionic strength, because $NaNO_3$ does not form any metal complexes and the rejection of $NaNO_3$ is similar to NaCN.

The rejection of zinc is also shown (Figure 13A) at various $CN_T/Zn_T$ ratios and constant ionic strength. The result clearly indicates that the rejection of $Zn_T$ increases at constant ionic strength as the $CN_T/Zn_T$ ratio increases; this is due to an increase in average negative charge, as indicated in Figure 6 by "$\bar{n}$" values. Figure 13B shows that as ionic strength (i.e., added $NaNO_3$) increases, rejection tends to decrease.

As stated in the Donnan equilibrium model and Nernst-Planck model, the rejection of simple salts decreases with an increase in concentration of the feed solution. This is again verified to be the same for zinc cyanide complex solution. The results are shown in Figure 14. Note also from this figure that the rejection of zinc at $CN_T/Zn_T$ = 40 is significantly lower at the high $Zn_T$ concentration level. This is due to the ionic strength effect.

For design purposes, a plot of log $C_f$ vs. log $C_i$ should give useful information. According to the Donnan equilibrium theory, the slope should lie between 1 and 2. Figure 14 shows the least squares lines for the two cases. The corresponding parameters, n and k, are shown in Table 3. The higher k value for metal rejection is seen in the case of $CN_T/Zn_T$ = 40, which is again due to much higher ionic strength even in the presence of higher average charge in the solution.

### C. Metal-EDTA Rejection Studies

The rejections of $Zn_T$, $Cd_T$, $Cu_T$, and $EDTA_T$ at various $L_T/M_T$ ratios and pH 4 were established. The rejection results for Zn and Cu are shown in Figures 15 and 16. At $EDTA_T/M_T$ = 1, $M_T$ rejection = $EDTA_T$ rejection, since the metal-containing species is primarily $M(EDTA)^{2-}$, which is a complex of the 1:1 type. It should also be noted that the Cd-EDTA

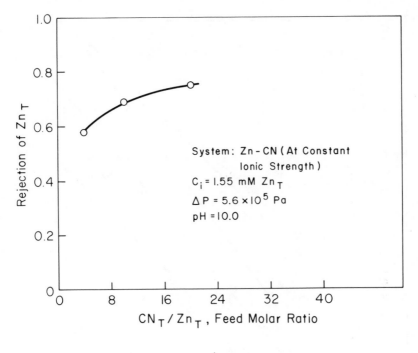

A

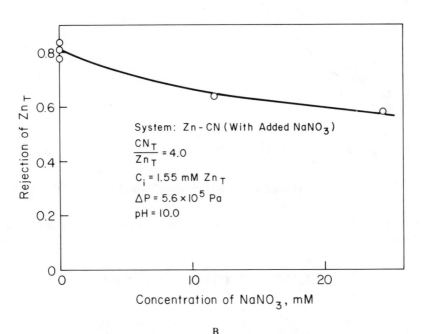

B

FIGURE 13.   Effect of ionic strength (NaNO₃) on $Zn_T$ rejection for Zn-CN system.

complexes showed lower rejection. This could be due to the orientation of the complexes inside the membrane pores. The metal rejection remains high at $EDTA_T/M_T$ between 1 and 2, because all the metals in the solution are in the highly rejected $M(EDTA)^{2-}$ complex form.

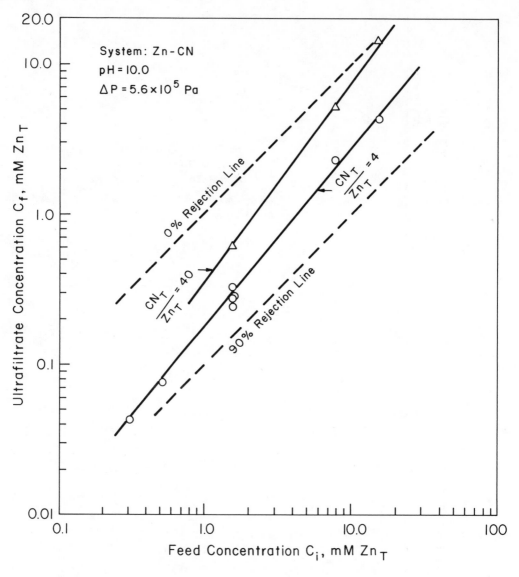

FIGURE 14.    Effect of feed $Zn_T$ concentration on ultrafiltrate zinc concentration for Zn-CN system.

### Table 3
### PARAMETERS TO FIT $C_f = kC_i^n$ FOR
### Zn-CN SYSTEM

| $CN_T/Zn_T$ | Rejection | k | n | Correlation coefficient |
|---|---|---|---|---|
| 4 | $Zn_T$ | 0.17 | 1.20 | 0.99 |
| 4 | $CN_T$ | 0.16 | 1.17 | 0.99 |
| 40 | $Zn_T$ | 0.34 | 1.34 | 0.99 |
| 40 | $CN_T$[a] | — | — | — |

[a]    Total cyanide rejection is negligible because of high free cyanide (90% of total cyanide).

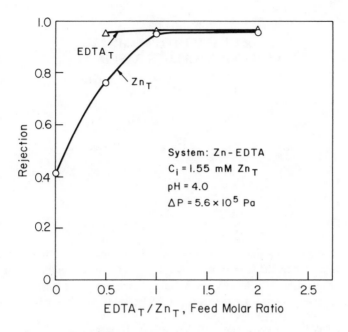

FIGURE 15.   Rejection behavior of $Zn_T$ and $EDTA_T$ as a function of feed molar ratio ($EDTA_T/Zn_T$) for Zn-EDTA system.

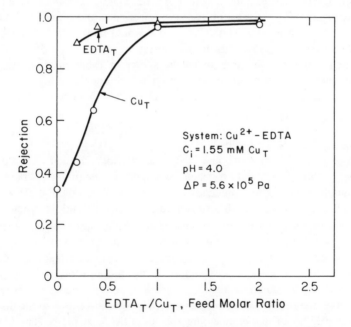

FIGURE 16.   Rejection behavior of $Cu_T$ and $EDTA_T$ as a function of feed molar ratio ($EDTA_T/Cu_T$) for $Cu^{2+}$-EDTA system.

The rejections of total metal and $M(EDTA)^{2-}$ were different at $EDTA_T/M_T < 1$ (metal species: $M(EDTA)^{2-}$ and $M^{2+}$) because of the presence of poorly rejecting free $M^{2+}$ in the solution. Thus, the total metal rejection was less than that of the $M(EDTA)^{2-}$ complexes. The rejection of $EDTA_T$ was lower at $EDTA_T/M_T < 0.5$. This is probably due to the fact

**Table 4**

**CALCULATED TOTAL METAL
REJECTION VALUES BASED ON
EXPERIMENTAL REJECTIONS OF
$M^{2+}$ AND $M(EDTA)^{2-}$**

| | | Rejection of $M_T$ | |
|---|---|---|---|
| Metal | $EDTA_T/M_T$ | Calculated | Experimental |
| Zn | 0.5 | 0.65 | 0.76 |
| Cd | 0.5 | 0.55 | 0.58 |
| Cu | 0.4 | 0.59 | 0.64 |
| Cu | 0.2 | 0.46 | 0.44 |

that $M(EDTA)^{2-}$ complexes show lower rejection in the presence of significant free metal ions. The rejection of free EDTA (primarily $H_2[EDTA]^{2-}$) was shown to be 0.8 at pH 4.0 (Figure 9). However, total EDTA rejection remained approximately constant at $1 < EDTA_T \leq 2$ in all three cases. Thus, the presence of $M(EDTA)^{2-}$ complexes may enhance the rejection of free EDTA ($H_2[EDTA]^{2-}$).

Total metal rejection at $EDTA_T/M_T < 1$ could be calculated if the rejection of free metal ($M^{2+}$) and the $M(EDTA)^{2-}$ complex is known, and in the absence of synergistic effects. For example, at a ratio of 0.4, the concentrations of $Cu(EDTA)^{2-}$ and $Cu^{2+}$ are 0.62 and 0.93 m$M$, respectively. Hence, if the rejection of $Cu(EDTA)^{2-}$ remains constant (R = 0.96, obtained from Figure 16) at a ratio of 1, $Cu^{2+}$ rejection is 0.34, and the calculated $Cu_T$ concentration in ultrafiltrate at $EDTA_T/Cu_T = 0.4$ would be 0.93 (1 − 0.34) + 0.62 (1 − 0.96) or 0.64 m$M$. Thus, $Cu_T$ rejection at 0.4 ratio is 1 − 0.65/1.55 = 0.59.

The calculated results for the three metals are shown in Table 4. Of course, as $EDTA_T/M_T \rightarrow 0$, calculated and experimental rejections must be equal. At higher ratios, the experimental values are higher than the calculated values, possibly due to enhanced rejection of $M^{2+}$ in the presence of $M(EDTA)^{2-}$.

## D. Metal-Oxalate Rejection Studies

The effects of pH, $OX_T/M_T$ ratio, and feed concentration variation were studied for the copper-oxalate system. For Cd and Zn, metal oxalate precipitate (in the pH range of interest) was formed, and hence ultrafiltration behavior was not studied. Solution pH was controlled in the pH 4 to 6 range so that $Cu(OH)_2$ precipitate formation could be avoided.

Rejections of total copper ($Cu_T$) in the presence of oxalate at three pH values are shown in Figure 17. The rejection of $Cu_T$ shows a significant increase as the pH of the solution increases. This is due to the fact that an increasing amount of charged species, $Cu(OX)_2^{2-}$, is converted from uncharged Cu(OX) in solution. In the presence of uncharged Cu(OX), it appears that $Cu(OX)_2^{2-}$ rejection is considerably less. Total oxalate ($OX_T$) rejection is also shown in the same figure. The sharp increase in $OX_T$ rejection is due to the dissociation of the $H(OX)^-$ at pH 3.5 to higher rejecting species, $OX^{2-}$, at pH 6. The $OX_T$ rejection behavior is similar to that observed in Figure 10.

Rejection of copper and oxalate at various $OX_T/Cu_T$ molar ratios at pH 4 is shown in Figure 18. The rejection of copper does not drop at high ratios indicating that the charge effect (see Figure 8) predominates over the ionic strength effect. The rejection of copper goes through a minimum at an $OX_T/Cu_T$ of about 1 because of the presence of a significant amount of uncharged Cu(OX) (zero rejection) in the solution (see Figure 5). At $OX_T/Cu_T < 1$, $Cu_T$ rejection increases because of an increase in $Cu^{2+}$ concentration. It should also

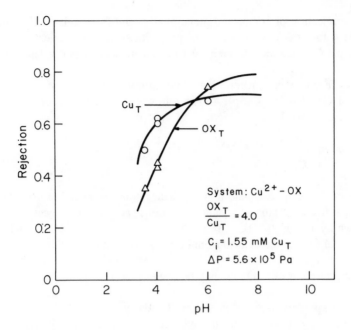

FIGURE 17.    Rejection behavior of $Cu_T$ and $OX_T$ as a function of pH for Cu-OX system.

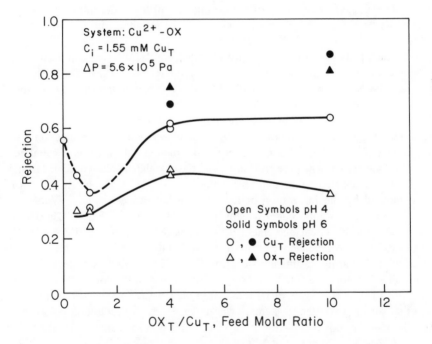

FIGURE 18.    Rejection behavior of $Cu_T$ and $OX_T$ as a function of feed molar ratio ($OX_T/Cu_T$) for $Cu^{2+}$-OX system at pH 4.0 and 6.0.

be noted that the rejection of $Cu_T$ is much higher at pH 6, particularly at ratio 10. This is because of the fact that $Cu_T$ is present primarily as $Cu(OX)_2^{2-}$.

The rejection of total oxalate is also shown in Figure 18. The total oxalate rejection is the sum of the rejection of complexed metal oxalate and free oxalate. The lower rejection of $OX_T$ at pH 4 for $OX_T/Cu_T > 4$ is principally due to the poor rejection of monovalent $H(OX)^{1-}$. At pH 6, since the predominant free oxalate species is divalent $OX^{2-}$, the total oxalate rejection is higher.

The effect of feed copper concentration on ultrafiltrate concentration of $Cu_T$ and $OX_T$ was investigated for $OX_T/Cu_T = 4$ at pH 4. The following correlations were obtained from the experimental rejection values:

- $Cu_T$ rejection: $C_f = 0.42 \, C_i^{0.90}$
- $OX_T$ rejection: $C_f = 0.56 \, C_i^{1.03}$

The power on $C_i$ of approximate unity indicates constant rejection over the entire concentration range. In contrast to total cyanide rejection (Table 3), $OX_T$ rejection is considerably lower as shown by a high "k" (k = 0.56) value. At $C_i = 1.55$ m$M$ the concentrations (in m$M$) of individual species (calculated from Figure 2 and 5) are $Cu(OX)_2^{2-} = 1.50$, $Cu(OX)$ = 0.05, and free oxalate = 3.15 [$H(OX)^- = 1.96$, $OX^{2-} = 1.19$]. The presence of a significant concentration of $H(OX)^-$ may be responsible for lower $OX_T$ rejection.

### E. Membrane Experiments at High Water Recovery

Since the rejection behavior of metals and complexing agents is a function of initial feed concentration, the overall solute removal would be dependent on the extent of water recovery. The fractional water recovery, r, is defined as the total volume of ultrafiltrate collected divided by the initial feed solution volume. The overall solute removal is defined as removal at any $r = 1 - C_{f,av}/C_i$, where $C_{f,av}$ = the average ultrafiltrate concentration. Experimental $C_{f,av}$ can easily be computed from the knowledge of the instantaneous $C_f$ and water flux vs. time data.

The relationships between $C_f$ and $C_i$ (for example, Table 3) have been used to correlate the experimental data in the case of negligible water recovery runs. The parameter "n" characterizes the system, and "k" is smaller for membranes with higher salt rejection. Thus, by knowing "n", one could predict the rejection behavior of similarly charged membranes simply by obtaining one $C_f$ value experimentally at a chosen $C_i$ from the negligible water recovery run on a different membrane batch. A membrane with higher rejection was deliberately selected for the high water recovery experiments. The fractional metal removals obtained from the experiment were compared with the calculated values obtained from known "n" and "k" values (from the negligible water recovery run). For batch water recovery, $C_{f,av}$ is computed by the integration of the material balance equation, $-Vdc - CRdV$, where V = volume of solution at any recovery, r, and $R = 1 - KC^{n-1}$. Zn-CN and Cu-OX systems exhibit excellent agreement between the experimental and calculated values (Figures 19 to 21). The dependence of metal removal at different initial metal concentrations is also calculated, and the results are shown in Figure 22 for Zn-CN system. The value of "n" for EDTA has not been found experimentally because of the concentration polarization problem with EDTA. However, metal removal in the Cu-EDTA system remains high compared to that for the two other systems, even under severe concentration polarization conditions. The metal removal behavior at all values of r was such that removal of M-EDTA > removal of M-CN > removal of M-OX.

## V. SUMMARY AND CONCLUSIONS

An extensive investigation was conducted with negatively charged, noncellulosic ultra-fultration membranes to establish the relative rejection behavior of complexed heavy metals under insignificant concentration polarization conditions. A transmembrane pressure of 5.6

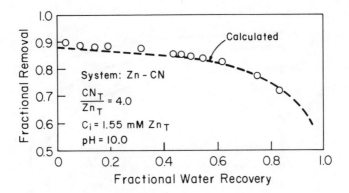

FIGURE 19. Fractional removal of $Zn_T$ as a function of water recovery for Zn-CN system at $CN_T/Zn_T = 4.0$.

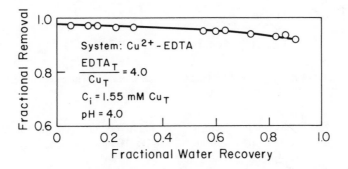

FIGURE 20. Fractional removal of $Cu_T$ as a function of water recovery for $Cu^{2+}$-EDTA system at $EDTA_T/Cu_T = 4.0$.

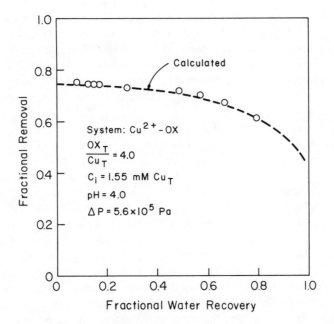

FIGURE 21. Fractional removal of $Cu_T$ as a function of water recovery for $Cu^{2+}$-OX system at $OX_T/Cu_T = 4.0$.

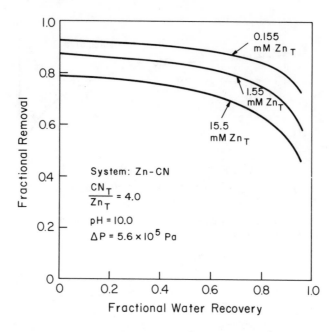

FIGURE 22.    Calculated fractional removals of $Zn_T$ as a function of water recovery for Zn-CN system at $CN_T/Zn_T = 4.0$.

$\times 10^5$ Pa provided maximum rejection. The negatively charged membranes used had a typical water flux of $13 \times 10^{-4}$ cm/sec at $5.6 \times 10^5$ Pa. The charge capacity (sulfonic acid groups) was between 300 and 400 m$M$.

The rejection dependence of the heavy metals ($Zn^{2+}$, $Cd^{2+}$, $Cu^{1+}$, and $Cu^{2+}$) and free complexing agents ($CN^-$, EDTA, and oxalate) was found to be a function of feed metal concentration, type of metal, complexing agent-to-metal feed molar ratio ($L_T/M_T$), pH, ionic strength, and transmembrane pressure difference (below $5.6 \times 10^5$ Pa). The effect of pH and feed $L_T/M_T$ ratio on rejection could be explained in terms of metal complex species distribution and the Nernst-Planck model. Extensive computer calculations were made to establish the distribution of various species in the solution. The effect of feed concentration on ultrafiltrate concentration was fit with power function. At a fixed $L_T/M_T$ ratio, the rejections of total metal and total complexing agents decreased with feed concentration for the case of the metal-cyanide system, while for the EDTA and oxalate systems, the rejection was approximately constant.

In addition to feed concentration dependence, metal rejection was also dependent upon the type of metal in the solution. For example, Figure 23A shows the relative rejection behavior of all three metals in M-CN systems. The average charge ($\bar{n} = Y - Z_m$) of species $[M^{Z_m+}(CN)_Y]^{-(Y-Z_m)}$ is shown in Figure 23B. Copper rejects much better than cadmium or zinc in M-CN systems. This is because of the higher average charge in the $Cu^{1+} - CN$ system at fixed $CN_T/M_T$ molar ratio. Cadmium rejects better than zinc above $CN_T/M_T = 20$ because of the higher average charge in the Cd-CN system. Cadmium exhibits lower rejection at lower ratios because of the presence of uncharged $Cd(CN)_2$ in the solution.

The rejection of metals was reduced in the presence of high free cyanide concentration in the solution ($CN_T/Zn_T > 12$). This was due to the ionic strength effect created in the solution at high ratios. In order to verify this postulate, constant ionic strength runs were performed with the Zn-CN system. Results clearly indicate that the $Zn_T$ rejection actually increases as $CN_T/Zn_T$ molar ratio increases in the absence of the ionic strength effect (as shown in Figure 24). Figure 24 also shows a calculated species distribution curve for Zn-CN systems. The formation of higher negatively charged species increases rejection.

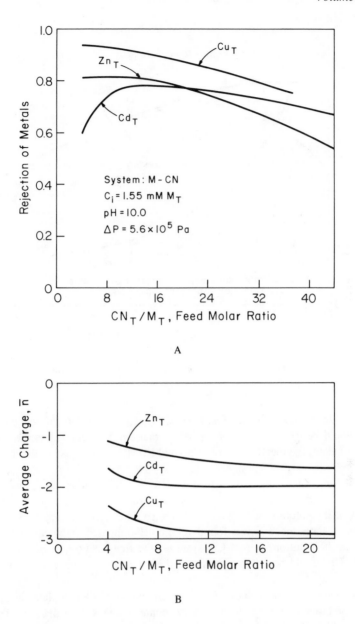

FIGURE 23.   Comparisons of metal rejection behavior with average charge ñ.

The relative metal rejection behavior in the presence of a complexing agent is compared in Figure 25. The ordinate indicates the relative charge in ultrafiltrate metal concentration in the presence of complexing agents. Knowing the metal rejection in the absence of complexing agent, one could use Figure 25 to predict the metal rejection that could be achieved at various $L_T/M_T$ ratios. The $Cu^{2+}$ salts were used for comparison in the M-EDTA and M-OX systems. The $Zn^{2+}$ salts were used for the M-CN system because of the nonavailability of $Cu(CN)_2$ salt. The $Zn^{2+}$-CN system would be expected to be similar to the $Cu^{2+}$-CN system. In all cases, the metal rejections show the trend: $R_{M\text{-}EDTA} > R_{M\text{-}CN} > R_{M\text{-}OX}$. Figure 25 also indicates that high metal rejections could be obtained at $L_T/M_T = 1.0$, pH $= 4$ to 10 with EDTA; at $L_T/M_T = 4$ to 6, pH $= 9$ to 10 with cyanide; and at $L_T/M_T = 10$ to 12,

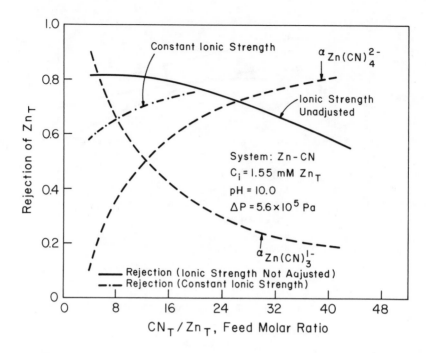

FIGURE 24.   Dependence of $Zn_T$ rejection on feed solution species distribution for Zn-CN system.

pH = 6 to 7 with oxalate. With EDTA, even at $L_T/M_T = 1$, the metal concentration in the ultrafiltrate would be 1/20 of the concentration that can be obtained without complexing agents. With oxalate, the metal rejection is poor at $L_T/M_T < 2$ because of uncharged Cu(OX) formation.

Free complexing agent rejection was found to be different in the presence of metal complexes. Higher rejections were found for free cyanide and free EDTA in the presence of metal ions in the solution at $L_T/M_T < 4$. Free oxalate rejection was poorer at pH = 4, but higher free oxalate rejection was found at high pH compared to the case where no metals were present in the solution. For M-CN systems, the free $CN^-$ rejection was negligible at $L_T/M_T > 10$. This could be advantageous for the selective concentration of metal cyanide complexes.

The approximate rejection values of various anions could be calculated from the experimental data, as shown in Table 5. The primary counter-ion is $Na^+$ in all cases. The following conclusions could be drawn from this table for the rejection of various species:

$$Cu(CN)_3^{2-} > Zn(CN)_4^{2-} > Cd(CN)_4^{2-}$$

$$Zn(CN)_4^{2-} > Zn(CN)_3^{1-}$$

$$Zn(EDTA)^{2-} \cong Cu(EDTA)^{2-} > Cd(EDTA)^{2-}$$

$$H(EDTA)^{3-} > H_2(EDTA)^{2-}$$

$$(OX)^{2-} > H(OX)^{1-}$$

$$M(CN)_4^{2-} > M(CN)_3^{1-} \gg CN^-$$

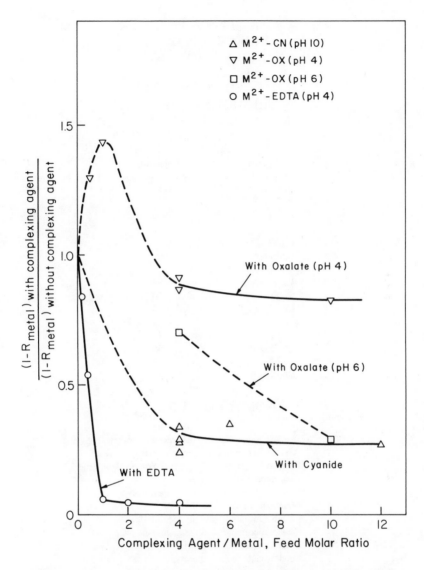

FIGURE 25.    Effect of three different complexing agents on relative metal rejection behavior.

$$M(EDTA)^{2-} \cong H(EDTA)^{3-} > H_2(EDTA)^{2-}$$

$$Cu(OX)^{2-} < (OX)^{2-}$$

Obviously, the species with higher charge density would be rejected better if similar species are compared. Higher species rejection was observed for larger species (low diffusion coefficient) provided that the charge densities were comparable. Selective separation from multimetal systems could be accomplished by adjusting the pH, complexing agent-to-metal molar ratio, ionic strength, transmembrane pressure, and feed complexing agent concentration.

## ACKNOWLEDGMENT

This study was supported by the Office of Water Research and Technology, Department of the Interior, under the provisions of Public Law 88-379, as Project Number B-059-KY.

## Table 5
### REJECTIONS OF VARIOUS ANIONS BY NEGATIVELY CHARGED COMPOSITE MEMBRANES OF $R_{CuCl2} = 0.34$ at 1.55 m$M$, $\Delta P = 5.6 \times 10^5$ Pa

| Species | Rejection |
|---|---|
| $Zn(CN)_3^{1-}$ | 0.81 |
| $Zn(CN)_4^{2-}$ | 0.90 |
| $Cu(CN)_3^{2-}$ | 0.94 |
| $Cu(CN)_4^{3-}$ | $0.95^+$ |
| $Cu(CN)_4^{2-}$ | 0.77 |
| $Zn(EDTA)^{2-}$ | 0.95 |
| $Cu(EDTA)^{2-}$ | 0.96 |
| $Cd(EDTA)^{2-}$ | 0.89 |
| $Cu(OX)^{2-}$ | 0.80 |
| $CN^-$ | 0.30 |
| $SO_4^{2-}$ | 0.81 |
| $H_2(EDTA)^{2-}$ | 0.80 |
| $H(EDTA)^{3-}$ | 0.96 |
| $OX^{2-}$ | 0.93 |
| $H(OX)^{1-}$ | 0.51 |

## NOMENCLATURE

| | |
|---|---|
| $C_f$ | Permeate concentration, m$M$ metal or complexing agents |
| $C_{f,av}$ | Average permeate concentration, m$M$ |
| $C_i$ | Feed concentration, m$M$ metal or complexing agents |
| $C_j$ | Concentration of $j^{th}$ ion, m$M$ |
| $C_{j(m)}$ | Concentration of $j^{th}$ ion in membrane phase, m$M$ |
| $C_m^*$ | Membrane charge capacity, m$M$ |
| $C_y$ | Co-ion concentration in the feed, m$M$ |
| $C_{y(m)}$ | Co-ion concentration in the membrane phase, m$M$ |
| $Cd_T$ | Total cadmium molar concentration, m$M$ |
| $CN_T$ | Total cyanide molar concentration, m$M$ |
| $Cu_T$ | Total copper molar concentration, m$M$ |
| $Dj_m$ | Diffusivity of $j^{th}$ ion in the membrane phase, cm$^2$/sec |
| $D_{(m)}$ | Diffusivity of counter-ion, cm$^2$/sec |
| $D_s$ | Diffusivity of solute, cm$^2$/sec |
| $D_y$ | Diffusivity of anion, cm$^2$/sec |
| $E$ | Equilibrium Donnan potential, volts |
| $EDTA_T$ | Total EDTA molar concentration, m$M$ |
| $F$ | Faraday constant |
| $J_j$ | Flux of $j^{th}$ ion, (mmol)/(cm$^2$·sec) |
| $J_s$ | Salt flux, (mmol)/(cm$^2$·sec) |
| $J_w$ | Water flux (in the presence of salt), cm/sec |
| $J_y$ | Co-ion (anion) flux, (mmol)/(cm$^2$·sec) |
| $K$ | Stepwise dissociation constant |
| $K^*$ | Salt distribution coefficient |
| $L_T$ | Total complexing agent concentration, m$M$ |
| $M_T$ | Total metal concentration, m$M$ |
| $\tilde{n}$ | Average charge |
| $OX_T$ | Total oxalate molar concentration, m$M$ |
| $\Delta P$ | Transmembrane pressure difference, Pa |
| $r$ | Fractional water recovery |
| $R$ | Membrane rejection |

| R' | Universal gas constant |
|---|---|
| $R_m$ | Resistance of ultrafiltration membrane to water flux, (Pa)/(cm/sec) |
| x | Distance coordinate in the membrane, cm |
| $Z_j$ | Charge on the $j^{th}$ ion |
| $Z_m$ | Charge on the metal cation |
| $Z_y$ | Charge on the anion |
| $Zn_T$ | Total zinc molar concentration, m$M$ |
| $\alpha_i$ | Fraction of metal-containing species in the solution |
| $\beta_i$ | Coupling coefficient of $j^{th}$ ion |
| $\beta$ | Overall stability constant |
| $\gamma$ | Activity coefficient of the salt in solution |
| $\gamma_j$ | Activity coefficient of the $j^{th}$ ion in solution |
| $\gamma_{j(m)}$ | Activity coefficient of the $j^{th}$ ion in the membrane phase |
| $\gamma_m$ | Activity coefficient of the salt in the membrane phase |

# REFERENCES

1. **Lonsdale, H. K.,** The growth of membrane technology, *J. Membr. Sci.,* 10, 81, 1982.
2. **Strathman, H.,** Membrane separation processes, *J. Membr. Sci.,* 9, 121, 1981.
3. **Sourirajan, S.,** Reverse Osmosis and Synthetic Membranes, Theory-Technology-Engineering, National Research Council of Canada, Ottawa, 1977.
4. **Bhattacharyya, D. and Grieves, R. B.,** Charged membrane ultrafiltration, in *Recent Developments in Separation Science,* Vol. 3, Li, N. N., Ed., CRC Press, Boca Raton, Fla., 1977, 261.
5. **Bhattacharyya, D., McCarthy, J. M., and Grieves, R. B.,** Charged membrane ultrafiltration of inorganic ions in single and multi-salt systems, *AIChE J.,* 20, 1206, 1974.
6. **Bhattacharyya, D., Moffitt, M., and Grieves, R. B.,** Charged membrane ultrafiltration of toxic metal oxyanions and cations from single- and multi-salt aqueous solutions, *Sep. Sci. Technol.,* 13, 449, 1978.
7. **Bhattacharyya, D., Schaaf, D. P., and Grieves, R. B.,** Charged membrane ultrafiltration of heavy metal salts: application to metal recovery and water reuse, *Can. J. Chem. Eng.,* 54, 185, 1976.
8. **Bhattacharyya, D., Garrison, K. A., Jumawan, A. B., and Grieves, R. B.,** Membrane ultrafiltration of non-ionic surfactant and inorganic salts from complex aqueous suspensions: design for water reuse, *AIChE J.,* 21, 1057, 1975.
9. **Bhattacharyya, D., Jumawan, A. B., and Grieves, R. B.,** Charged membrane ultrafiltration of heavy metals from nonferrous metal industry, *J. Water Pollut. Control Fed.,* 51, 176, 1979.
10. **Bhattacharyya, D., Farthing, S. S., and Cheng, C. S.,** Multiple-pass water reuse, *Environ. Prog.,* 1, 65, 1982.
11. **Gregor, H. P.,** Membranes in Separation Processes — a Workshop Symp., sponsored by the National Science Foundation, Case Western Reserve University, Cleveland, Ohio, 1973.
12. **Gyte, C. C. and Gregor, H. P.,** Poly(styrene sulfonic acid)-poly(vinyllidene fluoride) interpolymer ion-exchange membranes, ultrafiltration properties, *J. Polym. Sci.,* 14, 1855, 1976.
13. **Nomura, H., Seno, M., Takahashi, H., and Yamabe, T.,** Reverse osmosis by composite charged membranes, *Desalination,* 29, 239, 1979.
14. **Kimura, S. and Jitsuhara, I.,** Transport through charged ultrafiltration membranes, *Desalination,* 46, 399, 1983.
15. **Kamizawa, C.,** The permeation behavior of metal complex solutions through cellulose acetate membranes, *J. Appl. Polym. Sci.,* 22, 2867, 1978.
16. **Hopfenberg, H. B., Lee, K. L., and Wen, C. P.,** Improved membrane separations by selective chelation of metal ions in aqueous feeds, *Desalination,* 24, 175, 1978.
17. **Johnson, J. S.,** Polyelectrolytes in aqueous solutions — filtration, hyperfiltration, and dynamic membranes, in *Reverse Osmosis Membrane Research,* Lonsdale, H. K. and Podall, H. E., Eds., Plenum Press, New York, 1972, 379.
18. **Shor, A. J., Kraus, K. A., Smith, W. T., and Johnson, J. S.,** Salt rejection properties of dynamically formed hydrous zirconium (IV) oxide membranes, *J. Phys. Chem.,* 72, 2200, 1968.
19. **Dresner, L.,** Some remarks on the integration of the extended Nernst-Planck equations in the hyperfiltration of multicomponent systems, *Desalination,* 10, 27, 1972.
20. **Skoog, D. A. and West, D. M.,** *Fundamentals of Analytical Chemistry,* 2nd ed., Holt, Rinehart & Winston, New York, 1970.
21. **Ringborn, A. J.,** *Complexation in Analytical Chemistry,* Interscience, New York, 1963.

Chapter 7

# ADSORPTION OF GAS MIXTURES AND MODELING CYCLIC PROCESSES FOR BULK, MULTICOMPONENT GAS SEPARATION

**Wei-Niu Chen and Ralph T. Yang**

## TABLE OF CONTENTS

# I. INTRODUCTION

Separation of gas mixtures by "continuous chromatography" has long been an engineer's dream. Several ingenious approaches have been invented for this purpose, e.g., hypersorption,[1] temperature[2] and pressure[3] swings, and parametric pumping.[4] Among these only the temperature and pressure swings, which are cyclic, fixed-bed processes, have been widely accepted in industry. A combined pressure swing and parametric pumping process, pressure-swing parametric pumping,[5,6] has also been recently commercialized, mainly for small-scale oxygen enrichment from air. The majority of the commercial gas separation processes, however, is cryogenic in which one or more constituents are condensed and subsequently flashed off to the gas phase.[2] Although the cryogenic processes are well developed, their costs are escalating in parallel with the cost of energy, because they are energy-intensive processes. As a consequence, there has been an increasing interest in research and development on cyclic, fixed-bed processes which are less energy intensive.

The cyclic separation processes take advantage of the selective adsorption properties of a number of microporous adsorbents. The commercially used adsorbents are activated carbon, silica gel, alumina, and zeolites, the selection among which for a specific purpose is based mainly on adsorption/desorption capacity and selectivity. As reviewed by Keller and Jones, temperature swing is preferred for dilute mixtures, whereas pressure swing (PSA) should be used for "bulk" separation.[6] The PSA processes are among the most rapidly growing separation processes, with presently thousands of units already built world-wide for feed streams near or in excess of 1 million ft³/hr. However, study of the principles underlying the cyclic processes is still in its infancy.[7]

The most important information for modeling the cyclic processes is the equilibrium adsorption of gas mixtures. Since the composition of the mixture varies continuously in the bed and within each particle, one needs predictive equations or correlations for adsorption of each component from the mixture based on the single gas isotherms which are experimentally measured. Furthermore, for practical applications, the gas mixture is seldom binary, thus, one needs models for multicomponent mixtures.

In this chapter, we discuss the two interrelated subjects: adsorption of gas mixtures and modeling of cyclic separation processes. On each subject we present a brief state-of-the-art review, followed by new models developed in our laboratory.

# II. MODELS FOR ADSORPTION OF MIXTURES

The goal of modeling is to predict adsorption of mixtures based on single gas isotherms. A comprehensive review of all models published prior to 1962 is available by Young and Crowell.[8] The presently useful models, or the useful forms of the older models, were published after 1962. Some of the recent models have been reviewed by Danner and Choi.[9] Included in their review are the ideal adsorbed solution (IAS) theory,[10] the lattice solution model of Lee,[11] the two-dimensional gas model by Hoory and Prausnitz,[12] and the statistical thermodynamic approach developed for zeolites by Ruthven et al.[13] Only the IAS theory will be reviewed here as it will be used for comparison with other models. Sircar and Myers have developed a surface-potential theory for multilayer adsorption of mixtures, which is based on the IAS theory at high surface coverage using solution theory.[14]

The vacancy solution theory postulated by Dubinin and the dividing surface approach used by Lucassen-Reynders were used by Suwanayuen and Danner to formulate a four-parameter adsorption isotherm for single gases, which was extended to predict adsorption from binary mixtures, although extension to multicomponent mixtures can be made.[15] Using the four parameters from each single gas isotherm, iteration among four equations is necessary to predict the adsorbed phase composition for a binary system. Although iteration is needed for this theory as for all the aforementioned theories, the computation is relatively direct and simple.

## A. Langmuir Isotherm for Mixed Gases

This is a kinetic approach based on the concept of dynamic equilibrium due to Langmuir. The extension of the Langmuir isotherm to a binary gas mixture was made by Markham and Benton.[16] By equating the rate of condensation, $\alpha_i P_i(1 - \sum_{j=1}^{N} \theta_j)$, and the rate of evaporation, $v_i \theta_i$, for a mixture of N components, one may solve for the amount adsorbed for species i, $V_i$:

$$V_i = \frac{V_{mi} k_i P_i}{1 + \sum_{j=1}^{N} k_j P_j} \tag{1}$$

There is no problem in the algebraic derivation of Equation 1 using different values of $V_m$. However, it has been suggested by Innes and Rowley that a single $V_m$ value be used which, for a binary mixture,[8] may be taken as:

$$\frac{1}{V_m} = \frac{x_1}{V_{m1}} + \frac{x_2}{V_{m2}} \tag{2}$$

similar to the mixing rule as also used in the two-dimensional gas model.[12]

This model has been little tested. One reason for the lack of testing and, hence, application is that the Langmuir isotherm is seldom capable of fitting single gas adsorption data over a wide range of pressures. The model has been extended to multilayer adsorption of gas mixtures, as reviewed by Holland and Liapis.[17]

## B. Ideal Adsorbed Solution Theory

This theory assumes that the adsorbed phase forms an ideal solution and, hence, the partial pressure of an adsorbed component is given by the product of its mole fraction in the adsorbed phase and the pressure which it would exert as a pure adsorbed component at the same temperature and spreading pressure as those of the mixture,[10] i.e., according to Raoult's law:

$$Py_i = P_i^o x_i \tag{3}$$

where P is the total pressure, and $x_i$ and $y_i$ are mole fractions of component i in the adsorbed and gas phase, respectively. $P_i^o$ is the "vapor pressure" of pure component i adsorbed at the same temperature, T, and spreading pressure (or adsorbed volume — which is the potential theory approach to be discussed in the next section) as that of the adsorbed mixture.

The spreading pressure, $\pi_i$, can be calculated from the integrated Gibbs free energy equation at constant temperature:

$$\pi_i(P_i^o) = \frac{RT}{A} \int_0^{P_i^o} \frac{V_i(P)}{P} dP \tag{4}$$

using a single-gas isotherm for $V_i(P)$. By equating the spreading pressures and noting that

$$\Sigma x_i = \Sigma y_i = 1; \qquad \frac{1}{N_t} = \Sigma \frac{x_i}{N_i(P_i^0)} \tag{5}$$

one can calculate the amounts adsorbed of each component from a gas mixture.

The IAS method thus requires iteration among four or more equations. It has been used to model the performance of multicomponent liquid-phase adsorption in fixed beds.[18] However, it lends some insight concerning the origin of the "competitive adsorption" phenomenon.

A modification of the IAS theory, a real adsorbed solution theory, has been published which incorporates the activity coefficients of the components in the adsorbed phase by using the Wilson and UNIQUAC equations.[19]

## III. A UNIFIED POTENTIAL THEORY APPROACH FOR SINGLE AND MIXED GASES

### A. Polanyi and Extended Polanyi Potential Theories

Polanyi potential theory, presented in 1914,[8] precedes all modern theories of adsorption and is still regarded as fundamentally sound. The theory offers no explicit isotherm equation, but embraces a unique power of prediction that no other theory can offer.

Polanyi defines the adsorption potential, $\epsilon$, of any molecule within the attractive force field of the solid surface as the work required to remove the molecule to infinity from its location in the adsorbed phase. Thus, for 1 mol of adsorbate

$$\epsilon = \Delta F = RT\ln \frac{P_s}{P} \tag{6}$$

where $\delta F$ is the free energy change, $P_s$ is the saturated vapor pressure of liquid at adsorption temperature and $P$ is adsorption pressure. For any given gas-solid system, $\epsilon$ decreases continuously from the surface, thus, one may plot the accumulated volume of the adsorbed space, or the total amount adsorbed, vs. $\epsilon$. Such a plot is called characteristic curve of the system, which may be expressed as

$$W = f(\epsilon) \tag{7}$$

Polanyi made a postulation that the adsorption potential is independent of temperature, so that the characteristic curve should be the same at all temperatures.

The characteristic curve, Equation 7, must be obtained from an experimental isotherm by plotting the data in W against $\epsilon$. However, only one isotherm (at $T_i$) is needed, and since $\epsilon$ is assumed to be independent of T, adsorption isotherms at all temperatures may be predicted once the characteristic curve is constructed. The assumption that $\epsilon$ is independent of T is, indeed, an excellent one as voluminous data have shown; and it can also be theoretically justified.[20]

Besides the power of predicting adsorption at all temperatures from data at one temperature, it would be desirable to use the Polanyi theory to predict adsorption of other gases from data on one gas, on the same solid. In order to do so, Dubinin and collaborators assigned an "affinity coefficient" for each adsorbate, and the characteristic curve becomes[21]

$$W = f\left(\frac{\epsilon}{\beta}\right) \tag{8}$$

The characteristic curves for different adsorbates, the W vs. RT $\ln(P_s/p)$ curves, can be made to coalesce into a single curve if proper values of $\beta$ are assigned to the adsorbates.[8,21]

## Table 1
## CALCULATION FOR OBTAINING THE CHARACTERISTIC CURVES USING POTENTIAL THEORY

| Plotting A vs. B | Calculation of V | Calculation of $f_s$ or $P_s$ | Ref |
|---|---|---|---|
| $NV$ vs. $RT\ln\dfrac{P_s}{P}$ | As saturated liquid at $T < T_c$, as van der Waals const b at $T > T_c$ | Vapor pressure of liquid at $T < T_c$, 0.14 T/b for first trial at $T > T_c$ | 8 |
| $NV$ vs. $\dfrac{RT}{V}\ln\dfrac{P_s}{P}$ | Same as Polanyi[8] | Vapor pressure of liquid at $T < T_c$ Tr²Pc at $T > T_c$ | 21 |
| $N$ vs. $RT\ln\dfrac{KP_s}{P}$ | | Vapor pressure of liquid at $T < T_c$, extrapolation of Clapeyron eq at $T > T_c$ | 22 |
| $NV$ vs. $\dfrac{RT}{V}\ln\dfrac{f_s}{f}$ | Saturated liquid at gas pressure P | Vapor pressure of liquid at $T < T_c$, extrapolation of Antoine eq at $T > T_c$ | 23 |
| $NV$ vs. $\dfrac{RT}{V}\ln\dfrac{f_s}{f}$ | At saturated press at T (by eq of state and compression factor) | Vapor pressure of liquid at $T < T_c$, unspecified "extrapolation" at $T > T_c$ | 24 |
| $NV$ vs. $\dfrac{RT}{V}\ln\dfrac{f_s}{f}$ | Saturated liquid at normal boiling point | Vapor pressure of liquid at $T < T_c$ vapor pressure, extrapolation at $T > T_c$ | 25,26 |
| $NV$ vs. $\dfrac{RT}{V}\ln\dfrac{f_s}{f}$ | Saturated liquid at $T < T_b$, tangent to log V vs. log T plot at $T > T_b$ | Vapor pressure of liquid at $T < T_c$, using known value of V and T for P using eq state and compression factor | 27 |
| $NV$ vs. $\dfrac{RT}{V}\ln\dfrac{f_s}{f}$ | $V_{nbp}$ at $T < T_{nbp}$, $V_{nbp}$ $(T_c - T)/(T_c - T_{nbp})$ at $T_{nbp} < T \leq T_c$, b at $T \geq T_c$ | Reduced Kirchhoff eq for P at $T < T_c$, extrapolation of Kirchhoff eq for $T > T_c$ | 28 |

The value of β characterizes the attraction forces of the gas-solid system. Dubinin considered polarizability as the major important factor which, in turn, is proportional to the cube of molecular diameter. Thus, molar volume of the adsorbed molecules, $V_M$, was chosen as a measure of β. The characteristic curves can be unified as

$$W = f\left(\frac{RT}{V_M}\ln\frac{P_s}{P}\right) \tag{9}$$

This curve, in effect, becomes a correlation curve for all gases at all temperatures on a given solid. The function, however, remains unspecified.

Along the theme outlined in the above, many methods of plotting of the characteristic curves and many methods of computing parameters involved have been used in the literature. They are summarized in Table 1.

### B. Potential Theory Applied to Supercritical Conditions

The unified correlation curve, Equation 9, has had only limited success. The situation worsens when the adsorption temperature is above the critical temperature. Above the critical temperature, the adsorbed phase is ill-defined, since no condensation should take place thermodynamically. Consequently, as shown in Table 1, many methods of calculating the values of the molar volume of the adsorbed phase, $V_M$, and of $P_s$ (or $f_s$) have been suggested. The value of $V_M$ is also needed for calculating $f_s$ in the potential theory as will be discussed shortly.

The value of $V_M$ is, however, a real quantity, which should be measurable, in principle, under supercritical conditions. In the following, we present a method of calculating the value of $V_M$ under supercritical conditions based on data published in the literature. This value of $V_M$ may be considered as an experimental value.

Under supercritical conditions, adsorption of hydrocarbons on activated carbon follows the Langmuir isotherm, and the monolayer coverage, $V_m$, can be calculated. The surface area of the adsorbent, $S'$, can be independently measured by using the BET $N_2$ method, for example. Using these two experimental quantities, the molar volume of the adsorbed phase may be determined as:

$$V_M = \left( \frac{1}{f} \frac{22400S'}{N_o V_M} \right)^{3/2} N_o \left( \frac{cm^3}{gmol} \right)$$    (10)

where f is a packing factor, the value of which depends on the number of nearest neighbors. For 12 nearest neighbors in the bulk liquid and 6 on the plane — a common arrangement — the value of f is 1.091.

One may correlate the value of $V_M$, normalized against the molar volume of liquid at the normal boiling point, $V_{nbp}$ (which is a well-defined property) as a function of adsorption temperature and critical temperature.

Using the data of Reich et al.[28] for adsorption of $CH_4$, $C_2H_6$, $C_2H_4$, and $CO_2$ on BPL-type activated carbon above and below the critical temperatures, $V_M/V_{nbp}/T_c$ is plotted against the reduced temperature, $T_r$, as shown in Figure 1. Data on CO at high $T_r$ obtained in our laboratory[29] are also included in this figure.

As Figure 1 clearly shows, the adsorbed molar volume increases sharply at $T_r > 1$, or above the critical temperature, and, in fact, is approximately linear with $T_r$. Figure 1 will be used for calculating $V_M$ at temperatures above the critical, which in turn will be used to calculate $f_s$ for the potential theory in the next section.

## C. Unified Characteristic/Correlation Curve

The free energy equation, Equation 6, is for an ideal gas. For practical systems, however, the pressure is usually high and, hence, ideal gas behavior should not expected. For this reason, Lewis et al.[23] used fugacity to replace pressure in the free energy equation and the characteristic curves. As Table 1 indicates, fugacity has been used by all subsequent workers after Lewis et al.

Also, as indicated in Table 1, widely different opinions exist concerning the value of molar volume of the adsorbed phase, $V_M$, in correlating experimental data. Several extrapolation methods were suggested for $V_M$, each with limited success. Grant and Manes have used the saturated liquid molar volume at the normal boiling point for $V_M$,[25] $V_{nbp}$. This is a real and well-defined property and is chosen to reflect the affinity coefficient. We obtain the following characteristic curve:

$$W = f \left( \frac{T}{V_{nbp}} \ln \frac{f_s}{f} \right)$$    (11)

The use of $V_{nbp}$ to obtain the affinity coefficient is, needless to say, an approximation, since the affinity coefficient should be an indicator of the sum of all attractive forces of all contributing surface atoms for the adsorbate molecule, and it cannot be adequately represented by a single property such as $V_{nbp}$ nor any values of $V_M$ as calculated in Section B. On the other hand, the value of W, the total adsorbed volume, can be expressed exactly as:

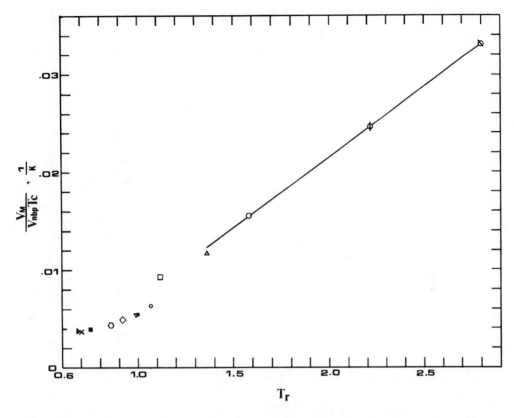

FIGURE 1. A generalized correlation for the experimental molar volume of the adsorbed phase with well-defined properties for adsorption data taken at Georgia Tech.[28] on BPL activated carbon for $CH_4$ at 301.4 K (○), 260.2 K (△), 212.7 K (□); $C_2H_4$ at 301.4 K (○), 260.2 K (◇), 212.7 (■); $C_2H_6$ at 301.4 K (▽), 260.2 K (◦), 212.7 K (▶); $CO_2$ at 301.4 K (●), 212.7 K (×); and data at SUNY at Buffalo[29] on PCB activated carbon for CO at 373.2 K (◌) and 296.2 K (◍).

$$W = NV_M(T) \tag{12}$$

where $V_M$ is strongly dependent on T as shown in Section B.

Thus, in our effort to find a unified characteristic/correlation curve for all gases on the same adsorbent, we use the following approaches:

1.  Affinity coefficient $= V_{nbp}$
2.  $W = NV_M(T)$

where the function $V_M(T)$ remains to be obtained using experimental data of various gases.

The experimental data of Reich et al. for four gases on BPL carbon are then plotted in the form of the characteristic curve, Equation 11, where $W = NV_{nbp}$, as shown in Figure 2. The fugacity was calculated using the Peng-Robinson equation of state.[30] The family of curves in Figure 2 can be unified by the use of Equation 12, i.e., by assigning a temperature dependence to $V_M$. Empirically, this dependence is found to be $W = V_{nbp}T_r^{0.6}$. Thus, we obtain the following unified correlation curve:

$$NV_{nbp}T_r^{0.6} = f\left(\frac{T}{V_{nbp}} \ln \frac{f_s}{f}\right) \tag{13}$$

The unified curve is shown in Figure 3.

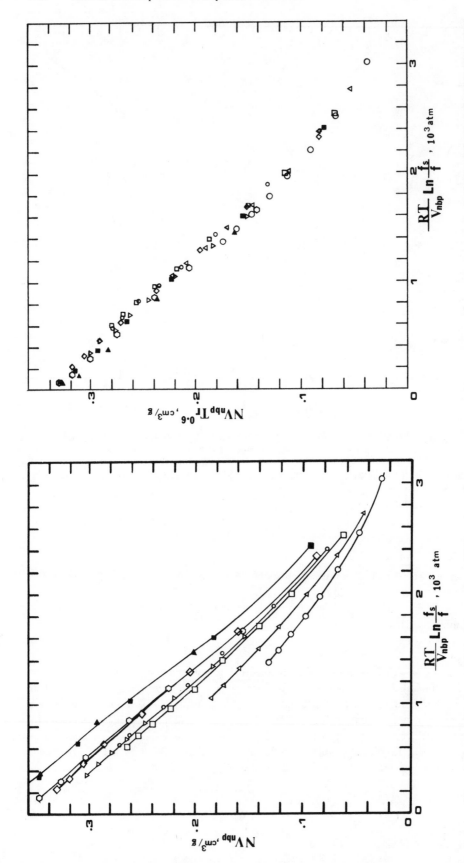

FIGURE 2.    The characteristic curves for adsorption on BPL activated carbon[28] for $CH_4$ at 301.4 K (○), 260.2 K (△), 212.7 K (□); $C_2H_4$ at 301.4 K (○), 260.2 K (◇), 212.7 K (■); $C_2H_6$ at 301.4 K (▽), 260.2 K (○), and 212.7 K (▲).

FIGURE 3.    The unified characteristic curve for adsorption on BPL activated carbon. The symbols are the same as in Figure 2.

## Table 2
## PREDICTION OF SINGLE GAS ADSORPTION BY UNIFIED
## CHARACTERISTIC CURVE ON BPL ACTIVATED CARBON AT 298.2 K

| Gas | P (atm) | Exptl $V^a$ (cm³/g) | $V_M$ (cm³/mol) | $f_s$ (atm) | f (atm) | $\epsilon$ (Eq 11) (atm) | W | Pred. $V^b$ (cm³/g) |
|---|---|---|---|---|---|---|---|---|
| $CO_2$ | 0.907 | 42.36 | 43.75 | 48.24 | 0.982 | 2870 | 0.044 | 32.04 |
| | 2.96 | 78.22 | 43.75 | 48.24 | 2.914 | 2069 | 0.107 | 77.92 |
| | 5.92 | 108.04 | 43.75 | 48.24 | 5.73 | 1570 | 0.162 | 117.9 |
| $CH_4$ | 0.987 | 20.62 | 108.7 | 131.3 | 0.985 | 3176 | 0.031 | 15.02 |
| | 3.95 | 46.51 | 108.7 | 131.3 | 3.91 | 2280 | 0.086 | 41.66 |
| | 6.91 | 61.64 | 108.7 | 131.3 | 6.80 | 1924 | 0.124 | 60.08 |
| CO | 0.987 | 8.970 | 127.5 | 184.4 | 0.986 | 3615 | 0.018 | 8.726 |
| | 3.95 | 26.90 | 127.5 | 184.4 | 3.94 | 2658 | 0.057 | 25.8 |
| | 6.91 | 37.21 | 127.5 | 184.4 | 6.88 | 2273 | 0.087 | 36.2 |
| $H_2$ | 1.97 | 1.51 | 128.4 | 251.5 | 1.98 | 4466 | 0.006 | 1.45 |
| | 3.95 | 2.86 | 128.4 | 251.5 | 3.96 | 3826 | 0.012 | 2.89 |
| | 6.91 | 4.45 | 128.4 | 251.5 | 6.94 | 3308 | 0.019 | 4.58 |

[a] Adsorption measured by Wilson and Danner.[31]
[b] Predicted from Figures 3 and 1 which are made from data on hydrocarbons on BPL carbon measured by Reich et al.[28]

In this correlation, $f_s$ is calculated as follows: below the critical temperature the adsorbed phase is assumed to be saturated liquid. The corresponding vapor pressure is used to calculate $f_s$ using the Peng-Robinson equation.[30] Above critical temperature conditions, Figure 1 is used to calculate $V_M$ from the adsorption temperature. A corresponding pressure is obtained which is then used to calculate $f_s$ using the Peng-Robinson equation. It should be noted, however, that extrapolation in Figure 1 to higher values of $T_r$ is not recommended.

The unified characteristic/correlation curve is further tested against the data of Wilson and Danner[31] on adsorption of $CO_2$, $CH_4$, CO, and $H_2$ on the same type (BPL) of activated carbon. Figure 3 is used to predict the adsorption of these four gases, at 298.2 K at pressures ranging from 1 to 7 atm. Thus, most of the experiments of Wilson and Danner were conducted at supercritical conditions. The predicted data are compared with their experimental data in Table 2. Some extrapolation of the curve in Figure 1 was performed for $H_2$.

The comparison shown in Table 2 is indeed encouraging and shows the predictive power of the potential-theory approach for predicting adsorption of other gases from data on one gas. The unified correlation curve is, of course, subject to further testing and improvement.

### D. Predicting Mixed-Gas Adsorption

Using the potential theory, adsorption of gas mixtures can be predicted from the single-gas isotherms, as illustrated by Lewis et al.[32] and by Grant and Manes.[33] Here we further use the unified correlation curve obtained in Section C, along with Figure 1 in Section B (for calculating $f_s$) for single gas adsorption, to predict the adsorption of mixtures. The calculation procedure, following Grant and Manes,[33] is as follows.

As in IAS theory, Raoult's law is assumed:

$$Py_i = P_i^o x_i \tag{3}$$

where $P_i^o$ is the "vapor pressure" of pure component i adsorbed at the same temperature T and adsorbate volume as that of the adsorbed mixture. The calculation begins with a set of assumed values of $x_i$ ($\Sigma x_i = 1$). A corresponding set of values of $(T/V_{nbpi})$ ln $(xf_s/f)$ i are calculated which should give the same adsorbate volume on the characteristic curves. Using Figure 3, the values of the potential function should be equal since the unified curve is used. The iteration is continued until the preceding is achieved. The total number of moles adsorbed is

$$N_t = \frac{W}{\Sigma x_i V_{Mi}(T)} ; \quad V_{Mi}(T) = V_{nbpi} T_{ri}^{0.6} \tag{14}$$

and the number of moles of component i adsorbed is

$$N_i = N_t x_i \tag{15}$$

The predicted results are shown in Tables 3 to 5. In the same tables the IAS, Langmuir, etc. methods are also shown and compared with the experimental data.

## IV. THE LOADING RATIO CORRELATION (LRC)

### A. The LRC Method

The loading ratio correlation is unique in that it is the only noniterative method for predicting adsorption from mixtures. This feature becomes a very important advantage in modeling gas separation processes where the adsorption (from mixtures) equations are coupled with the mass and energy balances as well as the rate expressions. The solution of the resultant set of equations represents a formidable computational problem.

The LRC method is a relatively little-known method, because it has been exclusively used for molecular sieve[34] and has rarely appeared in the literature. For molecular sieve adsorbents, the adsorption capacity is limited by the volume of intracrystalline cavities, and the amount of adsorbate required to fill the cavities is termed the maximum attainable loading. The amount adsorbed divided by the maximum loading is called loading ratio, hence, the term loading ratio correlation. The LRC method is based on the kinetic approach discussed in the preceding. As discussed in the Langmuir approach, a major problem is using the Markham-Benton equations[16] is that many adsorption systems do not follow the Langmuir isotherm; in that case the Freundlich isotherm can usually be used to fit the data. A more useful equation for fitting data is a hybrid of the Langmuir and Freundlich isotherms, viz.,

$$\theta = \frac{KP^{1/n}}{1 + KP^{1/n}} \tag{16}$$

The meaning of this equation can be interpreted in at least two ways. The first interpretation pertains to chemisorption, where each adsorbate molecule dissociates and occupies two site,[35] and n = 2 in that case. For physical adsorption, Sips[36] showed that Equation 16 leads to a site distribution, as a function of energy of adsorption, which is very similar to a Gaussian function. Therefore, Equation 16 is actually the Langmuir isotherm for a heterogeneous surface where the energy distribution follows a near-Gaussian distribution — not an uncommon situation.

Using Equation 16, Yon and Turnock have proposed the following noniterative equation for calculating adsorption from mixtures:[34]

$$\theta_i = \frac{K_p P_i^{1/n_i}}{1 + \sum_{i=1}^{N} K_i P_i^{1/n_i}} \tag{17}$$

## Table 3

### COMPARISON OF EXPERIMENTAL AND PREDICTED VALUES FOR ADSORPTION OF $CH_4(1)$-$C_2H_6(2)$ BINARY MIXTURES ON BPL ACTIVATED CARBON[a]

| T (K) | $Y_1$ (%CH$_4$) | P (kPa) | Experimental | | Unified potential | | | | IAS theory | | | | Markham-Benton | | | |
|---|---|---|---|---|---|---|---|---|---|---|---|---|---|---|---|---|
| | | | $n_1$ | $n_2$ | $n_1$ | %E | $n_2$ | %E | $n_1$ | %E | $n_2$ | %E | $n_1$ | %E | $n_2$ | %E |
| 301.4 | 0.267 | 2006 | 0.74 | 5.06 | 0.38 | 49.7 | 5.01 | 1.0 | 0.19 | 73.9 | 5.12 | −1.2 | 0.25 | 66.1 | 5.06 | 0.1 |
| | | 685 | 0.33 | 4.18 | 0.22 | 33.5 | 4.30 | 2.9 | 0.19 | 43.6 | 4.46 | −6.6 | 0.22 | 34.6 | 4.42 | −5.8 |
| | | 130 | 0.07 | 2.81 | 0.08 | −9.2 | 2.76 | −1.9 | 0.11 | −52.6 | 2.44 | 13.1 | 0.12 | −59.8 | 2.44 | 13.3 |
| | 0.499 | 1385 | 1.12 | 4.30 | 0.69 | 38 | 4.18 | 2.6 | 0.49 | 55.9 | 4.43 | −3.0 | 0.58 | 47.6 | 4.32 | −0.6 |
| | | 693 | 0.57 | 3.73 | 0.48 | 15.7 | 3.74 | −0.1 | 0.46 | 19.3 | 3.87 | −3.5 | 0.51 | 9.6 | 3.81 | −2.0 |
| | | 132 | 0.18 | 2.41 | 0.18 | 1.1 | 2.43 | −1.0 | 0.25 | −36.0 | 1.90 | 21.0 | 0.25 | −41.0 | 1.89 | 21.0 |
| | 0.745 | 1373 | 1.38 | 3.29 | 1.34 | 3.5 | 3.18 | 3.2 | 1.14 | 17.9 | 3.32 | −1.0 | 1.27 | 8.4 | 3.18 | 3.2 |
| | | 687 | 0.94 | 2.87 | 0.92 | 2.3 | 2.86 | 0.1 | 0.99 | −5.1 | 2.73 | 4.7 | 1.06 | −12.5 | 2.66 | 7.2 |
| | | 130 | 0.33 | 1.66 | 0.32 | 1.2 | 1.82 | −10.0 | 0.43 | −30.3 | 1.11 | 33.0 | 0.44 | 32.6 | 1.10 | 33.0 |
| 212.7 | 0.267 | 353 | 0.22 | 7.01 | 0.15 | 31.4 | 7.17 | −2.3 | 0.14 | 38.0 | 7.06 | −0.8 | 0.14 | 37.5 | 7.06 | −0.8 |
| | | 244 | 0.17 | 6.56 | 0.13 | 22.4 | 6.99 | 6.5 | 0.14 | 19.5 | 6.98 | −6.4 | 0.14 | 18.8 | 6.98 | −6.3 |
| | 0.499 | 695 | 0.56 | 7.27 | 0.48 | 13.1 | 7.07 | 2.7 | 0.37 | 32.8 | 6.88 | 5.4 | 0.38 | 32.3 | 6.88 | 5.4 |
| | | 348 | 0.38 | 6.52 | 0.36 | 7.2 | 6.72 | 3.0 | 0.37 | 4.6 | 6.75 | −3.5 | 0.37 | 3.7 | 6.75 | −3.5 |
| | 0.745 | 875 | 1.03 | 6.19 | 1.14 | 11.0 | 6.21 | 0.2 | 1.00 | 3.0 | 6.20 | 0.1 | 1.00 | 2.5 | 6.19 | 0.0 |
| | | 436 | 0.72 | 5.94 | 0.85 | 17.5 | 6.03 | 1.5 | 0.97 | −34.8 | 6.02 | −1.5 | 0.97 | −35.0 | 6.02 | −1.4 |

[a] Adsorption ($n_i$) is in mmol/g. E = percent error.

## Table 4
## COMPARISON OF EXPERIMENTAL AND PREDICTED VALUES FOR ADSORPTION OF $CH_4(1)$-$C_2H_4(3)$ BINARY MIXTURES ON BPL ACTIVATED CARBON[a]

| T (K) | $y_1$ (%CH₄) | P (kPa) | Experimental | | Unified potential | | | | IAS theory | | | | Markham-Benton | | | |
|---|---|---|---|---|---|---|---|---|---|---|---|---|---|---|---|---|
| | | | $n_1$ | $n_3$ | $n_1$ | %E | $n_3$ | %E | $n_1$ | %E | $n_3$ | %E | $n_1$ | %E | $n_3$ | %E |
| 301.4 | 0.260 | 2001 | 0.62 | 5.10 | 0.51 | 18.5 | 4.83 | 5.3 | 0.24 | 61.0 | 5.03 | 1.4 | 0.31 | 49.7 | 4.95 | 2.9 |
| | | 683 | 0.34 | 4.00 | 0.30 | 11.0 | 3.9 | 2.5 | 0.23 | 33.0 | 4.22 | -5.4 | 0.26 | 22.9 | 4.18 | -4.4 |
| | | 345 | 0.24 | 3.35 | 0.20 | 15.0 | 3.19 | 4.9 | 0.19 | 19.5 | 3.41 | -1.6 | 0.21 | 12.2 | 3.39 | -1.0 |
| | 0.536 | 2031 | 1.33 | 4.06 | 1.23 | 7.0 | 3.81 | 6.0 | 0.71 | 46.0 | 4.30 | -6.0 | 0.86 | 35.1 | 4.13 | -1.9 |
| | | 676 | 0.68 | 3.24 | 0.72 | -5.2 | 3.04 | 6.1 | 0.62 | 8.6 | 3.37 | -3.8 | 0.69 | -0.7 | 3.30 | -1.7 |
| | | 344 | 0.48 | 2.68 | 0.49 | -2.0 | 2.50 | 6.8 | 0.50 | -3.3 | 2.59 | 3.6 | 0.53 | -9.6 | 2.55 | 4.9 |
| | 0.765 | 1431 | 1.75 | 2.77 | 1.84 | -5.0 | 2.44 | 11.0 | 1.44 | 17.7 | 2.83 | -2.5 | 1.58 | 9.8 | 2.68 | 3.0 |
| | | 697 | 1.20 | 2.36 | 1.28 | -6.1 | 2.09 | 11.0 | 1.20 | -0.1 | 2.23 | 5.4 | 1.27 | -5.9 | 2.16 | 8.5 |
| | | 444 | 0.92 | 2.06 | 0.98 | -6.8 | 1.83 | 11.1 | 1.00 | -9.0 | 1.82 | 11.8 | 1.05 | -13.5 | 1.78 | 13.0 |
| 212.7 | 0.260 | 681 | 0.29 | 7.61 | 0.24 | 19.0 | 7.55 | 0.9 | 0.24 | 19.0 | 7.55 | 0.9 | 0.30 | -1.3 | 7.48 | 1.8 |
| | | 414 | 0.22 | 7.08 | 0.19 | 12.3 | 7.16 | -1.3 | 0.24 | -9.3 | 7.39 | -4.4 | 0.30 | -31.0 | 7.33 | -3.7 |
| | 0.536 | 1041 | 0.89 | 7.17 | 0.77 | 14.0 | 7.10 | 1.1 | 0.74 | 17.0 | 7.01 | 2.2 | 0.91 | -2.0 | 6.82 | 4.9 |
| | | 616 | 0.66 | 6.69 | 0.62 | 6.5 | 7.34 | -9.7 | 0.75 | 13.0 | 6.84 | -2.3 | 0.89 | -35.0 | 6.69 | 0.0 |
| | 0.765 | 896 | 1.44 | 5.92 | 1.46 | -1.4 | 5.72 | 3.3 | 1.78 | -23.0 | 5.68 | 4.0 | 2.04 | -41.0 | 5.40 | 8.7 |
| | | 443 | 1.17 | 5.49 | 1.13 | 2.9 | 5.49 | 0.1 | 1.75 | -49.6 | 5.31 | 3.0 | 1.93 | -65.0 | 5.12 | 6.8 |

[a] Adsorption ($n_i$) is in mmol/g. E = percent error.

**Table 5**

**COMPARISON OF EXPERIMENTAL AND PREDICTED VALUES FOR ADSORPTION OF $C_2H_6(2)$-$C_2H_4(3)$ BINARY MIXTURES ON BPL ACTIVATED CARBON[a]**

| T (K) | $Y_2$ (%$C_2H_6$) | P (kPa) | Experimental | | Unified potential | | | | IAS theory | | | | Markham-Benton | | | |
|---|---|---|---|---|---|---|---|---|---|---|---|---|---|---|---|---|
| | | | $n_2$ | $n_3$ | $n_2$ | %E | $n_3$ | %E | $n_2$ | %E | $n_3$ | %E | $n_2$ | %E | $n_3$ | %E |
| 301.4 | 0.240 | 1982 | 1.61 | 4.32 | 1.82 | −12.5 | 3.66 | 15.0 | 1.56 | 3.2 | 3.86 | 10.0 | 1.60 | 0.9 | 3.82 | 11.0 |
| | | 1341 | 1.47 | 3.89 | 1.77 | −20.0 | 3.40 | 12.0 | 1.52 | −2.8 | 3.72 | 4.4 | 1.55 | −4.8 | 3.69 | 5.1 |
| | | 737 | 1.35 | 3.36 | 1.65 | −22.0 | 2.94 | 12.6 | 1.40 | −4.2 | 3.42 | −1.8 | 1.42 | −5.7 | 3.40 | −1.1 |
| | | 308 | 1.11 | 2.73 | 1.40 | −25.0 | 2.25 | 17.1 | 1.14 | −2.0 | 2.75 | −0.8 | 1.14 | −2.8 | 2.74 | −0.5 |
| | 0.472 | 1133 | 2.86 | 2.51 | 3.03 | −6.0 | 2.03 | 19.2 | 2.77 | 3.3 | 2.39 | 4.6 | 2.80 | 2.1 | 2.36 | 5.9 |
| | | 550 | 2.53 | 2.05 | 2.70 | −7.1 | 1.66 | 18.0 | 2.48 | 1.9 | 2.13 | −4.0 | 2.50 | 1.1 | 2.11 | −3.0 |
| 212.7 | 0.240 | 405 | 2.53 | 5.25 | 2.93 | −16.0 | 4.82 | 8.2 | 2.62 | −3.7 | 4.95 | 5.6 | 3.00 | −18.5 | 4.54 | 13.4 |
| | | 140 | 2.32 | 4.40 | 2.75 | −18.0 | 3.98 | 9.4 | 2.59 | 11.6 | 4.55 | −3.3 | 2.83 | −21.8 | 4.29 | 2.4 |
| | 0.682 | 343 | 5.81 | 1.61 | 5.99 | −3.0 | 1.46 | 13.2 | 5.74 | 1.2 | 1.60 | 5.4 | 5.99 | −3.2 | 1.33 | 21.3 |
| | | 137 | 5.27 | 1.40 | 5.61 | −6.5 | 1.23 | 11.8 | 5.60 | −6.3 | 1.46 | −4.4 | 5.77 | −9.5 | 1.28 | 8.4 |

[a] Adsorption $n_i$ is in mmol/g. E = percent error.

where

$$\theta_i = \frac{V_i}{V_{im}}$$

The value of $V_{im}$ is the amount adsorbed at monolayer coverage for species i from a single gas.

Equation 17 can be derived from Equation 16 using the kinetic approach of Markham and Benton for a differential increment of energy of adsorption and integrating over the entire energy range in the same manner as used by Roginskii.[8] Thus, Equation 17 does have a theoretical basis.

For the correlation of data rather than predictive purposes, Equation 17 can be adjusted to the following form where the $\eta_i$ describe the adsorbed phase interactions between different species:

$$\theta_i = \left(\frac{K_i P_i^{1/n_i}}{\eta_i}\right) \Big/ \left[1 + \sum_{i=1}^{N} (K_i P_i^{1/n_i}/\eta)\right] \qquad (18)$$

Equation 17 provides a noniterative method for calculating adsorption from mixtures. Satisfactory correlation using this method has been shown for molecular sieves.[34] Equation 18 can be used to further improve data correlation.

The LRC method is used here for PCB-type activated carbon for $H_2$-$CH_4$ mixtures over a wide temperature and pressure range, under supercritical condition. These data are obtained in our laboratory.[37] The predicted data are shown in Table 6, where they are also compared with that from other methods and the experimental data.[37]

## B. Comparison of Various Models

From Tables 3 to 6, it is clear that all the models considered here are roughly equivalent in their predictive abilities, and all are quite satisfactory. It is interesting to note, however, that for a few specific data points, all methods fail rather equally badly.

This is hard to explain since the three methods all derive from different origins. The only point in common is that the interactions among the adsorbed molecules have not been included in these models, except in Equation 18 which we have not used in this comparison. The LRC method is the recommended technique for modeling the performance of cyclic separation processes, due to the fact that it yields results comparable with the other methods and is much simpler to apply.

## V. MODELING OF BULK, MULTICOMPONENT GAS SEPARATION BY CYCLIC PROCESSES

Mainly because of the escalation costs of energy, the commercial use of cyclic separation processes using fixed beds has also been increasing. For gas separation, pressure swing adsorption (PSA) is more desirable than temperature swing adsorption (TSA), because of shorter cycle times and, hence, higher throughputs. A significant number of commercial units has been installed primarily for hydrogen purification and air drying.[38,52,53] Other applications include separation of straight chain hydrocarbons and oxygen enrichment from air.[3,52,53] Other potential PSA applications are separation of ammonia plant purge gas, hydrodealkylation plant purge gas, cryogenic plant recycle gas, and refinery fuel gas.

Although cyclic separation processes have been widely used in industry, theoretical understanding is still in a primitive stage. Theoretical developments for PSA have been ex-

## Table 6
### COMPARISON OF EXPERIMENTAL DATA ON ADSORPTION OF $H_2(1)$-$CH_4(2)$ MIXTURE ON PCB ACTIVATED CARBON WITH LRC AND IAS PREDICTIONS[a]

| T (K) | P (kPa) | $y_1$ | $y_2$ | Experimental | | | | LRC | | | | IAS | | | |
|---|---|---|---|---|---|---|---|---|---|---|---|---|---|---|---|
| | | | | $n_1$ | $n_2$ | $x_1$ | $x_2$ | $n_1$ | $n_2$ | $x_1$ | $x_2$ | $n_1$ | $n_2$ | $x_1$ | $x_2$ |
| 295 | 738 | 0.91 | 0.09 | 0.187 | 0.638 | 0.227 | 0.773 | 0.27 | 0.644 | 0.243 | 0.757 | 0.207 | 0.633 | 0.246 | 0.754 |
| 295 | 1138 | 0.90 | 0.10 | 0.253 | 0.915 | 0.217 | 0.783 | 0.286 | 1.019 | 0.219 | 0.781 | 0.287 | 0.981 | 0.225 | 0.775 |
| 375 | 1613 | 0.94 | 0.06 | 0.241 | 0.232 | 0.510 | 0.490 | 0.190 | 0.262 | 0.420 | 0.580 | 0.191 | 0.255 | 0.428 | 0.572 |
| 480 | 896 | 0.83 | 0.17 | 0.044 | 0.111 | 0.284 | 0.716 | 0.039 | 0.103 | 0.275 | 0.725 | 0.039 | 0.102 | 0.276 | 0.724 |
| 480 | 807 | 0.91 | 0.09 | 0.037 | 0.056 | 0.398 | 0.602 | 0.039 | 0.050 | 0.438 | 0.562 | 0.039 | 0.049 | 0.446 | 0.554 |

[a] The amount adsorbed ($n_1$) is in mmol/g.

tensively reviewed by Wankat[3] and, more recently, by Cheung and Hill.[7] In all of the published models, (1) linear adsorption isotherms are used, thus, limiting the models to adsorption of dilute species and (2) local equilibrium between the bulk gas phase and the intraparticle absorbed phase is assumed, implying no mass transfer limitations. The only exception to (2) is the model by Chihara and Suzuki[39] where the linear-driving-force approach is used. In the equilibrium models for PSA, the method of characteristics can be used to derive simple algebraic equations to estimate the steady-state product concentration, the effects of parameter changes on products, and provide a bound for real system operation. These models include those of Shendalman and Mitchell,[40] Weaver and Hamrin,[41] Fernandez and Kenney,[42] and Hill et al.[7,43] for one adsorbate, and Chan et al.,[44] Nataraj and Wankat,[45] etc. for two or more adsorbates. Another approach involving the numerical solution of mass and energy balance equations, plus the mass transfer rate equations, provides better approximations at the expense of increased computational time; e.g., see Chihara and Suzuki[39] and Carter and Wyszynski.[46] All the preceding examples were developed for dilute systems and, thus, linear isotherms were used. The model by Sircar and Kumar[47] included mass transfer rates (linear driving force), but involved one adsorbate which obeyed the Langmuir isotherm, even though the method of characteristics was used.

Multicomponent, bulk gas separation, i.e., for gases containing high concentrations of adsorbates (more than 10 wt % according to Keller[52]), is in its infancy from the point of view of both commercial application and theoretical understanding. However, its potential applications are quite important. One of the important feasible applications is in the separation of fuel gases in advanced coal gasification and liquefactions processes. In the designs of advanced coal gasification processes, the costs of gas separation amount to about one third or more of the total plant costs using cryogenic processes. Here, gas separation involves removal of the sour gases ($H_2S$ and $CO_2$) and separating the high-Btu gas (over 90% $CH_4$) and hydrogen (and CO). All of these components except $H_2S$ are present at high percentages. The feasibility of separating this mixture into three product streams by cyclic processes has recently been demonstrated by using temperature swing in the authors' laboratory.[48] The published theoretical models are not applicable to multicomponent, bulk gas separation because of the two limitations outlines in the foregoing. Equilibrium, multicomponent models have indeed been developed by Hill and co-workers[44] and by Wankat and co-workers.[45] However, the use of independent, noninterfering linear isotherms limits the use of their models to dilute systems.

## A. Temperature Swing Cycle

To circumvent the two limitations described above, in what follows we present a model which (1) incorporates the loading ratio correlation equations and (2) allows for mass transfer limitations. Such a model may then be used to interpret the results for bulk gas separation for multicomponent mixtures. The model is used in this case for separating a mixture containing 50/50 of $H_2/CH_4$ by a temperature swing cycle which can be readily extended to more components, since the adsorption of both $H_2$ and $CH_4$ are accounted for simultaneously.

The following assumptions and simplications are made for the model:

1.  The ideal gas law applies. (The compressibility factor for the gas mixture was calculated to be 0.99 under our experimental conditions of 34 atm and 25°C.)
2.  The axial pressure gradient across the bed is neglected.
3.  Inside the pores, instantaneous equilibrium exists between the gas phase and the adsorbed phase.
4.  Plugflow conditions apply, i.e., longitudinal dispersion along the bed is neglected.
5.  The temperature is assumed to vary uniformly in the bed, i.e., temperature is equilibrated instantaneously between gas and solid.

Mass balances for both components $CH_4$ (A) and $H_2$ (B) in the packed-bed column at system pressure, P, and temperature, T, are given by:

$$\alpha \frac{\partial C_A}{\partial t} + \frac{\partial u C_A}{\partial z} - S_A = 0 \tag{19}$$

and

$$\alpha \frac{\partial C_B}{\partial t} + \frac{\partial u C_B}{\partial z} - S_B = 0 \tag{20}$$

The quantity S is the molar flux through the exterior surfaces of the particles in a unit volume of bed. Converting to mole fraction, $y_A$, noting that $y_B = 1 - y_A$, and rearranging, we obtain

$$\frac{\partial u}{\partial z} = (S_A + S_B) \frac{RT}{P} + \alpha \frac{1}{T} \frac{\partial T}{\partial t} \tag{21}$$

$$\alpha \frac{\partial y_A}{\partial t} + u \frac{\partial y_A}{\partial z} + y_A(S_A + S_B) \frac{RT}{P} - \frac{S_A RT}{P} = 0 \tag{22}$$

We further assume that the bed is composed of spherical particles of uniform radius, a.

$$S_A = \frac{3\rho_B}{\rho_p} \frac{1}{a} [N_{Ar}]_{r=a} \tag{23}$$

$$S_B = \frac{3\rho_B}{\rho_p} \frac{1}{a} [N_{Br}]_{r=a} \tag{24}$$

The mass balances for A and B in the pores of a spherical particle at axial location z are given by:

$$\epsilon \frac{\partial C_A^*}{\partial t} + \rho_p \frac{\partial q_A}{\partial t} + \frac{1}{r^2} \frac{\partial}{\partial r} (r^2 N_{Ar}) = 0 \tag{25}$$

$$\epsilon \frac{\partial C_B^*}{\partial t} + \rho_p \frac{\partial q_B}{\partial t} + \frac{1}{r^2} \frac{\partial}{\partial r} (r^2 N_{Br}) = 0 \tag{26}$$

By using the following volume-average quantities[49] (for both A and B):

$$\bar{C} = \frac{3}{a^3} \int_0^a C_A^* r^2 dr \tag{27}$$

$$\bar{q}_A = \frac{3}{a^3} \int_0^a q_A r^2 dr \tag{28}$$

and noting

$$N_{Ao} = [N_{Ar}]_{r=a} = y_{As}^* (N_{Ao} + N_{Bo}) - CD_e \left[ \frac{\partial y_A^*}{\partial r} \right]_{r=a} \tag{29}$$

From Equations 25 and 26 and using the volume-average quantities:

$$N_{Ao} + N_{Bo} = - \frac{a\rho_p}{3} \left( \frac{\partial \bar{q}_A}{\partial t} + \frac{\partial \bar{q}_B}{\partial t} \right) + \epsilon \frac{P}{RT^2} \frac{\partial T}{\partial t} \left( \frac{2}{3} \right) \tag{30}$$

From Equations 29 and 30 we have

$$N_{Ao} = [N_{Ar}]_{r=a} = - \frac{a}{3} \rho_p y_{As}^* \left( \frac{\partial \bar{q}_A}{\partial t} + \frac{\partial \bar{q}_B}{\partial t} \right)$$

$$+ \epsilon \frac{P}{RT^2} \frac{\partial T}{\partial t} y_{As}^* \left( \frac{a}{3} \right) - CD_e \left[ \frac{\partial y_A^*}{\partial r} \right]_{r=a} \tag{31}$$

Expressing Equation 25 in volume-average quantities:

$$\frac{\partial \bar{C}_A}{\partial t} = - \frac{3}{a^3 \epsilon} a^2 [N_{Ar}]_{r=a} - \frac{\rho_p}{\epsilon} \frac{\partial \bar{q}_A}{\partial t} \tag{32}$$

or

$$\frac{\partial \bar{y}_A}{\partial t} = \frac{3 \bar{D}_e}{a \epsilon} \left[ \frac{\partial y_A^*}{\partial r} \right]_{r=a} - \frac{\rho_p}{\epsilon} \frac{RT}{P} \frac{\partial \bar{q}_A}{\partial t}$$

$$+ \frac{y_{As}^* \rho_p}{\epsilon} \frac{RT}{p} \left( \frac{\partial \bar{q}_A}{\partial t} + \frac{\partial \bar{q}_B}{\partial t} \right) + (\bar{y}_A - y_{As}^*) \frac{1}{T} \frac{\partial T}{\partial t} \tag{33}$$

In order to simplify the equation further, we assume a parabolic concentration profile as first used by Chao and co-workers:[50]

$$y_A^* = K_o + K_2 r^2 ,$$

$$K_o = y_{As}^* - K_2 a^2 ,$$

$$K_2 = (y_{As}^* - \bar{y}_A) \frac{5}{2a^2} \tag{34}$$

Equation 25 can now be converted to:

$$\frac{\partial \bar{y}_A}{\partial t} = \frac{15 \, D_e}{a^2 \epsilon} (y_{As}^* - \bar{y}_A) - \frac{\rho_p}{\epsilon} \frac{RT}{P} \frac{\partial \bar{q}_A}{\partial t} + \frac{y_{As}^* \rho_p}{\epsilon}$$

$$\left( \frac{\partial \bar{q}_A}{\partial t} + \frac{\partial \bar{q}_B}{\partial t} \right) \frac{RT}{P} + (\bar{y}_A - y_{As}^*) \frac{1}{T} \frac{\partial T}{\partial t} \tag{35}$$

The quantities $\bar{q}_A$ and $\bar{q}_B$ must be related to $\bar{y}_A$ and $\bar{y}_B$ via adsorption isotherms. The LRC equations are

$$\bar{q}_A = \bar{q}_A(T, p, \bar{y}_A) = \frac{V_{mA} k_A \bar{y}_A}{1 + k_A \bar{y}_A + k_B (1 - y_A)^n} \tag{36}$$

$$\bar{q}_B = \bar{q}_B(T, p, \bar{y}_B) = \frac{V_{mB} k_B \bar{y}_A (1 - \bar{y}_A)^n}{1 + k_A \bar{y}_A + k_B (1 - y_A)^n} \tag{37}$$

where

$$V_{mA} = k_3 T^{k_4} \text{ and } \quad k_A = k_5 e^{k_6/T} ,$$

$$V_{mB} = k_7 T^{k_8} \text{ and } \quad k_B = k_9 e^{k_{10}/T} , \text{ and}$$

$$n = a + bT \tag{38}$$

Equations 23 and 24 can also be expressed in the volume-average quantities as

$$S_A = -\frac{P}{RT} \frac{15}{a^2} \frac{\rho_B}{\rho_p} D_e(y_A - \bar{y}_A) - \rho_B y_A \left( \frac{\partial \bar{q}_A}{\partial t} + \frac{\partial \bar{q}_B}{\partial t} \right)$$

$$+ \frac{\epsilon \rho_B}{\rho_p} y_A \frac{P}{RT^2} \frac{\partial T}{\partial t} \tag{39}$$

$$S_B = -\frac{P}{RT} \frac{15}{a^2} \frac{\rho_B}{\rho_p} D_e(y_A - \bar{y}_A) - \rho_B(1 - y_A) \left( \frac{\partial \bar{q}_A}{\partial t} + \frac{\partial \bar{q}_B}{\partial t} \right)$$

$$+ \frac{\epsilon \rho_B}{\rho_p} (1 - y_A) \frac{P}{RT^2} \frac{\partial T}{\partial t} \tag{40}$$

Equations 36 to 38 can now be substituted into Equation 35 and Equation 35 is solved simultaneously with Equations 21 and 22 in which $S_A$ and $S_B$ are obtained from Equations 39 and 40. The model is, thus, complete with proper boundary and initial conditions.

The boundary conditions for the $H_2/CH_4$ separation example are

$$t = 0, \quad z > 0; \quad \bar{y}_A = 0.5, \quad y_A = 0.5 \tag{41}$$

$$t > 0, \quad z = 0; \quad y_A = 0.5 \tag{42}$$

**Method of solution** — Standard numerical methods may be used to obtain solutions for Equations 21, 22, and 35. Equation 35 is first integrated by Euler method to obtain the value of $\bar{y}_A$ at a new time step. With the value of $\bar{y}_A$, $\partial \bar{q}_A/\partial t$ and $\partial \bar{q}_B/\partial t$ can be evaluated by differentiating Equations 36 and 37. Using the values of $S_A$ and $S_B$ from Equations 39 and 40, Equation 22 is then solved by numerical integration with a proper boundary condition, e.g., at a fixed u at the exit end of column. Finally, Equation 21 is solved by the Crank-Nicolson method, which has been found both stable and convergent for a wide range of parameter values. Because central differences for the distance derivative are averaged at the beginning and end of each time step, a mixed order or error is not involved. The packed bed is divided into 20 cells for computation. The number of time steps is 160.

The model has been applied to the data obtained in our laboratory on temperature swing separation of $H_2$-$CH_4$ mixture using PCB-activated carbon.[51] In our previous model[51] the adsorption of hydrogen was neglected and no assumption was made on the functional form of the concentration profile within the particle. The experimental conditions are shown in Table 7 along with the input parameters used in this model. The LRC parameters are not shown but are taken from the data used in Table 6. The computed results using this model are shown for one heating half-cycle and one cooling half-cycle in Figure 4. The results are only slightly better than the previous model neglecting $H_2$ adsorption,[51] because the adsorption of $H_2$ is, indeed, quite small. The model is now being used for cyclic separation — both temperature and pressure swings — for multicomponent gas mixtures containing adsorbate of similar adsorptivities.

**Table 7**
## EXPERIMENTAL CONDITIONS FOR TEMPERATURE SWING SEPARATION USING PCB ACTIVATED CARBON AND INPUT PARAMETERS FOR MODELING

|  | Heating half-cycle | Cooling half-cycle |
|---|---|---|
| Packed-bed length, L | 40.0 cm | 40.0cm |
| Initial temperature, $T_o$ | 298 K | 523 K |
| Final temperature, $T_f$ | 523 K | 298 K |
| System pressure, P | 34 atm | 34 atm |
| Heating or cooling rate, $T_r$ | 4 K/min | −7.0 K/min |
| Inlet concentration of $CH_4$, y in | 0.50 | 0.50 |
| Product flow rate, G | 600 cm³/min | 300 cm³/min |
| Particle size, a | 0.05 cm | 0.05 cm |
| Effective diffusivity | $1.5 \times 10^{-5}$ cm²/min | $4.0 \times 10^{-5}$ cm²/min |
| At initial conditions, $D_{e_\rho}$ | $(2.5 \times 10^{-7}$ cm²/sec) | $(6.6 \times 10^{-7}$ cm²/sec) |

## B. Pressure Swing Adsorption

Pressure swing adsorption processes are widely used by industry and are rapidly growing in their applications. Two state-of-the-art reviews on PSA are available; they are, however, not on the modeling aspects.[52,53]

The basic steps involved in PSA processes are (1) pressurization, (2) high-pressure adsorption, (3) cocurrent depressurization, (4) counter-current blowdown, and (5) low-pressure purge. The pressurization step may be done by using either the feed gas mixture or the effluents from the depressurization and blowdown steps, and in the latter case it is called pressure equalization. The five steps may be arranged and synchronized by using four beds.[52,53] In a commercial operation, two to five of the above steps are used depending on the requirements of product recovery and product purity.

Most of the fundamental studies, i.e., modeling, in the published literature were on cycles involving two to three steps. Although the nonequilibrium models should be better than the equilibrium models, a direct comparison has not been made.

A one-column apparatus has been built and operating in this laboratory for studying PSA for separating fuel gases, such as from coal gasification processes. Using the five steps with a cycling time of 5 to 10 min, a 50/50 $H_2/CH_4$ can be separated into 92% $CH_4$ and 96% $H_2$ with product recoveries of 96% for $CH_4$ and 81% for $H_2$; the latter is used as the purge gas.

A major difference between bulk gas separation and purification processes, e.g., air drying and hydrogen purification, is in the temperature variation in the bed. In air drying, the bed temperature varies within about ±5°C.[38,39] In bulk separation, such as separation of 50/50 $H_2/CH_4$, we have found that the temperature in the bed varies typically up to ±25 to 30°C, at all locations in the bed. Because the heat transfer rate between the column and the ambient is much lower than the heating and cooling rates in the bed due to adsorption and desorption, the process may be assumed as adiabatic. In fact, the large temperature variation in the bed is the basis for a new design of the columns, whereby two or more columns are arranged in a shell-and-tube heat-exchanger configuration.[54] In this configuration the heat of adsorption is used to provide the heat for desorption, and the separation is substantially more efficient.

As a consequence of the large temperature variation in the bed for bulk separation, the model must incorporate the energy balance equations. The model presented in Section A has been modified by adding the energy balance equations to account for the temperature variation. The detailed results on PSA at this laboratory will be published, shortly, elsewhere.

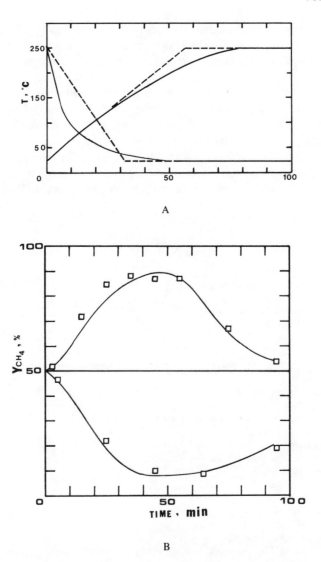

FIGURE 4.   Temperature-swing separation of $H_2/CH_4$ mixture under conditions shown in Table 7. (A) Temperature: experimental (solid line) and used in model (dashed line); B. effluent concentration: experimental ($\square$)[51] and theoretical (solid line).

## VI. SUMMARY AND CONCLUSION

First, a review has been made on the existing theories of adsorption from mixed gases, i.e., the prediction of mixed gas adsorption from single gas isotherms. A detailed discussion is given on three specific theories: the ideal adsorbed solution theory, the potential theory, and the loading ratio correlation.

Second, the Polanyi potential theory can be used, in principle, to predict, for a given gas-solid system, the adsorption isotherms at all temperatures from the isotherm at any specific temperature. In this paper, a method is suggested for predicting adsorption of all gases based on adsorption from a specific gas on the same solid. Critical temperatures are used in this method. A discussion is also given on the extension of the potential theory for predicting mixed gas adsorption from single gas data.

Third, a comparison of the three theories with experimental data showed that there is no distinct advantage for any theory over the others in predictive power. Despite the different origins of these theories, their predictive power appears to be consistent for any given adsorption system, i.e., they succeed or fail to the same extent for the same system. However, the loading ratio correlation is the only noniterative method and is the simplest to use. It is recommended that this method be used in modeling of the dynamics of adsorbers.

Fourth, a critical review is made on the published models for the performance of cyclic gas separation processes, primarily, pressure swing adsorption. For bulk, multicomponent gas separation, the published models are not applicable. A detailed temperature-swing model is presented which incorporates the loading ratio correlation equations and mass transfer rates with a specific attention to pore diffusion. Our ongoing work on PSA is also briefly described.

## NOMENCLATURE

| | |
|---|---|
| a | Average radius of sorbent particle, cm |
| b | A constant related to the net enthalpy, $\delta H$, of adsorption according to Langmuir theory |
| C | Concentration in bulk flow, mol/$\ell$ |
| De | Effective diffusivity, $cm^2$/min or $cm^2$/sec |
| $k_i$, K | Constants |
| n | Constant |
| N | Molar flux in radial direction, $mol/cm^2$/sec; number of adsorbates; or number of moles of adsorbates |
| $N_o$ | Avogadro number |
| p | Pressure of adsorbate, atm |
| P | Total pressure, atm |
| q | Number of mole of sorbate adsorbed per gram of solid |
| r | Radial distance, cm |
| R | Gas constant |
| S | Overall rate of sorption per unit volume of bed, mol/min/$\ell$ of bed |
| S' | Surface area, $cm^2$/g |
| t | Time, min or sec |
| T | Solid or bed temperature, K |
| $T_c$ | Critical temperature, K |
| $T_r$ | Reduced temperature |
| u | Superficial velocity, cm/min |
| V | Volume adsorbed per gram of sorbent, $cm^3$ (STP)/g or molar volume of adsorbed phase (in potential theory), $cm^3$/gmol |
| $V_m$ | Volume adsorbed corresponding to monolayer coverage, $cm^3$ (STP)/g |
| $V_M$ | Molar volume of the adsorbed molecules, $cm^3$/gmol |
| $V_{nbp}$ | Molar volume of the adsorbed phase for saturated liquid at normal boiling point, $cm^3$/gmol |
| W | Volume of the adsorbed phase, $cm^3$/g |
| x | Mole fraction in the adsorbed phase |
| y | Mole fraction in the gas phase |
| z | Axial distance along the bed, cm |

## Greek Letters

| | |
|---|---|
| $\alpha$ | Interparticle void fraction |
| $\beta$ | Affinity coefficient |
| $\epsilon$ | Intraparticle void fraction; or potential function in the potential theory |
| $\rho_B$ | Bed density, $g/cm^3$ |
| $\rho_p$ | Particle density, $g/cm^3$ |
| $\eta$ | Interaction parameter between adsorbates |
| $\theta$ | Particle density, $g/cm^3$; or surface coverage |

## Subscripts

| | |
|---|---|
| A | $CH_4$ |
| B | $H_2$ |
| i | Species |
| m | Monolayer coverage based on single adsorbate |
| o and s | At surface of particle |
| r | In the radial direction |
| s | Saturated state |
| t | Total |

## Superscripts

| | |
|---|---|
| — | Volume—average quantities |
| * | In the pores |

## ACKNOWLEDGMENTS

Portions of this work were sponsored by the Department of Energy, Morgantown Energy Technology Center, under Contract DE-AC-83MC20183. We are also grateful to Dr. M. J. Desai and Mr. S. J. Doong for performing most of the work on modeling cyclic processes.

## REFERENCES

1. **Berg, C. and Bradley, W. E.,** Hypersorption — new fractionating process, *Petrol. Eng.,* 18, 115, 1947.
2. **King, C. J.,** *Separation Processes,* 2nd ed., McGraw-Hill, New York, 1980, chap. 4.
3. **Wankat, P. C.,** Cyclic separation techniques, in *Percolation Processes, Theory and Applications,* Rodrigues, A. E. and Tondeur, D., Eds., Alphen aan den Rign, The Netherlands, 1981, 443.
4. **Sweed, N. H.,** Parametric pumping, in *Recent Developments in Separation Science,* Vol. 1, Li, N. N., Ed., CRC Press, Cleveland, 1972, chap. 3.
5. **Turnock, P. H. and Kadlec, R. H.,** Separation of nitrogen and methane via periodic adsorption, *AIChE J.,* 17, 335, 1971.
6. **Keller, G. E., II and Jones, R. L.,** A new process for adsorption separation of gas streams, *ACS Symp. Ser.,* 135, 275, 1980.
7. **Hill, F. B. and Cheung, H. C.,** Separation of Helium Methane Mixtures by Pressure Swing Adsorption, paper 48c, AIChE Meet., Houston, March 27 to 31, 1983.
8. **Young, D. M. and Crowell, A. D.,** *Physical Adsorption of Gases,* Butterworths, London, 1962, chap. 11.
9. **Danner, R. P. and Choi, E. C. F.,** Mixture adsorption equilibria of ethane and ethylene on 13X molecular sieves, *Ind. Eng. Chem. Fundam.,* 17, 248, 1978.
10. **Myers, A. L. and Prausnitz, J. M.,** Thermodynamics of mixed gas adsorption, *AIChE J.,* 11, 121, 1965.
11. **Lee, A. K. K.,** Lattice theory correlation for binary gas adsorption equilibria on molecular sieves, *Can. J. Chem. Eng.,* 51, 688, 1973.
12. **Hoory, S. E. and Prausnitz, J. M.,** Monolayer adsorption of gas mixtures on homogeneous and heterogeneous solids, *Chem. Eng. Sci.,* 22, 1025, 1967.
13. **Ruthven, D. M., Loughlin, K. F., and Holborow, K. A.,** Multicomponent sorption equilibrium in molecular sieve zeolites, *Chem. Eng. Sci.,* 28, 701, 1973.
14. **Sircar, S. and Myers, A. L.,** Surface potential theory of multilayer adsorption from gas mixtures, *Chem. Eng. Sci.,* 28, 489, 1973.
15. **Suwanayuen, S. and Danner, R. P.,** Vacancy solution theory of adsorption from gas mixtures, *AIChE J.,* 26, 76, 1980.
16. **Markham, E. C. and Benton, A. F.,** Adsorption of gas mixtures by silica, *J. Am. Chem. Soc.,* 53, 497, 1931.

17. **Holland, C. D. and Liapis, A. I.,** *Computer Methods for Solving Dynamic Separation Problems,* McGraw-Hill, New York, 1983, chap. 11.
18. **Wang, S. C. and Tien, C.,** Further work on multicomponent liquid phase adsorption in fixed beds, *AIChE J., 28,* 565, 1982.
19. **Costa, E., Sotelo, J. L., Calleja, G., and Marron, C.,** Adsorption of binary and ternary gas mixtures on activated carbon, *AIChE J., 27,* 5, 1981.
20. **Manes, M.,** The Polanyi adsorption potential theory and its applications to adsorption from water solutions to activated carbon, in *Activated Carbon Adsorption,* Suffet, I. H. and McGuire, M. J., Eds., Ann Arbor Science, Ann Arbor, 1980.
21. **Dubinin, M. M.,** Adsorption of vapors on active charcoals in relation to the physical properties of the adsorbate, *D. Acad. Sci. U.S.S.R. (Int. Ed.), 55,* 137, 1947.
22. **Toth, J.,** Gas-(Dampf-) adsorption an festen oberflachen inhomoger aktivitat, *Acta Chim. Hung., 30,* 415, 1962.
23. **Lewis, W. K., Gilliland, E. R., Chertow, B., and Cadogan, W. P.,** Pure gas isotherms — adsorption equilibria, *Ind. Eng. Chem., 42,* 1326, 1950.
24. **Maslan, F. D., Altman, M., and Aberth, E. R.,** Prediction of gas — adsorbent equilibria, *J. Phys. Chem., 57,* 106, 1953.
25. **Grant, R. J., Manes, M., and Smith, S. B.,** Adsorption of normal paraffins and sulfur compounds on activated carbon, *AIChE J., 3,* 403, 1962.
26. **Grant, R. J. and Manes, M.,** Correlation of some gas adsorption data extending to low pressures and supercritical temperatures, *Ind. End. Chem. Fundam., 3,* 221, 1964.
27. **Cook, W. H. and Basmadjian, D.,** Correlation of adsorption equilibria of pure gases on activated carbon, *Can. J. Chem. Eng., 42,* 146, 1964.
28. **Reich, R., Ziegler, W. T., and Rogers, K. A.,** Adsorption of methane, ethane and ethylene gases and their mixtures and carbon dioxide on activated carbon at 212-301 K and pressures to 35 atmospheres, *Ind. Eng. Chem. Proc. Des. Dev., 19,* 336, 1980.
29. **Ritter, J. A.,** Equilibrium Adsorption of Gas Mixtures of $H_2$, $CO^{oh}$, $CH_4$, $CO_2$ and $H_2S$ on Activated Carbon, M.S. thesis, Department of Chemical Engineering, State University of New York at Buffalo, Buffalo, N. Y., 1984.
30. **Peng, D. Y. and Robinson, D. B.,** A new two-constant equation of state, *Ind. Eng. Chem. Fundam., 15,* 59, 1976.
31. **Wilson, R. J. and Danner, R. P.,** Adsorption of synthesis gas mixture components on activated carbon, *J. Chem. Eng. Data, 28,* 14, 1983.
32. **Lewis, W. K., Gilliland, E. R., Chertow, B., and Cadogan, W. P.,** Adsorption equilibria — hydrocarbon mixtures, *Ind. Eng. Chem., 42,* 1319, 1950.
33. **Grant, R. J. and Manes, M.,** Adsorption of binary hydrocarbon gas mixtures on activated carbon, *Ind. Eng. Chem. Fundam., 5,* 490, 1966.
34. **Yon, C. M. and Turnock, P. H.,** Multicomponent adsorption equilibria on molecular sieves, *AIChE Symp. Ser., 67(117),* 3, 1971.
35. **Langmuir, I.,** The adsorption of gases on plane surfaces of glass, mica and platinum, *Am. Chem. Soc., 40,* 1361, 1918.
36. **Sips, R.,** On the structure of a catalyst surface, *J. Chem. Phys., 16,* 490, 1948.
37. **Saunders, J. T.,** Adsorption of $CH_4$ and $H_2$ from Single and Mixed Gases on Carbonaceous Sorbents, M.S. thesis, Department of Chemical Engineering, State University of New York at Buffalo, Buffalo, N.Y., 1982.
38. **Skarstrom, C. W.,** Heatless fractionation of gases over solid adsorbents, in *Recent Developments in Separation Science,* Vol. 2, Li, N. N., Ed., CRC Press, Cleveland, 1973.
39. **Chihara, K. and Suzuki, M.,** Simulation of nonisothermal pressure swing adsorption, *J. Chem. Eng. Jpn., 16,* 53, 1983.
40. **Shendalman, L. H. and Mitchell, J. E.,** A study of heatless adsorption in the model system $CO_2$ in He, *Chem. Eng. Sci., 27,* 1449, 1972.
41. **Weaver, K. and Hamrin, C. E., Jr.,** Separation of hydrogen isotopes by heatless adsorption, *Chem. Eng. Sci., 29,* 1873, 1974.
42. **Fernandez, G. F. and Kenney, C. N.,** Modelling of the pressure swing air separation process, *Chem. Eng. Sci., 38,* 827, 1983.
43. **Hill, F. B., Wong, Y. W., and Chan, Y. N. I.,** A temperature swing process for hydrogen isotope separation, *AIChE J., 28,* 1, 1982.
44. **Chan, Y. N., Hill, F. B., and Wong, Y. H.,** Equilibrium theory of a pressure swing adsorption process, *Chem. Eng. Sci., 36,* 243, 1981.
45. **Nataraj, S. and Wankat, P. C.,** Multicomponent pressure swing adsorption, in *Recent Advances in Adsorption and Ion Exchange,* Vol. 78, MA, Y. H., Ausikaitis, J. P., LeVan, M. D., and Sweed, N. H., Eds., *American Institute of Chemical Engineers,* New York, 1982.

46. **Carter, J. W. and Wyszynski, M. L.,** The pressure swing adsorption drying of compressed air, *Chem. Eng. Sci.,* 38, 1093, 1983.
47. **Sircar, S. and Kumar, R.,** Adiabatic adsorption of bulk binary gas mixtures: analysis by constant pattern model, *Ind. Eng. Chem. Proc. Des. Dev.,* 22, 271, 1983.
48. **Wang, S. S. and Yang, R. T.,** Multicomponent separation by cyclic processes, *Chem. Eng. Commun.,* 20, 183, 1983.
49. **Sheth, A. C. and Dranoff, J. S.,** Adsorption of ethylene on 4A molecular sieve particles, *Chem. Eng. Prog. Symp. Ser.,* 69(134),76,1973.
50. **Liaw, C. H., Wang, J. S., Greenkorn, R. A., and Chao, K. C.,** Kinetics of fixed bed adsorption — a new solution, *AIChE J.,* 25, 376, 1979.
51. **Tsai, M. C., Wang, S. S., and Yang, R. T.,** A pore diffusion model for cyclic separation: temperature swing separation of hydrogen and methane at elevated pressures, *AIChE J.,* 29, 966, 1983.
52. **Keller, G. E., II,** Gas adsorption processes: state of the art, in *Industrial Gas Separations,* Whyte, T. E., Jr., Yon, C. M., and Wagener, E. H., Eds., ACS Symp. Ser. 223, American Chemical Society, Washington, D. C., 1983, 145.
53. **Cassidy, R. T. and Holmes, E. S.,** Twenty-Five Years of Progress in "Adiabatic" Adsorption Processes, paper presented at the AIChE Meet., November 1, 1983, Washington, D. C.
54. **Yang, R. T.,** A New Design for Bulk Gas Separation by Pressure Swing Adsorption with Internal Heat Exchange, U.S. Patent, pending.

Chapter 8

# ROTATING-DISK THIN-LAYER CHROMATOGRAPHY: THEORY AND PRACTICE

**R. J. Laub and D. L. Zink**

## TABLE OF CONTENTS

# I. INTRODUCTION

Liquid chromatography (LC) in its many forms has, since the turn of this century, enjoyed considerable success as a separations technique. The well-known advantages attributed to column, thin-layer, and paper modes of LC are indeed impressive and include practicality, high sample capacity, and simplicity of operation. However, a major drawback of these methods is that in the absence of an external applied force, the linear velocity of mobile phase through the porous support/stationary-phase bed generally is slow. In those instances where capillary rise is the force responsible for elutriation, the time of development of bands or spots may, in fact, become interminably long. Moreover, while the renaissance in "high-performance" column LC has gone far toward eliminating this problem, the attendant simplicity of the more classical techniques is thereby lost.

Centrifugal force has on occasion been applied to conventional paper and thin-layer chromatography (TLC) systems in efforts to reduce the time of analysis, the history of which has been recounted by Laub and Zink.[1] Since publication of that report, several other studies of the topic have come to light.[2-11] For example, Heftmann and colleagues[4,6-9] utilized a scaled-down version of the TLC apparatus first described in 1947 by Hopf,[12] which consisted of a rotating drum of adsorbent with which 100-g quantities of solutes could readily be separated, detected, and collected. Two patterns of design of such instruments have since emerged. In one type, the top of a rotated TLC plate is left exposed to the atmosphere within a chamber which may or may not contain the vapor of a second solvent.[7,8,10] In contrast, the apparatus considered here and employed previously by Laub and Zink[1] makes use of a layer of sorbent 1 mm to 1 cm thick which is sandwiched between two parallel plates, the adsorbent being retained at the periphery with a porous spacing ring (the system thus amounts to a hybrid of TLC and column chromatography). Although neither type of apparatus has achieved widespread popularity, each offers several practical advantages as a result of unique patterns of elution, band-spreading, and flow, the more interesting aspects of which are here presented and discussed.

# II. EXPERIMENTAL

## A. Instrumentation

The rotating-disk thin-layer chromatographic (RDTLC) system was an Hitachi® Model CLC-5 which consists of a plate chamber and motor housing, a flow-through detector, a fraction collector, and electronic controls. The heart of the device is the separations disk which is shown in Figure 1. The unit consists of a lower plate with a socket which rests on the spindle of a direct-drive induction motor; an upper tempered-glass plate with a metal border that attaches to the lower plate with set screws; a reservoir cap; and a porous stainless-steel spacing washer which determines the thickness of the adsorbent layer (here 2 mm unless otherwise noted). Solvent is continuously introduced through the reservoir cap via a syphon, and eluent passing from the edge of the plate is funneled from the housing base by gravity through the detector to the fraction collector.

## B. Materials

All solvents were reagent-grade and, except as noted, were recycled by rotary evaporation and dried over molecular sieves. The silica adsorbent was Baker-Grade Silica 7 of 325 mesh without binder. The spacer filler (see later) was Davidson silica of 100 to 200 mesh. The solutes (dyes) were tetraphenylcyclopentadienone (student preparation); amaplast yellow AGB, amaplast red LB, amaplast red-violet PB, and amaplast orange LFP, graciously supplied by the American Color and Chemical Corporation; fatty orange and indophenol from the Fotodyne Corporation; and fat red 7B from Pfaltz and Bauer.

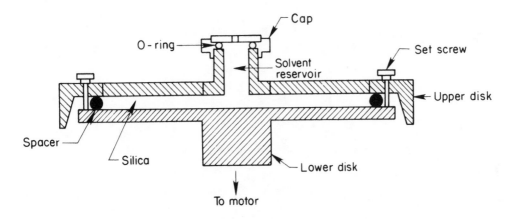

FIGURE 1.    Schematic representation of the rotating-disk assemblage; cf. text.

## C. Procedures

Mechanical balancing of the empty plate was carried out first in order to ensure uniform (slurry-packed) distribution of the adsorbent as well as vibration-free operation at 1000 rpm (16.7 Hz, the maximum speed of the motor at hand). The pores (*circa* 20 μm) of the spacer then required filling with coarse dry silica in order to contain finer-mesh materials within the confines of the disk: approximately 10 mℓ of the former were therefore added to the reservoir of the clean dry disk at rest, the cover replaced, and the motor turned quickly to maximum speed. After 2 to 3 min rotation, the speed was slowly reduced to full stop (in order not to collapse the dry silica boundary) and the plugging ring examined for uniformity. Gaps in the silica dictated repetition of the procedure, whereas if none were present the plate was once again taken to 1000 rpm for an additional 3 to 5 min in order to ensure firm compaction. Thin slurries of dehydrated (120°C; *in vacuo* overnight) silica in dichloromethane were then added to the plate at 400 rpm in a continuous stream. When the disk was filled with slurry to the edge of the solvent reservoir, the speed was increased to 1000 rpm to compact the adsorbent. The solvent thereby became visible as a sharply defined ring anterior to the silica boundary. When the solvent boundary came to within $1^1/_2$ cm of the edge of compacted adsorbent the plate speed was reduced to 400 rpm and more slurry was added. Three or four repetitions of the procedure were usually required in order to fill the plate completely, following which an additional 200 mℓ of solvent were passed through the system at 1000 rpm to ensure uniform packing of the silica sandwich. Thereafter, the packing did not collapse even upon prolonged standing.

Plate speeds of 300 to 500 rpm were found to yield the best results for injection of mixtures of solutes: the solvent front was allowed to approach to within 5 mm of the edge of the silica and the solute solution was then deposited onto the bottom of the reservoir with a syringe. The closer the solvent was to the adsorbent, the narrower was the resultant applied band of solutes. The reservoir was next rinsed several times with additional solvent until the solute mixture was washed completely onto the adsorbent. However, since the plate speed was slow, little further development occurred; that is, the solutes were concentrated at the inner edge of the silica in a narrow (1 to 2 mm) ring. The reservoir was then refilled with pure solvent and the speed adjusted slowly to that desired (otherwise, cracks developed in the adsorbent). For solvents other than that employed as the slurry medium, gentle concentration gradients of mobile phase were required to displace one liquid with the other; abrupt (20%) changes in composition led to channeling, pocket formation, and eventual cracking of the silica plate.

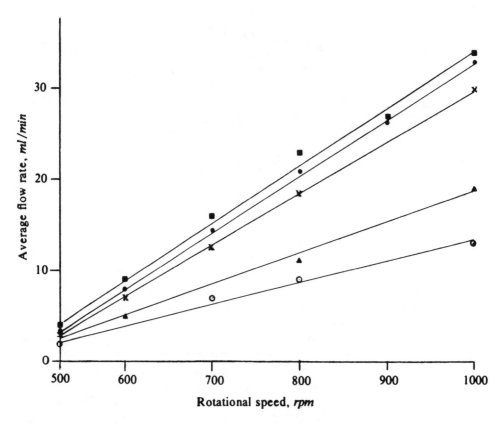

FIGURE 2.    Volume flow rate as a function of plate rotational speed for the solvents: ■, toluene; ●, dichloro-methane; X, chloroform; ▲, *iso*-octane; and ◕, 80:20 v/v xylenes:chloroform.

## III. RESULTS AND DISCUSSION

### A. Mobile-Phase Flow Rate

Shown in Figure 2 are plots of volume flow rate against plate rotational speed for toluene, dichloromethane, chloroform, and *iso*-octane solvents at room temperature (20 to 25°C). There appears to be little question as to the approximate linearity of the data, although there is a considerable difference in the flow rates that could be achieved with these materials at 1000 rpm. For example, toluene eluted at the rate of ca. 34 m$\ell$ min$^{-1}$, whereas the maximum for *iso*-octane was only 18 m$\ell$ min$^{-1}$. The bottom line represents an 80:20 (v/v) blend of xylenes/chloroform which yielded only 12 m$\ell$ min$^{-1}$. The properties (25°C) of the pure solvents are listed in Table 1, but cast little light on the magnitudes of the respective flow rates. For example, the densities of toluene and chloroform differ by nearly a factor of 2, while the viscosity and surface tension of the latter liquid are close to half those of the former. Nonetheless, the two solvents exhibit nearly the same flow characteristics under the influence of centrifugal force. Moreover, while the wetting properties of these two mobile phases would be expected to come into play, the surface tension, dipole moment, and dielectric permittivity data failed to correlate in any apparent manner with flow rate. Perhaps the most surprising result was that the mixed solvent gave the slowest of all flow rates, whereas each of the pure components fell closer to toluene than to *iso*-octane. Plots of flow

## Table 1
## PHYSICAL PROPERTIES OF PURE SOLVENTS AT 25°C[13]

| Solvent | $M/\text{g mol}^{-1}$ | $\rho/\text{g cm}^{-3}$ | $\eta/\text{cS}^a$ | $\gamma/\text{dyne cm}^{-1b}$ | $\mu/D^c$ | $\epsilon^d$ |
|---|---|---|---|---|---|---|
| Toluene | 92.134 | 0.86230 | 0.64947 | 27.93 | 0.38[e] | 2.379 |
| | | | | | 0.34[f] | |
| Dichloromethane | 84.94 | 1.31630 | 0.37835 | 27.21 | 1.59[f] | 8.93[g] |
| Chloroform | 119.389 | 1.47985 | 0.3635 | 14.58 | 1.04[e] | 4.639 |
| | | | | | 1.14[f] | |
| *iso*-Octane | 114.224 | 0.68777 | 0.69405 | 18.33 | — | 1.936 |

[a]   Viscosity.
[b]   Surface tension.
[c]   From Reference 17.
[d]   Dielectric permittivity at $10^5$ Hz.
[e]   Gas-phase.
[f]   Benzene, carbon tetrachloride solvents.
[g]   From Reference 18.

rate against mobile-phase composition at constant plate speed, hence, must exhibit minima for this system; those for cybotactic (e.g., *tert*-butyl alcohol/water) to interactive (e.g., chloroform/acetone) combinations of solvents would therefore be of some interest for further study in this regard.

### B. Retentions

Figure 3 illustrates the variation of retention time as a function of rotational speed for five solute/solvent systems. All exhibited an exponential decrease as the latter was increased. Plate speed clearly had the largest effect (i.e., most concaved curves) on solutes which were more strongly retained; the retention times for those which eluted fastest, broadly speaking, varied with flow rate nearest to inverse linearity. These must, of course, approach true linearity as the capacity factor $k'$ $(t_R - t_A)/t_A$ approaches zero, i.e., as the solute partition coefficient $K_R$ (ratio of stationary-phase/mobile-phase solute concentrations $C_i^S/C_i^M$) becomes negligible.

In contrast to retention times, the linear velocity of solutes, $dx/dt$, at constant rotational speed appeared to be a linear function of the distance traveled, x. Figure 4 illustrates band velocity plotted against migration distance for tetraphenylcyclopentadienone and indophenol solutes with dichloromethane solvent. Each point corresponds to elution of a solute band for a time at 1000 rpm, whereupon the plate was stopped and the distance traveled was gauged with a vernier caliper. Despite the scatter, it is nonetheless difficult to envisage other than linear regression through each set of data.

In general, the linear velocity of mobile phase in "modern" column LC is constant along the length of the bed, while in contrast in conventional TLC and paper chromatography, the linear velocity profile is one of exponential decrease. However, superposition of centrifugation upon such systems apparently results in a linear profile, the slope of which presumably depends not only upon the solute partition coefficient, but also upon the magnitude of the applied force (here proportional to the rotational speed). The mobile-phase flow profile in RDTLC thus portends utilization of the entire system at the optimum linear velocity, i.e., at maximum separation efficiency (however, see Giddings[14] and Sternberg[15]).

Retentions in LC are of course also a function of the type and degree of activation of the adsorbent and the composition of the carrier liquid. The various situations commonly en-

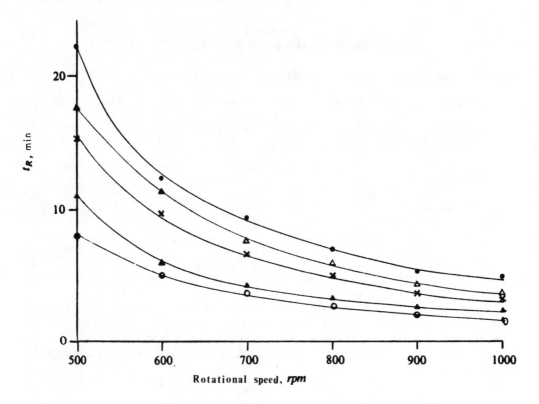

FIGURE 3.   Variation of retention time, $t_R$, as a function of plate rotational speed for the solute/solvent systems: ●, amaplast red-violet PB/chloroform; △, indophenol/dichloromethane; X, fat red/toluene; ▲, amaplast yellow AGB/dichloromethane; and ○, amaplast red BL/chloroform.

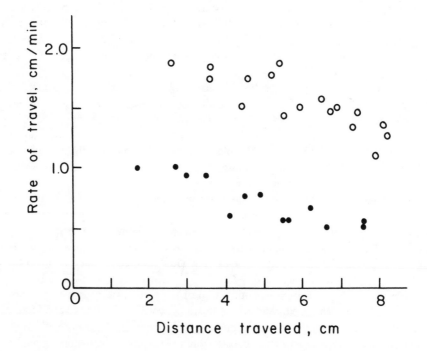

FIGURE 4.   Plots of rate of travel with distance traveled for the solutes: ○, tetraphenyl-cyclopentadienone and ●, indophenol with dichloromethane solvent (1000 rpm plate speed).

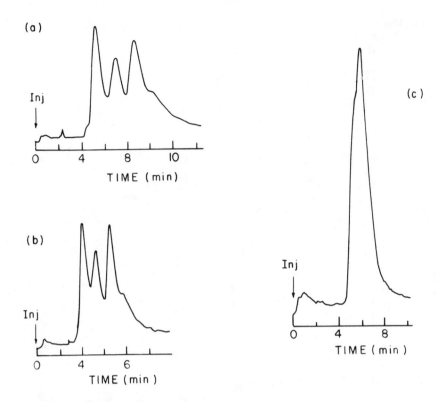

FIGURE 5.    Chromatograms of amaplast yellow AGB (first peak), indophenol, and Sudan II with dichloromethane eluent and silica adsorbent. (a) Silica heat-activated, dichloromethane dried over molecular sieves; (b) silica water-saturated, dry dichloromethane; (c) silica heat-activated, dichloromethane water-saturated.

countered with silica are illustrated in Figure 5. The first chromatogram (a) was obtained with carefully dehydrated adsorbent and dichloromethane eluent, while the second (b) was that found for dry mobile phase but with water-washed silica (dried only at 60°C). The separation obtained is comparable to that in (a), but the elution times are substantially reduced. The effect of moisture saturation of the carrier is shown in chromatogram (c), where resolution has been lost completely.

## C. Band Spreading

The number of theoretical plates N was found to increase (decreasing plate height H) with increasing rotational speed for all solute/solvent systems studied here. Representative plots of H against speed are shown in Figure 6, where in no case was there observed a minimum usually associated with curves of the van Deemter type. Toluene solvent provided noticeably higher efficiency than did the other eluants, although there was some dependence of H upon identity of the solute. However, as pointed out by Laub and Zink,[1] the curves may well be specious, because of the very large post-plate volume extant in the system. Since higher rotational speeds (with concomitant higher volume flow rates) result in more efficient washing of the solutes from the collection basin into the detector, H would be expected to decrease as the speed is increased. Further, since toluene yielded the highest flow rate, H was smallest for this solvent. (Reese and Scott[16] have discussed at length extra-column band dispersion in liquid chromatography.) Harrison[7] has also shown that the connection volume between the edge of the plate and the detector can be reduced to a milliliter or so, in which case efficiencies of at least an order of magnitude greater than those achieved here can be anticipated.

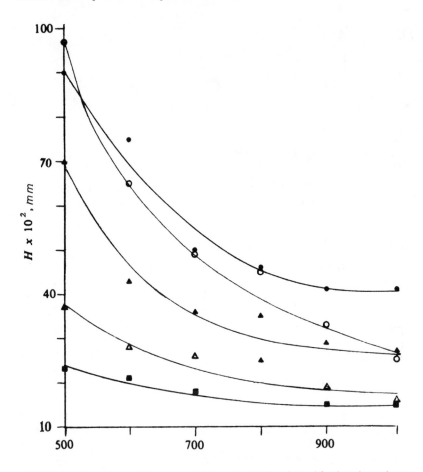

FIGURE 6.    Plots of theoretical plate height H against rotational speed for the solute/solvent systems: ●, amaplast yellow AGB/dichloromethane; ○, indophenol/dichloromethane; ▲, amaplast red-violet PB/chloroform; △, amaplast red BL/chloroform; and ■, fat red/toluene.

The mode of band spreading of a spot (as opposed to a ring) of solute dye from the inner edge to the outer periphery of the plate provided further insight into the system efficiency.* Upon leaving the origin, spots traversed a straight line to the edge of the plate and not, as expected originally, as a logarithmic spiral. Furthermore, the spots spread into arcs of finite width which were symmetric about the path of migration, viz.,

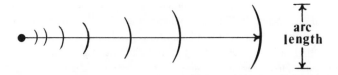

Plots of arc length as a function of migration distance are shown in Figure 7 for tetra-phenylcylcopentadienone and indophenol solutes with dichloromethane solvent. Both the symmetry and degree of development of the arcs appeared to be independent of the solute and, in addition, to be a linear function of migration distance (the same experimental difficulties were encountered in these measurements as in those for Figure 4). In retrospect,

* We thank R. P. W. Scott for suggesting this experiment to us.

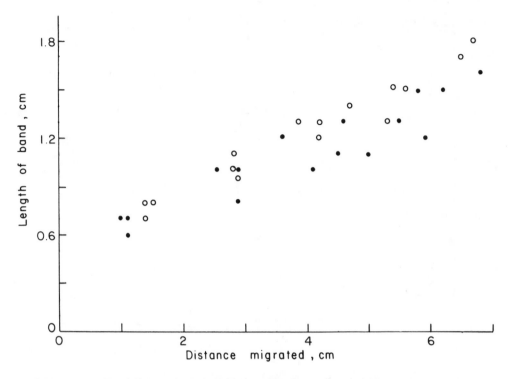

FIGURE 7.    Plots of arc length (cf. text) against migration distance for tetraphenylcyclopentadienone ○ and indophenol ● solutes with dichloromethane solvent. Plate speed: 1000 rpm.

symmetric spreading of solute molecules as well as the linearity of arc length with distance traveled are expected as a result of normal forces of dispersion. That is, molecular diffusion is faster than the maximum linear carrier velocity possible with the system. On the other hand, because a solute ring (or spot) is eluted into a bed of adsorbent of continuously increasing volume on passing from the center to the outer boundary of the plate, molecules in the front of the eluting band move with a linear velocity which is slower than those at the back. The result, despite the excessive extra-system mixing volume with the unit used here, is that band "tailing" (diffuse back edge) is reduced considerably and the observed chromatographic peaks (e.g., Figure 5) were very nearly symmetric.

## D. Sample Capacity

In the original work by Hopf,[12] hectogram amounts of samples were easily resolved, whereas the slightly smaller apparatus of Heftmann et al.[6] provided separations of up to 5 g each of several amino acids. RDTLC thus has been used from an historical standpoint as a preparative-scale tool. A particular advantage of the device employed in this work is that the plate thickness is controlled by the width of interchangeable spacing rings, such that adsorbent layers of 1 mm can be used for analytical separations, while those of 5 or 10 mm find utility in a small (100 mg)-scale preparative mode (the desirability of an instrument of versatility of 1 mm to 20 cm is self-evident and has been expressed and discussed by us previously[1]). An additional advantage inherent in the technique is that solutes can easily be concentrated at the inner edge of the adsorbent, since band development does not commence until the rotational speed of the plate is increased beyond 500 rpm. Figure 8 (toluene solvent; 2-mm spacer) provides illustration of each of these features, where two sets of solutions of fat red 7B were prepared such that the total mass of injected solute was either 0.05 or 3.125 mg. The number of theoretical plates obtained for each set confirmed

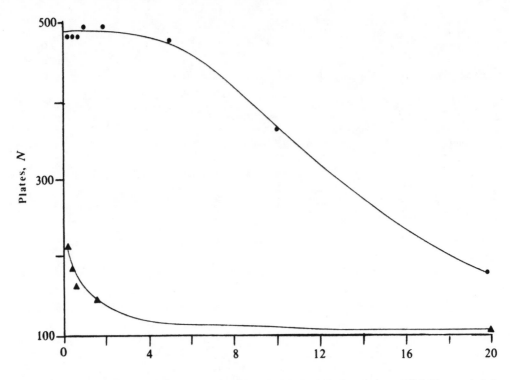

FIGURE 8.    Plots of theoretical plates N as a function of injection volume for solutions of ● 0.05 mg and of ▲ 3.125 mg fat red 7B solute in toluene (toluene eluent; 1000 rpm plate speed).

that the system efficiency deteriorates badly under conditions of solute overload. However, uniform efficiency was found for the samples containing the smaller mass of solute even when injected in up to as much as 8 m$\ell$ of solvent. RDTLC would therefore seem to offer the potential for simultaneous trace-constituent concentration, analytical separation, and preparative-quantity isolation, regarding which we hope soon to report further study and evaluation.

## ACKNOWLEDGMENTS

We gratefully acknowledge support of this work provided by the Department of Energy Office of Basic Energy Sciences.

## REFERENCES

1.  **Laub, R. J. and Zink, D. L.,** Rotating-disk thin-layer chromatography, *Am. Lab.,* 13(1), 55, 1981.
2.  **Morr, C. V., Nielson, M. A., and Coulter, T. A.,** Centrifugal Sephadex® procedure for fractionation of concentrated skim milk, whey, and similar biological systems, *J. Dairy Sci.,* 50, 305, 1967.
3.  **Ribi, E., Filz, C. J., Goode, G., Strain, S. M., Yamamoto, K., Harris, S. C., and Simmons, J. H.,** Chromatographic separation of steroid hormones by centrifugation through columns of microparticulate silica, *J. Chromatogr. Sci.,* 8, 577, 1970.
4.  **Heftmann, E., Krochta, J. M., Farkas, D. F., and Schwimmer, S.,** The chromatofuge, an apparatus for preparative rapid radial column chromatography, *J. Chromatogr.,* 66, 365, 1972.
5.  **Heftmann, E.,** Liquid chromatography — past and future, *J. Chromatogr. Sci.,* 11, 295, 1973.

6. **Finley, J. W., Krochta, J. M., and Heftmann, E.,** Rapid preparative separation of amino acids with the chromatofuge, *J. Chromatogr.,* 157, 435, 1978.
7. **Harrison, S.,** U.S. Patent 4,139,458, February 1979.
8. **Derguini, F., Balogh-Nair, V., and Nakanishi, K.,** A versatile synthesis of retinoids via condensation of the side-chain to cyclic ketones, *Tetrahedron Lett.,* 4899, 1979.
9. **Nes, W. D., Heftmann, E., Hunter, I. R., and Walden, M. K.,** Determination of solasodine in fruits of *Solanum khasianum* by a combination of chromatofuge and high-pressure liquid chromatography, *J. Liq. Chromatogr.,* 3, 1687, 1980.
10. **Hostettmann, K., Hostettmann-Kaldas, M., and Sticher, O.,** Rapid preparative separation of natural products by centrifugal thin-layer chromatography, *J. Chromatogr.,* 202, 154, 1980.
11. **Korzum, B. P. and Brody, S.,** Centrifugally accelerated thin-layer chromatography, *J. Pharm. Sci.,* 53, 454, 1964.
12. **Hopf, P. P.,** Radial chromatography in industry, *Ind. Eng. Chem.,* 39, 938, 1947.
13. **Dreisbach, R. R.,** *Physical Properties of Chemical Compounds,* Vol. 1, American Chemical Society, Washington, D.C., 1955; Vol. 2, 1959; Vol. 2, 1961.
14. **Giddings, J. C.,** Role of column pressure drop in gas chromatographic resolution, *Anal. Chem.,* 36, 741, 1964.
15. **Sternberg, J. C.** Effect of pressure gradient on chromatographic column efficiency, *Anal. Chem.,* 36, 921, 1964.
16. **Reese, C. E. and Scott, R. P. W.,** Microbe columns — design, construction, and operation, *J. Chromatogr. Sci.,* 18, 479, 1980.
17. **McClellan, A. L.,** *Tables of Experimental Dipole Moments,* Vol. 2, Rahara Enterprises, El Cerrito, Calif., 1974.
18. **Mellan, I.,** *Industrial Solvents Handbook,* 2nd ed., Noyes Data Corp., Park Ridge, N.J., 1977.

Chapter 9

# OLEFIN SEPARATION BY FACILITATED TRANSPORT MEMBRANES

**R. D. Hughes, J. A. Mahoney, and E. F. Steigelmann**

## TABLE OF CONTENTS

# I. INTRODUCTION

Membranes have long been considered as an attractive alternative to conventional separation methods, such as distillation. Membrane separations, because of their potential for both low capital costs and high energy efficiency, should have an inherent advantage over conventional separation methods. To date, however, few membrane separations are of commercial importance. This is because membrane systems are uneconomical because of slow transfer rates and low selectivities in separations. However, using the principle of facilitated transport or carrier-mediated diffusion, it is possible to devise highly selective and permeable membranes. This paper reviews the development of one such membrane — a facilitated transport membrane for olefin separation.

# II. SCOPE

This paper presents the development of a membrane system for separating olefins from paraffins. Specifically, the purpose of this study was to develop a membrane for separating ethylene and propylene from light paraffin gases. Conventionally, this separation is done by distillation and is quite expensive.

A highly selective membrane was developed using the principles of facilitated transport. Silver ion was used as the carrier for olefins in the membrane. Silver ion forms labile complexes with olefins, but not with paraffins. When incorporated into a membrane, silver ion then transports olefins, but not paraffins, across the membrane.

The laboratory and pilot plant scale developments of the silver ion-based olefin selective membrane are the primary focus of this paper. First, however, a review of some of the literature and principles of facilitated transport will be presented. This will be followed by a discussion of the development of the olefin selective membrane. The paper is concluded with a summary of the advantages and limitations of the olefin membrane.

# III. FACILITATED TRANSPORT PROCESS

## A. Description

Facilitated transport is the active rather than passive transport of permeant molecules across a membrane. Active transport is achieved with a carrier species. The carrier must react with permeant molecules to form a labile complex. By doing so, the carrier increases the permeant concentration in the membrane. When constrained within a membrane, the carrier shuttles the permeant between the membrane boundaries. Permeant will be transported in the direction of higher to lower permeant concentrations. When contacted with a feed mixture containing only one component that the carrier can react with, only the transport of that one component will be "facilitated" across the membrane. The net effect of facilitated transport is to augment the flux of certain permeant components across a membrane producing both high flux and high selectivity.

The process of facilitated transport is best understood with a general example. A feed mixture of "A" and "C" are contacted with a membrane containing carrier "B". A labile complex "AB" is formed between "A" and "B" (Equation 1). By contrast, "C" does not react with "B" (Equation 2).

$$A + B \rightleftarrows AB \tag{1}$$

$$C + B \not\rightarrow BC \tag{2}$$

"A" and not "C" will be preferentially absorbed into the membrane. "A" is thus preferentially transported across the membrane.

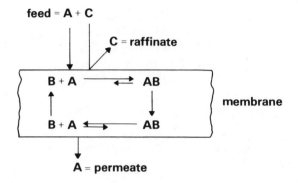

FIGURE 1. Mechanism for facilitated transport of component A by carrier B. Component C is rejected at the membrane surface.

The facilitated transport of "A" across the membrane is depicted in Figure 1. At the top, membrane surface "A" reacts with "B" to form "AB". "AB" diffuses across the membrane where it dissociates back to "A" and "B". "A" is released from the membrane and "B" diffuses back to the top membrane surface. Here "B" can react with "A" to repeat the process.

## B. Requirements
The driving force for facilitated transport is a concentration gradient of permeant across the membrane. In gas phase separations the driving force is normally a partial pressure difference of the permeant gas. This driving force is produced by using elevated feed pressures at the top, relative to the bottom, membrane surface. Using reduced pressures or purge gases at the bottom membrane surface also increases the driving force.

Facilitated transport membranes must meet three other criteria to be useful. These are

1. The formation of "AB" must be fast and reversible (Equation 1).
2. Both "B" and "AB" must be mobile in the membrane.
3. "B" must be stable in the separation environment.

If Equation 1 is not reversible, then "A" will not be released from the membrane. If the formation of "AB" is slow compared to the diffusion rate of free "A", then the flux of "A" due to facilitated transport will be negligible. If either "B" or "AB" are immobile, then the facilitated diffusion of "A" is also limited. In fact, if "B" or "AB" is immobile, "A" can only be facilitated across the membrane by a slow site-jumping or "bucket-brigade" mechanism. Finally, if "B" is unstable in the separation environment, its ability to continually shuttle "A" is limited. The membrane simply would not maintain its facilitated transport capability.

## C. Literature
Scholander et al.[1] reported the first work on facilitated transport. He studied the transport of oxygen through aqueous hemoglobin solutions. Since then, reviews by Schultz et al.[2,3] and Smith et al.[4] have extensively examined the mathematical expressions that describe facilitated transport. In 1979, Kimura et al.[5] summarized potential industrial applications of facilitated transport. More recently, Halwachs and Schügerl[6] and Way et al.[7] have reviewed facilitated transport in liquid membranes.

Facilitated transport systems for $O_2$, $CO_2$, and $H_2S$ have been widely studied. Ward and Robb[8] and Winneck et al.[9] used basic solutions to transport $CO_2$ across a membrane. Roughton[10]

**Table 1**
**EQUILIBRIUM CONSTANTS**
**FOR FORMATION OF THE**
**SILVER-OLEFIN COMPLEX**
**WITH ACYCLIC OLEFINS**

| Olefin | $K_{Ag-O\ell}$[a] | Ref.[b] |
|---|---|---|
| Ethylene | 98 | 26 |
| Propylene | 97 | 26 |
| 1-Butene | 141 | 25 |
| *cis*-2-Butene | 72 | 25 |
| *trans*-2-Butene | 29 | 23, 25 |
| 2-Methylpropene | 61 | 22, 25 |
| *cis*-2-Pentene | 112 | 22, 23 |
| *trans*-2-Pentene | 62 | 22, 23 |
| 2-Methyl-2-butene | 13 | 22 |
| 1-Hexene | 860 | 22 |

$$ {}^aK_{Ag-O\ell} = \frac{[Ag - olefin]^+}{[Ag^+][olefin]} $$

[b]   All measurements were reported with aqueous 1 $M$ AgNO$_3$.

and Wittenberg[11] studied O$_2$ and CO$_2$ transport in physiological systems. Otto and Quinn[12] and Donaldson and Quinn[13] have extensively studied the mechanism of CO$_2$ transport in bicarbonate solutions. Enns[14] improved CO$_2$ transport in bicarbonate solutions by adding the enzyme carbonic anhydrase. Matson et al.[15] used hot carbonate solutions to remove H$_2$S from a low quality coal gas.

Facilitated transport systems for both NO and CO have been reported. Smith et al.[16] and Hughes and Steigelmann[17] used cuprous ion to transport CO. Ward[18] has used ferrous ion to transport NO. Also, Ward[19] and Bdzil et al.[20] electrically induced the transport of NO by ferrous ion.

Numerous examples of the facilitated transport of gaseous permeants have been reported. All these examples demonstrate that facilitated transport offers a means of producing both high flux and highly selective membranes. The remainder of this paper presents the development and testing of a facilitated transport membrane for olefins.

## IV. OLEFIN MEMBRANE SYSTEM

### A. Silver-Olefin Chemistry

Silver ion forms a one-to-one complex with olefins (Equation 3).

$$ Ag^+ + O\ell \rightleftarrows Ag\text{-}O\ell^+ \tag{3} $$

where $O\ell$ = an olefin. Lucas and co-workers[21-26] studied the equilibrium of olefins with aqueous solutions of AgNO$_3$ at 25°C. Table 1 presents the equilibrium constants for silver-olefin complexes. These data show the general nature of the Ag$^+$ olefin reaction. In particular, note that ethylene and propylene have identical equilibrium constants. Other soluble silver salts react with olefins, also. Quinn and Glew[27] and Featherstone and Sorrie[28] studied the formation of silver-olefin complexes with AgClO$_4$ and AgBF$_4$. Beverwijk et al.[29] has published a comprehensive review of organosilver chemistry.

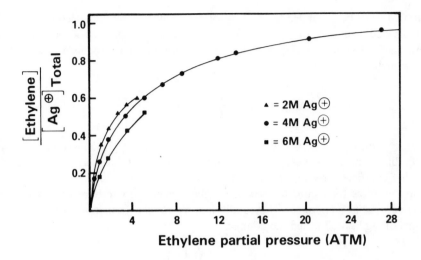

FIGURE 2. The solubility of ethylene in AgNO$_3$ solutions. Plot of the (ethylene/Ag$^+$) ratio vs. ethylene pressure for Ag$^+$ solutions of 2, 4, and 6 $M$.

Silver-olefin complexes are both labile and mobile in aqueous solutions. Thus, Ag$^+$ is a potential carrier for olefins in a membrane. The source of Ag$^+$ should not matter. However, because of its low cost and high stability relative to other silver salts, AgNO$_3$ was chosen as the Ag$^+$ source for development of an olefin selective membrane. AgNO$_3$ is also highly soluble in water. Solutions up to 8.5 $M$ in Ag$^+$ can easily be made from AgNO$_3$ at 25°C.

## B. Ethylene Solubility

The solubility of ethylene in AgNO$_3$ solutions at 25°C was measured as a function of ethylene pressure and Ag$^+$ concentration. These measurements should indicate how olefin pressure and Ag$^+$ concentration should affect membrane performance. Ethylene pressures up to 28 atm and Ag$^+$ concentrations of 2, 4, and 6 $M$ were used to construct the isotherms presented in Figure 2. The data show the moles of ethylene per mole of total Ag$^+$ (ethylene/Ag$^+$) as a function of total ethylene pressure. As expected, in going to higher olefin pressure, the ethylene/Ag$^+$ ratio increases asymptotically to a value of unity. As the ethylene pressure increases, the incremental amount of the Ag$^+$-ethylene complex formed decreases. Also, as the total Ag$^+$ content of the solution increases, the ethylene/Ag$^+$ ratio decreases at any ethylene pressure. Ag$^+$ at higher concentrations becomes less efficient on a per-mole basis at reacting with ethylene. Undoubtedly, this is caused by reduced Ag$^+$ activity at the higher concentrations.

## C. Transport Equations

The steady-state flux rate of a permeant gas across a membrane is a function of membrane area (A), membrane thickness (L), and the partial pressure difference (ΔP) of the gas across the membrane. The equation describing steady-state flux is the integrated form of Fick's Law (Equation 4):

$$\text{flux} = AK \frac{\Delta P}{L} \tag{4}$$

The proportionality constant, K, is the "permeability" coefficient. Equation 4 satisfactorily describes permeant flux in virtually all passive transport membranes.

Equation 4 must be expanded to account for permeant flux by facilitated transport. Two terms are needed — one for transport of the free permeant species and a second for permeant transported as the complex. For $Ag^+$ transport of olefins, the expanded rate expression becomes Equation 5.

$$\text{flux}_{(O\ell)} = AK_1 \frac{\Delta P_{O\ell}}{L} + AK_2 \frac{\Delta C_{Ag\text{-}O\ell}}{L} \tag{5}$$

where $O\ell$ = olefin, $Ag\text{-}O\ell$ = silver-olefin complex, $C_{Ag\text{-}O\ell}$ = concentration of the silver-olefin complex (moles/liter), $K_1$ = permeability coefficient for free olefin, and $K_2$ = permeability coefficient for the silver-olefin complex.

The driving force for facilitated transport is the concentration difference of the complex ($\Delta C_{Ag\text{-}O\ell}$) across the membrane. The equilibrium expression (Equation 6) relates the concentrations of $Ag^+$, olefin, and the complex.

$$K_{Ag\text{-}O\ell} = \frac{[Ag\text{-}O\ell]}{[Ag^+][O\ell]} \tag{6}$$

Combining Equation 6 and the $Ag^+$ mass balance expression (Equation 7) and solving for the concentration of the complex ($[Ag\text{-}O\ell]$) yields Equation 8.

$$[Ag^+]_{total} = [Ag^+] + [Ag\text{-}O\ell] \tag{7}$$

$$[Ag\text{-}O\ell] = K_{Ag\text{-}O\ell} \frac{[O\ell][Ag^+]_{total}}{1 + K_{Ag\text{-}O\ell}[O\ell]} \tag{8}$$

The olefin concentration at either membrane surface is proportional to the olefin partial pressure ($P_{O\ell}$) as given in Equation 9 (Henry's Law).

$$[O\ell] = K'_{O\ell}P_{O\ell} \tag{9}$$

Combining Equations 5, 8, and 9 yields the expression for olefin flux through an $Ag^+$ membrane (Equation 10):

$$\text{flux}_{(O\ell)} = AK_1 \frac{\Delta P_{O\ell}}{L} + \frac{AK_{Ag\text{-}O\ell}K_2K'_{O\ell}}{L}[Ag^+]_{total}$$

$$\left( \frac{(P_{O\ell})_T}{1 + K_{Ag\text{-}O\ell}K'_{O\ell}(P_{O\ell})_T} - \frac{(P_{O\ell})_B}{1 + K_{Ag\text{-}O\ell}K'_{O\ell}(P_{O\ell})_B} \right) \tag{10}$$

where $(P_{O\ell})_T$ = olefin partial pressure at the top membrane surface, and $(P_{O\ell})_B$ = olefin partial pressure at the bottom membrane surface.

Because facilitated transport is much faster than passive transport, the passive transport term may be neglected. Rewriting Equation 10 and combining constants gives Equation 11.

$$\text{flux}_{(O\ell)} \approx A \frac{K_3}{L}[Ag^+]_{total} \left( \frac{(P_{O\ell})_T}{1 + K_4(P_{O\ell})_T} - \frac{(P_{O\ell})_B}{1 + K_4(P_{O\ell})_B} \right) \tag{11}$$

where $K_3 = K_{Ag\text{-}O\ell}K_2K'_{O\ell}$ and $K_4 = K_{Ag\text{-}O\ell}K'_{O\ell}$. This expression is similar to the one derived by Schultz.[2] It describes the steady-state flux of a volatile permeant through a liquid membrane that contains a carrier for the permeant. It is valid when the rate of formation of the carrier-permeant complex is fast compared to the diffusion rate of the complex.

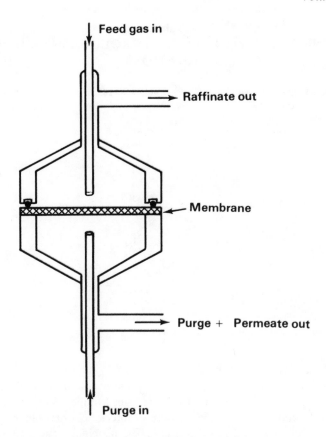

FIGURE 3.   Membrane test cell for flat films. This cell uses a purge
to remove the permeate from the cell.

## D. Membrane Configuration

Facilitated transport membranes require that a carrier be constrained within a membrane. Incorporating the carrier in the micropores of a porous membrane is a simple method of doing this. Practical membranes must also be configured in a viable physical form. The olefin membrane was developed using both flat film and hollow fiber forms. The $Ag^+$ carrier was either incorporated into micropores of the membrane or retained atop a porous barrier film. Flat films were used for initial development. However, final development was done using the hollow fiber form, because fibers are the most practical and least costly membrane configuration.

## V. OLEFIN MEMBRANE DEVELOPMENT

## A. Flat Film Tests
### 1. General Considerations

Flat films are the most convenient configuration for laboratory tests of membranes. A modified Millipore® cell (Figure 3) was used to measure the effects of olefin pressure, $Ag^+$ concentration, and membrane thickness on membrane performance. This cell allows for a continuous flow of a feed mixture to the top of the membrane. A purge stream can be supplied to the bottom of the membrane. The purge stream lowers the olefin permeant partial pressure at the bottom membrane surface. The olefin driving force can be controlled by adjusting the feed and purge pressures and rates.

## Table 2
### ETHYLENE PERMEATION RATES THROUGH IMMOBILIZED Ag⁺ SOLUTIONS

| Ag⁺ solution | Nominal thickness, L (cm × 10²) | Permeability × 10³ [cm³ (NTP)-cm/cm²-min-atm][a] | $\left(\dfrac{\text{Permeability in Ag}^+}{\text{permeability in water}}\right)^{b}$ |
|---|---|---|---|
| 1 *M* AgNO₃ | 1.65 | 5.61 | 48 |
| | 3.30 | 6.16 | 53 |
| | 4.95 | 4.79 | 40 |
| 2 *M* AgNO₃ | 1.65 | 7.98 | 69 |
| | 3.30 | 7.20 | 62 |
| | 4.95 | 7.39 | 64 |
| 4 *M* AgNO₃ | 1.65 | 10.2 | 88 |
| | 3.30 | 9.39 | 81 |
| | 4.95 | 9.71 | 84 |
| | 8.25 | 9.26 | 80 |
| 6 *M* AgNO₃ | 1.65 | 9.62 | 83 |
| | 3.30 | 6.75 | 58 |
| | 4.95 | 5.61 | 48 |

[a]    Permeabilities are corrected for a tortuosity of 1.4 and a porosity of 70% in the immobilizing filters.
[b]    Ratio of the ethylene permeability in Ag⁺ to that in water.

### 2. Ethylene Flux in Porous Filters

Initial experiments were designed to determine the maximum possible flux of ethylene through an Ag⁺-containing membrane. Cellulose ester Millipore® filters were impregnated by soaking in a solution of Ag⁺. Surface tension forces hold the solution in the pores of the filters. The filters are fixed in the test cell. The membrane thickness was adjusted by stacking filters atop one another in this cell.

The Millipore® filters had a nominal thickness of 165 μm, a tortuosity of 1.4, and a porosity of 70%. The porosity was determined from the weight uptake when the filters were filled with water. The tortuosity was estimated from measurements of the ethylene flux through the filters filled with water compared to that in water alone. Values of $1.85 \times 10^{-5}$ cm²/sec and $0.466 \times 10^{-5}$ mol/cm³ have been reported for the diffusivity and solubility, respectively, of ethylene in water.[30] Others have also reported a tortuosity of 1.4 for similar filters.[16,31-33]

Steady-state permeation rates were measured for ethylene in membranes containing 1, 2, 4, and 6 *M* Ag⁺. The ethylene feed gas was saturated with water vapor to minimize water loss from the membrane. Hydrogen peroxide (∼1 wt %) was added to Ag⁺ solution to minimize reduction of Ag⁺ to Ag°. A helium stream presaturated with water vapor was used as the purge. The ethylene flux was determined from measurements of stream flow rates and compositions. Results are given in Table 2 for both measured and corrected ehtylene fluxes.

The ratio of the ethylene flux rate in the Ag⁺ solution to that in water gives a measure of facilitated transport due to Ag⁺ (see Table 2). This value represents a minimum facilitation factor, because aqueous solutions at comparable ionic strengths (but not containing Ag⁺) would be less permeable to ethylene than pure water. Clearly, Ag⁺ facilitates ethylene transport by at least a factor of 40.

### 3. Barrier Film Tests

A barrier film system was used to determine the selectivity of Ag⁺ to ethylene over methane and ethane. A test cell (Figure 4) contained a solution supported atop the barrier

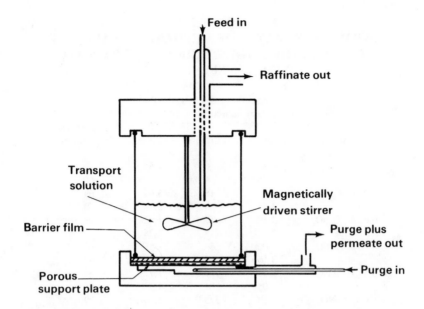

FIGURE 4.   Membrane cell for testing transport solutions. The cell uses a porous Teflon®
barrier film to support the transport solution. The solution is stirred during the test.

film. A porous Teflon® barrier (Gore-Tex®, W.L. Gore and Assoc.) was used to support a
1- to 2-cm-deep solution of Ag⁺. Surface tension forces are too large to allow the aqueous
solution to penetrate the 0.02-μ$M$ pores in the Gore-Tex® at pressures up to 6 atm. Gore-
Tex® is an ideal barrier film. Although it retains the transport solution, it provides little
resistance to olefin permeation.

A feed mixture of methane, ethane, and ethylene (∼1/3 each) was supplied to the Ag⁺
solution in the test cell. The solution was stirred to minimize the diffusion resistance through
the solution. The bottom side of the membrane was purged with helium. Measurements
were made over at least a 4-hr period, although the system reached equilibrium within 30
min. The results are given in Table 3. Ethylene flux is clearly enhanced over that of methane
and ethane. Up to about 1 $M$ Ag⁺, an increase in Ag⁺ concentration increased the ethylene
flux rate. Increased Ag⁺ also decreased the methane and ethane flux rates, presumably,
because of the "salting-out" effect. The practical utility of Ag⁺ transport of ethylene is
clearly demonstrated in producing 99 + % ethylene from a stream containing only 33%
ethylene.

## B. Hollow Fiber Membranes
### 1. Permeator Construction
Commercial hollow fibers designed for reverse osmosis applications were chosen for
development of the olefin membrane. These cellulose ester fibers are anisotropic with a
dense skin and porous substructure. The porous substructure gives the fiber strength and
also serves as a container for the Ag⁺ transport solution. The dense outer skin helps to retain
the Ag⁺ solution in the fibers, yet should be a minimal resistance to permeant flux.

Hollow fiber units were prepared by potting from 1 to 16 fibers into stainless steel tubes.
The fibers were taken from Dow RO4K® (Dow Chemical Corp.) permeators and were used
as received. These fibers have 0.023- and 0.081-cm inside and outside diameters, respectively. The configuration of the laboratory scale permeator is shown in Figure 5. A typical
permeator was 61 cm long with 35.5 cm of active fiber length. The active membrane area,
based on the logarithmic mean diameter of the fibers, was 25.5 cm².

**Table 3**

**PERMEATION RATES OF METHANE, ETHANE, AND ETHYLENE THROUGH STIRRED SOLUTIONS ATOP POROUS TEFLON® FILMS**

| Solution | Permeation rates [cm³ (NTP)/cm²-min-atm] × 10³ᵃ | | | Mol % ethylene in permeateᵇ |
|---|---|---|---|---|
| | Methane | Ethylene | Ethane | |
| Water | 3.37 | 8.57 | 2.64 | 59 |
| | 3.15 | 7.98 | 2.28 | 60 |
| 0.01 *M* AgNO₃ | 3.42 | 15.6 | 3.01 | 69 |
| | 3.65 | 18.1 | 3.24 | 70 |
| 0.05 *M* AgNO₃ | 2.19 | 17.4 | 1.92 | 79 |
| | 2.37 | 18.1 | 2.28 | 81 |
| 0.10 *M* AgNO₃ | 3.00 | 54.7 | 2.55 | 90 |
| | 3.15 | 86.1 | 2.83 | 93 |
| 0.50 *M* AgNO₃ | 1.92 | 129.0 | 1.73 | 97 |
| | 2.05 | 201.0 | 1.92 | 98 |
| 1.0 *M* AgNO₃ | 2.10 | 373.0 | 1.69 | 99+ |
| | 2.32 | 483.0 | 2.05 | 99+ |
| 2.0 *M* AgNO₃ | 1.28 | 201.0 | 1.09 | 99+ |
| | 1.23 | 255.0 | 1.05 | 99+ |
| 4.0 *M* AgNO₃ | 0.64 | 134.0 | 0.50 | 99+ |
| 6.0 *M* AgNO₃ | 0.44 | 268.0 | 0.30 | 99+ |
| | 0.34 | 298.0 | 0.22 | 99+ |

ᵃ  The feed gas was a mixture of ∼1/3 each of the three gases. The first row of data for each different solution was collected at a total feed pressure of 2.4 atm and the second at 3.75 atm.

ᵇ  Purge-free content.

Water contained in the reverse osmosis fibers was exchanged for $Ag^+$ solution. This was done by pumping the $Ag^+$ solution both through the fiber bores and over the outside (shell side of the fibers). After 2 hr, the exchange of $Ag^+$ solution for water was complete. Once drained of the excess solution, the permeator was ready for use. This simple procedure was found adequate for impregnating hollow fibers with $Ag^+$ solution.

## 2. Laboratory Tests

Hollow fiber permeators were evaluated in tests similar to those described for flat films. An olefin-containing feed gas was supplied to each permeator and both the olefin flux and selectivity were measured. Selectivity is a measure of membrane specificity for olefins. It is defined by Equation 12.

$$\text{selectivity} = S = \left( \frac{(X_{O\ell})_P}{(1 - X_{O\ell P})} \right) \left( \frac{(1 - X_{O\ell})_F}{(X_{O\ell})_F} \right) \tag{12}$$

where $(X_{O\ell})_F$ = mole fraction of olefin on the feed side of the fibers, and $(X_{O\ell})_P$ = mole fraction of olefin on the permeate side of the fibers.

The olefin flux was not corrected for either the olefin driving force or the fiber thickness. The olefin driving force decreases along the length of the fiber in a permeator. Because olefins permeate many times faster than paraffins, the raffinate stream is much lower in olefins than the feed stream. To account for this, the "true" olefin driving force reported in this work was determined from an average of the olefin content at the feed and raffinate ends of the permeator.

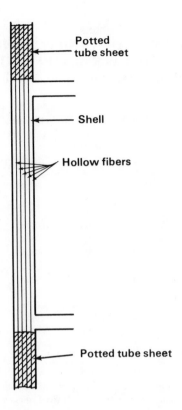

FIGURE 5.  Laboratory-scale hollow fiber permeator.

The feed gas was supplied to either the shell or bore side of the fibers with a nitrogen or helium purge supplied to the other side. Stream pressures were adjusted to minimize the pressure differential across the fiber wall. This reduces the stress on the fibers and minimizes loss of $Ag^+$ solution from the fiber. Tests were made using $Ag^+$ concentrations of 2, 4, and 6 $M$, feed pressures up to 15 atm, and temperatures of 21 to 38°C.

The results from the hollow fiber permeators are summarized in Tables 4 and 5. Several effects are obvious. An increase of the $Ag^+$ concentration in the fibers increased selectivity, primarily by reducing the flux of the paraffins. As the olefin feed partial pressure increased, the selectivity decreased. This decrease occurred because at higher olefin feed pressures less silver-olefin complex was formed per unit increase in olefin pressure. However, the paraffin solubility in $Ag^+$ solution, while quite low, increased linearly with increased feed pressure. In short, increased feed pressures produce a linear increase in the flux of paraffins, but a nonlinear increase in the flux of olefins. The permeators were found equally selective with the feed gas either on the bore or the shell side of the fibers.

Figure 6 shows the performance of one permeator over a 168-day test period. This permeator was operated with the feed on the shell side at an average ethylene driving force of 2 to 3 atm. The ethylene flux rate ranged from 0.089 to 0.147 cm³ (NTP)/cm²-min and the selectivity exceeded 100 for the entire run. These data demonstrate long-term operability of the olefin permeator.

The final phase of laboratory development was to test larger permeators, propylene as the olefin, and a liquid as the purge. Larger permeators were constructed containing up to 400 fibers and an active membrane area of up to 730 cm². Distribution problems of the feed gas in the permeators should show up in the larger permeators, if there are any. A feed stream of propylene and propane was used. It was chosen both to demonstrate selectivity

## Table 4
### EFFECT OF Ag⁺ CONCENTRATION AND FEED DIRECTION ON THE SELECTIVITY AND ETHYLENE PERMEATION RATES THROUGH Ag⁺ IMPREGNATED HOLLOW FIBERS

| Days on stream | Feed direction[a] | Av ethylene pressure (atm)[b] | [$Ag^+$] (mol/ℓ) | Temp (°C) | Mol % ethylene in permeate[c] | Ethylene permeation rate [cm³ (NTP)/cm²-min] | S |
|---|---|---|---|---|---|---|---|
| 8 | i | 3.3 | 2 | 23 | 94.5 | 0.090 | 35 |
| 21 | i | 3.3 | 2 | 38 | 90.5 | 0.118 | 18 |
| 37 | i | 3.3 | 2 | 29 | 93.4 | 0.104 | 24 |
| 60 | o | 1.6 | 2 | 38 | 97.1 | 0.084 | 67 |
| 66 | o | 1.6 | 2 | 29 | 97.8 | 0.089 | 89 |
| 75 | o | 1.6 | 2 | 21 | 97.8 | 0.081 | 87 |
| 21 | o | 2.2 | 4 | 21 | 98.4 | 0.168 | 140 |
| 109 | o | 3.5 | 4 | 32 | 98.5 | 0.086 | 140 |
| 116 | o | 3.2 | 4 | 43 | 98.2 | 0.097 | 115 |
| 6 | i | 3.3 | 6 | 23 | 98.2 | 0.120 | 105 |
| 14 | i | 3.3 | 6 | 29 | 98.2 | 0.103 | 105 |
| 22 | i | 3.3 | 6 | 38 | 98.4 | 0.102 | 115 |
| 37 | o | 1.6 | 6 | 38 | 98.7 | 0.083 | 160 |
| 45 | o | 1.6 | 6 | 29 | 98.8 | 0.084 | 165 |
| 59 | o | 1.6 | 6 | 23 | 99.1 | 0.101 | 220 |

[a]   The feed was a 1/3-each mixture of methane, ethylene, and ethane. It was supplied either to the inside (i = bore side) or the outside (o = shell side) of the fibers.

[b]   The ethylene pressure was determined from the average of the ethylene contents of the feed and raffinate streams.

[c]   Purge-free content.


## Table 5
### EFFECT OF TEMPERATURE AND ETHYLENE PRESSURE ON THE SELECTIVITY AND ETHYLENE PERMEATION RATES THROUGH HOLLOW FIBERS CONTAINING 6 $M$ Ag⁺

| Av ethylene pressure (atm)[a] | Temp (°C) | Mol % ethylene in permeate[b] | Ethylene permeation rate [cm³ (NTP)/cm²-min] | S |
|---|---|---|---|---|
| 1.3 | 21 | 98.7 | 0.105 | 150 |
| 1.2 | 29 | 98.6 | 0.103 | 140 |
| 1.3 | 38 | 98.8 | 0.104 | 160 |
| 1.2 | 38 | 98.7 | 0.101 | 160 |
| 1.3 | 38 | 98.9 | 0.137 | 170 |
| 2.4 | 38 | 98.4 | 0.197 | 125 |
| 2.6 | 38 | 98.6 | 0.224 | 140 |

[a]   The feed gas was a 1/3-each mixture of methane, ethylene, and ethane. It was supplied to the shell side of the fibers. The average olefin pressure was determined from the average ethylene content of the feed and raffinate streams.

[b]   Purge-free content.

of the permeators to a second olefin and because the separation of propylene from propane has commercial importance. A liquid purge is desired to make the final recovery of the olefin gas from the purge (a requirement in any commercial process) easier. Hexane was chosen as the liquid purge because it is inexpensive and easily separated from ethylene or

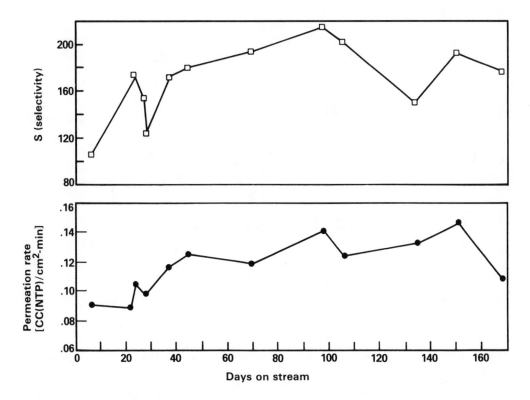

FIGURE 6.    Performance of a hollow fiber module vs. time. The top plot shows ethylene selectivity vs. days of operation, and the bottom plot shows ethylene permeation rate vs. days on stream.

propylene. A small pilot plant was built to accomodate the larger permeators and the hexane purge (Figure 7). This unit contained a distillation train to separate propylene from the hexane purge, thereby allowing the hexane to be recycled.

Data from two runs using nitrogen and hexane purges are given in Table 6 and 7, respectively. Both purge materials proved effective in the permeators. The propylene flux was comparable at 0.1 to 0.2 cm³ (NTP)/cm²-min. Selectivity to propylene was, however, higher with the nitrogen purge. It was found that with a liquid purge the purge must be on the shell side of the fibers. Large pressure drops developed along the length of the fiber when the liquid was pumped through the small fiber bore. The large pressure drop produced a large pressure imbalance across the fiber wall and accelerated failure of the fibers.

In all permeators a decline in performance, primarily in the propylene flux, was found with time. This decline was caused by loss of water or Ag⁺ from the fibers. Permeator performance was effectively restored by periodic regeneration of the units. This was done by circulating fresh Ag⁺ through the fibers for a 2-hr period.

Table 6 shows a typical response from one regeneration cycle. This simple procedure typically restored the olefin flux to at least 70% of its original value.

The performance decline with time, while reversible, makes it difficult to compare performances of different permeators. It was clear that all the permeators were selective to propylene. Also, even though comparisons between permeators are not straightforward, no major feed or purge distribution problems were found with the larger laboratory permeators or with the liquid purge.

## 3. Model Development

A simple mathematical model was needed to allow meaningful comparisons of different

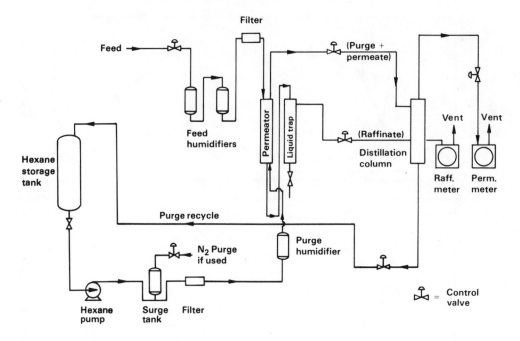

FIGURE 7.   Pilot plant for evaluating laboratory-scale hollow fiber permeators. Note that either a liquid hexane or a nitrogen gas purge may be used.

## Table 6
### PERMEATION RATES OF PROPYLENE THROUGH HOLLOW FIBERS IMPREGNATED WITH 6 $M$ Ag$^+$ (NITROGEN GAS PURGE)

| Days on stream | Av propylene pressure (atm)[a] | Mol % propylene in permeate[b] | Propylene permeation rate [cm³ (NTP)/cm²-min] | S |
|---|---|---|---|---|
| 1 | 2.9 | 99.6 | 0.203 | 230 |
| 3 | 2.6 | 99.6 | 0.185 | 245 |
| 9 | 2.5 | 99.6 | 0.206 | 285 |
| 14 | 2.4 | 99.7 | 0.195 | 350 |
| 17 | 2.5 | 99.7 | 0.166 | 390 |
| 22 | 2.7 | 98.7 | 0.103 | 95 |
| 24 | 2.6 | 98.6 | 0.097 | 87 |

**Fibers Reimpregnated with AgNO₃ Solution**

| | | | | |
|---|---|---|---|---|
| 28 | 2.7 | 99.4 | 0.116 | 215 |
| 34 | 2.9 | 99.6 | 0.162 | 295 |
| 45 | 3.0 | 99.7 | 0.109 | 315 |
| 62 | 3.0 | 99.7 | 0.158 | 280 |
| 76 | 3.0 | 99.7 | 0.136 | 270 |
| 93 | 2.9 | 99.6 | 0.106 | 285 |
| 104 | 3.2 | 99.7 | 0.101 | 240 |
| 110 | 3.2 | 99.6 | 0.098 | 225 |

[a]   The feed was an equimolar mixture of propane and propylene at a total pressure of 7.8 atm.
[b]   Purge-free content.

## Table 7
## PERMEATION RATES OF PROPYLENE THROUGH HOLLOW
## FIBERS IMPREGNATED WITH 6 $M$ Ag$^+$ (HEXANE LIQUID PURGE)

| Days on stream | Av propylene pressure (atm)[a] | Mol % propylene in permeate[b] | Propylene permeation rate [cm³ (NTP)/cm²-min] | S |
|---|---|---|---|---|
| 2 | 2.8 | 99.4 | 0.244 | 230 |
| 4 | 3.0 | 99.1 | 0.203 | 140 |
| 6 | 3.2 | 98.8 | 0.148 | 93 |
| 9 | 3.3 | 98.5 | 0.119 | 73 |
| 15 | 3.7 | 98.7 | 0.116 | 66 |
| 20 | 3.8 | 99.3 | 0.064 | 115 |
| 23 | 3.8 | 99.2 | 0.067 | 100 |

[a]  The feed was an equimolar mixture of propane and propylene at a total pressure of 7 atm.
[b]  Purge-free content.

permeators. The model should be able to predict performance of hollow fiber permeators operated under varying conditions. Starting with Equation 11, developed for predicting olefin flux in flat films, two simplifications were made. First, because only one type of fiber was used in this study, the membrane thickness (L) was the same in all permeators and constant for all tests. Second, at Ag$^+$ concentrations $\geq 2$ $M$, the olefin flux was independent of Ag$^+$ concentration. All permeators were run at $>2$ $M$ in Ag$^+$ concentration, and, therefore, [Ag$^+$]$_{total}$ was also constant for all tests. Including both L and [Ag$^+$]$_{total}$ as constants with K$_3$ of Equation 11 gives a new constant (K$_4$) in Equation 13.

$$\text{flux}_{O\ell} \approx AK_4 \left( \frac{(P_{O\ell})_F}{1 + K_2(P_{O\ell})_F} - \frac{(P_{O\ell})_P}{1 + K_2(P_{O\ell})_P} \right) \tag{13}$$

where $(P_{O\ell})_F$ = olefin partial pressure on the feed side of the fibers, $(P_{O\ell})_P$ = olefin partial pressure on the permeate side of the fibers, and where

$$K_4 = \frac{K_3}{L} [Ag^+]_{total}$$

A second equation was developed for the flux of the uncomplexed species through the fibers (Equation 14). This empirical expression gave the best fit of the data.

$$\text{flux}_{(u)} = AK_5[(P_u)_F^{K_6} - (P_u)_P^{K_7}] \, e^{-K_8[Ag^+]_{total}} \tag{14}$$

where $(P_u)_F$ = partial pressure of the uncomplexed permeant on the feed side, and $(P_u)_P$ = partial pressure of the uncomplexed permeant on the permeate side.

A nonlinear least-squares regression analysis was used to fit over 2000 data points to Equations 13 and 14. From that analysis the "best values" found for the constants K$_2$ and K$_4$ to K$_8$ were

- K$_2$ = 0.392 atm$^{-1}$
- K$_4$ = 0.167 cm³ (NTP)/cm²-min-atm
- K$_5$ = 3.48(10$^{-3}$) cm³ (NTP)/cm²-min-atm
- K$_6$ = K$_7$ = 0.682
- K$_8$ = 0.197 $\ell$/mol

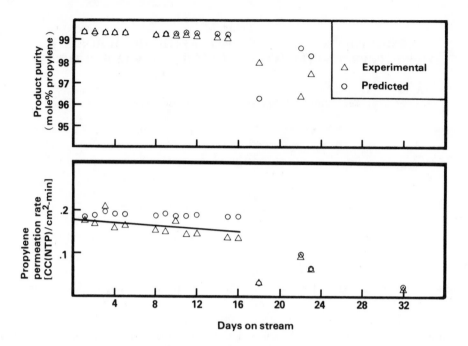

FIGURE 8.    Predicted and actual performance of a laboratory-scale hollow fiber permeator vs. time. The top plot shows the mol % propylene in the permeate vs. days on stream. The bottom plot shows the propylene permeation rate vs. days on stream.

Both Equations 13 and 14 were incorporated into a mathematical model for hollow fiber permeators. In the model, the permeator is first subdivided into 50 equal axial segments. The feed stream entering the first permeator segment is adjusted to account for the permeant in that segment. This gives a new feed composition to use for the second segment. This process is repeated for all 50 segments. Integration both inside and outside the fibers yields predicted raffinate and permeate streams from the permeator.

The model and flux expressions have some obvious limitations. First, it was assumed that all nonfacilitated permeants (paraffins) permeate the fibers at the same rate. While this assumption is incorrect, it was used for the sake of simplicity. Even so, the prediction of the total paraffin flux was quite good. Second, neither flux expression contains a temperature term. This reflects the observation that in the range of 16 to 43°C there was no significant temperature effect on permeant flux.

The actual and predicted performance of a typical permeator is shown in Figure 8. Excellent agreement was found for predictions of propylene purity. However, the propylene flux rate was overpredicted by the model. Even so, this simple model proved adequate for predicting the relative magnitude and direction of change of permeator performance due to changes in feed rates, pressures, and/or compositions.

## C. Field Test Development

The final phase of development of the olefin membrane was to demonstrate the performance of a commercial-sized permeator in a commercial environment. A commercial-sized permeator was prepared by modifying a large reverse osmosis permeator. Modification consisted of constructing and machining a second epoxy tube sheet on the fiber bundle. This allowed an open pathway through the fiber bores from one end of the permeator to the other. An outlet was also constructed on the permeator shell. The end result was a radial flow permeator of the design shown in Figure 9. Flow on the bore side of the fibers would be axial in the permeator, but the flow on the shell side would be radial and across the fibers.

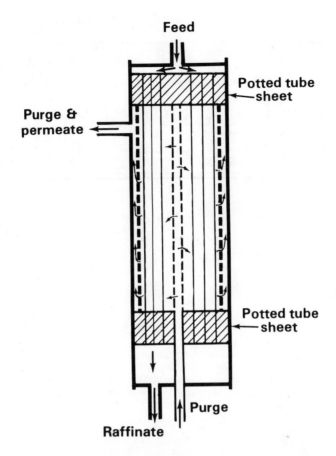

FIGURE 9.   Diagram with flow patterns of a radial flow hollow fiber
permeator. The center tube with slits is a distributor tube for the purge
liquid. The outermost heavy dashed lines are porous spacers where the
purge plus permeate collect. The purge and permeate discharge from
the port on the shell of the vessel. The solid lines running from one
potted tube sheet to the other represent hollow fibers.

The modified commercial permeator contained over 130,000 fibers. The final active length
was 40 to 66 cm providing an active membrane area of 22 to 37 m². Each permeator was
impregnated with 6 *M* AgNO₃ solution in the same manner (but, of course, on a larger
scale), as was done with the smaller permeators. The commercial-sized permeator was a
factor of 300 to 500 times larger than the largest laboratory unit.

A test stand (Figure 10) was built to accomodate one large permeator. The test stand also
had other special features worth noting. It had its own distillation train to recover the purge
(hexane) from the permeate (propylene). A catalytic unit was included to remove hydrogen
(a reducing agent for Ag⁺) from the feed. The test stand also had its own dedicated analytical
instrumentation to automatically measure stream compositions, rates, and pressures.

The permeators were tested at a commercial polypropylene plant. They were used to
recover propylene from the vent gas stream from the polymerization process. This stream
contained ~75 mol % propylene with the remainder being methane, ethane, propane, hydrogen, and traces of higher hydrocarbons.

The test stand was interfaced with the polypropylene plant as shown in Figure 11. The
vent gas stream from the recycle drum was (1) treated to remove hydrogen, (2) saturated
with water vapor, (3) filtered to remove particulates, and then (4) supplied to the bore side

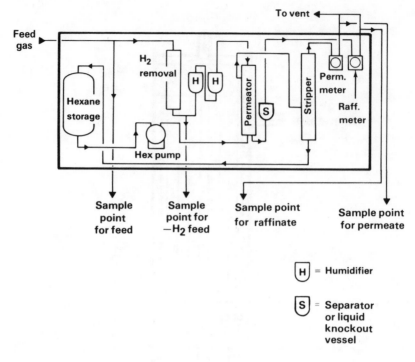

FIGURE 10.    Test stand for a commercial-sized permeator. Note the hydrogen removal unit on the feed gas stream. A stripper (distillation train) separates the permeate from the purge.

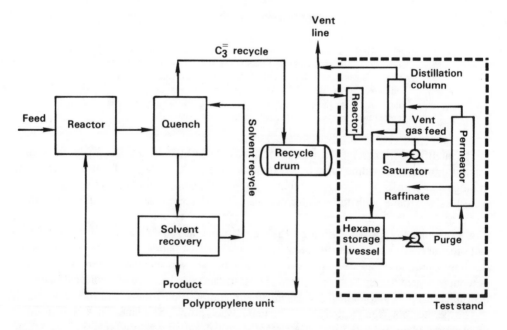

FIGURE 11.    Interface of the test stand with the polypropylene plant. The feed to the test stand is the overhead stream from the recycle drum. The recycle drum is used to control the level of impurities (hydrogen and paraffins) that build up in the reactor.

of the fibers. Hexane was pressurized, filtered, and supplied as the purge to the shell side of the fibers. Permeator performance was assessed from measurements of stream compositions and rates.

## Table 8
### PERFORMANCE OF COMMERCIAL-SIZED HOLLOW FIBER MEMBRANE MODULES

| Run no. | Days on stream | Total feed pressure (atm) | Feed rate (m³/hr) | Purge rate (ℓ/hr) | Propylene permeation rate [cm³ (NTP)/cm²-min] | | Mol % propylene in purge[a] | |
|---|---|---|---|---|---|---|---|---|
| | | | | | Measured | Predicted | Measured | Predicted |
| 1 | 7 | 4.8 | 3.2 | 41 | 0.0878 | 0.0761 | 98.6 | 98.8 |
| 2 | 14 | 5.0 | 2.4 | 43 | 0.0534 | 0.078 | 98.5 | 99.0 |
| | 16 | 7.1 | 4.1 | 71 | 0.0730 | 0.134 | 97.7 | 99.4 |
| 3 | 2 | 6.8 | 2.4 | 74 | 0.074 | 0.081 | 96.3 | 98.4 |
| | 9 | 4.8 | 4.1 | 77 | 0.073 | 0.115 | 98.7 | 99.5 |
| | 14 | 4.8 | 2.4 | 75 | 0.053 | 0.061 | 98.0 | 98.4 |
| 4 | 4 | 7.3 | 2.9 | 116 | 0.165 | 0.170 | 99.3 | 99.5 |
| | 10 | 6.5 | 2.9 | 116 | 0.125 | 0.165 | 96.6 | 99.4 |
| 5 | 4 | 6.8 | 3.2 | 111 | 0.155 | 0.178 | 99.4 | 99.8 |
| | 13 | 5.4 | 3.2 | 90 | 0.099 | 0.170 | 98.9 | 99.7 |

[a]  Purge-free content.

The field tests were designed to determine the effects of process variables on the performance of a large-scale permeator. Feed pressures of 5 to 13 atm and rates of 1.4 to 4.2 m³/hr were used. Purge rates were varied between 35 and 115 ℓ/hr. In all tests, every effort was made to minimize the pressure differential across the fiber wall. Data from a number of these tests are summarized in Table 8. These results clearly show operability of the commercial-sized permeator over a wide range of conditions. High purity propylene was produced at rates comparable to that found with the laboratory units.

Figures 12 and 13 show permeator performance as a function of time. In both these figures the predicted membrane performance is shown for at least part of each test. It is clear that (as with the laboratory units) the model overpredicts propylene product purity. This occurred mainly because of incorrect weighting of the flux of certain paraffins through the fibers. The model also overpredicts the propylene flux in the larger permeators. With time, the differences between predicted and measured values became even more significant. A loss of water and/or $Ag^+$ from the fibers caused deterioration of performance in the large units. This deactivation was not accounted for in the model. Thus, significant differences between predicted and measured values were found with time with the larger permeators.

The most severe problem encountered in the operation of the larger permeators was the decline of performance with time. The high heat of dissolution of propylene into the hexane purge caused temperature rises of 5 to 17°C in the permeator. Because of the near-adiabatic nature of the permeator vessel, this temperature rise could not be moderated. The increase in temperature accelerated the loss of water from the fibers. With laboratory-scale units the total heat produced was not significant and was easily dissipated to the atmosphere. Thus, with laboratory units the decline of performance with time was quite small.

Periodic regeneration of the large permeators was successful in restoring their performance. Regeneration was performed by pumping fresh $Ag^+$ solution through the fibers. The dotted lines in Figure 13 show the points where regeneration was carried out. In most cases, the performance was restored somewhat.

Two permeators were operated for longer than 60 days. In both tests, the permeators deactivated with time. Part of the deactivation was reversible. In the longest test, one permeator processed over 50,800 kg of feed gas producing >6300 kg of 98+ mol % propylene product at an average flux of 0.095 cm³ (NTP)/cm²-min.

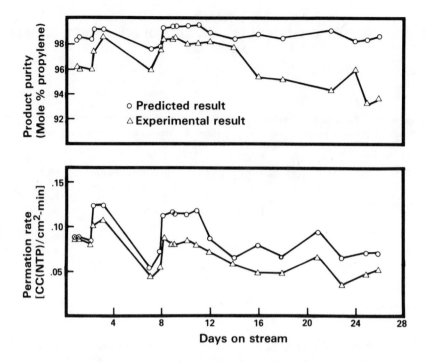

FIGURE 12.   Predicted vs. actual performance of a commercial-sized olefin permeator with time. The top plot shows the mol % propylene in the permeate stream vs. days on stream. The bottom plot gives the propylene permeation rate with time.

## VI. SUMMARY

Silver ion-impregnated membranes are selective to olefins over paraffins. Silver acts to facilitate olefin transport across the membrane. The rate-limiting step to olefin transport is diffusion through the $Ag^+$ transport solution. Permeability coefficients of 5.5 to 10.0 $\times$ $10^{-3}$ cm$^3$ (NTP)-cm/cm$^2$-min-atm were measured for ethylene through immobilized $Ag^+$ solutions. LeBlanc et al.[34] reported a permeability of only 1 $\times$ $10^{-3}$ cm$^3$-cm/cm$^2$-min-atm for ethylene through $Ag^+$ immobilized in an ion-exchange membrane. The large difference between these two values is probably due to lowered $Ag^+$ mobility and, thus, ethylene transport in the ion exchange membrane.

Both ethylene and propylene are transported by $Ag^+$ at comparable rates. This is not surprising since the equilibrium constant (see Table 1) for formation of the silver-olefin complex is identical for both olefins. At low olefin pressures (up to 1 atm) the permeability of olefins increases linearly with olefin pressure. However, at higher pressures the olefin flux approaches a limiting value. This is due to saturation of the $Ag^+$ in the membrane with olefins. To account for the phenomena a rate equation was developed that uses the silver-olefin equilibrium expression.

Lower olefin fluxes were found for hollow fibers impregnated with $Ag^+$ than with flat films of immobilized $Ag^+$ solutions. This was caused by the dense surface skin on the hollow fibers used in this study. The dense skin reduced the availability and mobility of the $Ag^+$ solution at the membrane surface and, thus, reduced the olefin transport rate.

Olefin flux was independent of $Ag^+$ concentration when the $Ag^+$ concentration was >2 $M$. However, increased $Ag^+$ did reduce paraffin permeability and, therefore, improved the membrane's selectivity to olefins.

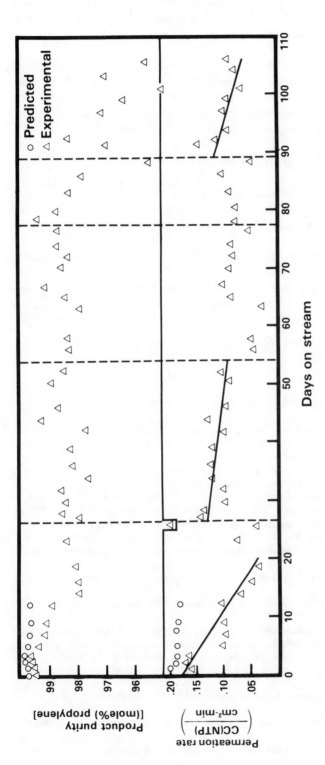

FIGURE 13.    Performance of a commercial-sized olefin permeator with time. The top plot gives the mol % propylene in the permeate with time. The bottom plot shows the propylene permeation rate vs. days on stream. The vertical dotted lines represent times when the permeator was regenerated. The solid lines on the bottom plot show the decline in performance with time. Predicted data points are also shown for the early part of the test. Note that this test lasted 108 days.

The operation of the olefin membrane was demonstrated with a commercial-sized permeator. The major limitation found with the commercial permeator was deactivation due to loss of water from the fibers. Periodic regeneration was partially effective in restoring permeator performance. Both laboratory and commercial-sized hollow fiber permeators had comparable selectivity and permeability to olefins.

Several limitations of the olefin membrane must be pointed out. First, the membrane cannot tolerate poisons to $Ag^+$, such as acetylenes, $H_2$, or $H_2S$. Second, decline of performance due to water and/or $Ag^+$ loss can become significant. Periodic regeneration can minimize this problem. Finally, a minimum pressure differential across the fiber wall is necessary to maintain optimum performance and maximize permeator lifetime.

## VII. FUTURE DIRECTIONS

The development of the olefin membrane was a technical success. However, when compared to distillation, the olefin membrane is uneconomical in producing polymer-grade olefins (99.95 + wt % olefin). This results primarily because the olefin membrane cannot produce polymer-grade olefin with one membrane stage. At least two stages would be needed. A two-stage olefin membrane system is simply too expensive in both capital and operating costs for producing polymer-grade olefins.

No further research is currently planned on the $Ag^+$-based olefin membrane. The technology has been proven. Further development awaits the identification of a suitable and economic application.

## REFERENCES

1. **Scholander, P. F.**, Oxygen transport through hemoglobin solutions, *Science*, 131, 585, 1960.
2. **Schultz, J. S., Goddard, J. D., and Suchdeo, S. R.**, Facilitated transport via carried-mediated diffusion in membranes. I, *AIChE J.*, 20, 417, 1974.
3. **Schultz, J. S.**, Carrier-mediated transport in liquid-liquid membrane systems, in *Recent Developments in Separation Science*, Vol. 3B, Li, N. N., Ed., CRC Press, Cleveland, 1977, 243.
4. **Smith, D. R., Lander, R. J., and Quinn, J. A.**, Carrier-mediated transport in synthetic membranes, in *Recent Developments in Separation Science*, Vol. 3B, Li, N. N., Ed., CRC Press, Cleveland, 1977, 225.
5. **Kimura, S. G., Matson, S. L., and Ward, W. J.**, Industrial applications of facilitated transport, in *Recent Developments in Separation Science*, Vol. 5, Li, N. N., Ed., CRC Press, Cleveland, 1979, 11.
6. **Halwachs, W. and Schügerl, R.**, The liquid membrane technique — a promising extraction process, *Int. Chem. Eng.*, 20, 519, 1980.
7. **Way, J. D., Noble, R. D., Flynn, T. M., and Sloan, E. D.**, Liquid membrane transport: a survey, *J. Membrane Sci.*, 12, 239, 1982.
8. **Ward, W. J. and Robb, W. L.**, Carbon dioxide-oxygen separation: facilitated transport of carbon dioxide across a liquid film, *Science*, 156, 1481, 1967.
9. **Winneck, J., Marshall, R. D., and Schubert, F. H.**, An electrochemical device for carbon dioxide concentration. I. System design and performance, *Ind. Eng. Chem. Proc. Des. Dev.*, 13, 59, 1974.
10. **Roughton, F. J. W.**, Transport of oxygen and carbon dioxide, in *Handbook of Physiology*, Section 3: Respiration, Vol. 1, Fenn, W. O. and Rahn, H., Sect. Eds., American Physiological Society, Washington, D.C., 1964, 767—825.
11. **Wittenberg, J. B.**, The molecular mechanism of hemoglobin-facilitated oxygen diffusion, *J. Biol. Chem.*, 241, 115, 1966.
12. **Otto, N. C. and Quinn, J. A.**, The facilitated transport of carbon dioxide through bicarbonate solutions, *Chem. Eng. Sci.*, 26, 949, 1971.
13. **Donaldson, T. L. and Quinn, J. A.**, Carbon dioxide transport through enzymatically active synthetic membranes, *Chem. Eng. Sci.*, 30, 103, 1975.
14. **Enns, T.**, Facilitation by carbonic anhydrase of carbon dioxide transport, *Science*, 155, 44, 1967.
15. **Matson, S. L., Herrick, C. S., and Ward, W. J.**, Progress on the selective removal of $H_2S$ from gasified coal using an immobilized liquid membrane, *Ind. Eng. Chem. Proc. Des. Dev.*, 16, 370, 1977.

16. **Smith, D. R. and Quinn, J. A.,** The facilitated transport of carbon monoxide through cuprous chloride solutions, *AIChE J.,* 26, 112, 1980.

17. **Hughes, R. D. and Steigelmann, E. F.,** Process for Separating Carbon Monoxide, U.S. Patent 3,823,529, 1974.

18. **Ward, W. J.,** Analytical and experimental studies of facilitated transport, *AIChEJ,* 16, 405, 1970.

19. **Ward, W. J.,** Electrically induced carrier transport, *Nature (London),* 227, 162, 1970b.

20. **Bdzil, J., Carlier, C. C., Frisch, H. L., Ward, W. J., and Breiter, M. W.,** Analysis of potential difference in electrically induced carrier transport systems, *J. Phys. Chem.,* 77, 846, 1973.

21. **Ebenz, W. F., Welge, H. J., Yost, D. M., and Lucas, H. J.,** The hydration of unsaturated compounds. IV. The rate of hydration of isobutene in the presence of silver ion. The nature of the isobutene-silver complex, *J. Am. Chem. Soc.,* 59, 45, 1937.

22. **Winstein, S. and Lucas, H. J.,** The coordination of silver ion with unsaturated compounds, *J. Am. Chem. Soc.,* 60, 836, 1938.

23. **Lucas, H. J., Moore, R. G., and Pressman, D.,** The coordination of silver with unsaturated compounds. II. Cis- and trans-2-pentene, *J. Am. Chem. Soc.,* 65, 227, 1943.

24. **Lucas, H. J., Billmeyer, F. W., Jr., and Pressman, D.,** The coordination of silver ion with unsaturated compounds. III. Mixtures of trimethylethylene and cyclohexane, *J. Am. Chem. Soc.,* 65, 230, 1943.

25. **Hepner, F. R., Trueblood, K. N., and Lucas, H. J.,** Coordination of silver ion with unsaturated compounds. IV. The butenes, *J. Am. Chem. Soc.,* 74, 1333, 1952.

26. **Trueblood, K. N. and Lucas, H. J.,** Coordination of silver ion with unsaturated compounds. V. Ethylene and propylene, *J. Am. Chem. Soc.,* 74, 1338, 1952.

27. **Quinn, H. W. and Glew, D. N.,** Coordination compounds of olefins with solid complex silver salts, *Can. J. Chem.,* 40, 1103, 1962.

28. **Featherstone, W. and Sorrie, A. J. S.,** Silver-hydrocarbon complexes, *J. Chem. Soc.,* p. 5235, 1964.

29. **Beverwijk, C. D. M., Van Der Kerk, G. J. M., Leusink, A. J., and Noltes, J. G.,** Organosilver chemistry, *Organomet. Chem. Rev. A,* 5, 215, 1970.

30. **Duda, J. L. and Vrentas, J. S.,** Laminar liquid jet diffusion studies, *AIChE J.,* 14, 286, 1968.

31. **Keller, K. H. and Friedlander, S. K.,** Diffusivity measurements of human methemoglobin, *J. Gen. Physiol.,* 49, 681, 1966.

32. **Wittenberg, J. B.,** Myoglobin-facilitated oxygen diffusion: role of myoglobin in oxygen entry into muscle, *Physiol. Rev.,* 50, 559, 1970.

33. **Colton, C. K., Stroeve, P., and Zahka, G. J.,** Mechanism of oxygen transport augmentation by hemoglobin, *J. Appl. Physiol.,* 35, 307, 1973.

34. **LeBlanc, O. H., Jr., Ward, W. J., III, Matson, S. L., and Kimura, S. G.,** Facilitated transport in ion-exchange membranes, *J. Membr. Sci.,* 6, 339, 1980.

Chapter 10

# SEPARATION OF EUROPIUM WITH LIQUID SURFACTANT MEMBRANES

**Yu Jianhan, Jiang Changyin, and Zhu Yongjun**

## TABLE OF CONTENTS

# I. INTRODUCTION

The factors controlling the separation of europium ion with liquid surfactant membranes have been investigated. A layered structure model for high-viscosity liquid surfactant membranes is suggested. Equations describing the relationship between the effective membrane thickness, the time, and other factors are derived and verified experimentally. Results show that under certain conditions the decreasing concentration of europium ion in the external phase is proportional to the square root of time, the acidity of the internal phase, and the carrier concentration in the membrane phase. This model might be used for the description of mass transfer processes in liquid surfactant membranes.

The liquid membrane separation process developed by Li,[1] due to its significant potential in various fields, has been receiving increased attention. The liquid surfactant membrane technique combines membrane separation with solvent extraction and has its own unique characteristics. Although many aspects of mass transfer through the liquid membrane have been studied, there is still a lack of knowledge on mass transfer dynamics in these systems. Marr et al.[2] have suggested a layered structure model for high-viscosity liquid surfactant membranes and have proven that for such a system the effective membrane thickness is proportional to the square root of the mass transfer time. In fact, the effective membrane thickness is related to many other factors. The present study examines this kind of model in detail. In the experimental work, the membrane phase consists of di(2-ethylhexyl) phosphoric acid (HDEHP, carrier), kerosene (solvent), Span-80 (surfactant), and polybutadiene (additive). The resultant emulsion phase has a high viscosity (100 cp at 25°C).

# II. THEORETICAL

## A. Model

The complexation and the decomplexation reaction between europium ion and the HDEHP carrier is as follows:

$$Eu^{+3} + 3H_2A_2 \rightleftharpoons Eu(HA_2)_3 + 3H^+ \tag{1}$$

where $H_2A_2$ is the dimer form of HDEHP. The equilibrium constant for Equation 1 is

$$K = \frac{[Eu(HA_2)_3] \cdot [H^+]^3}{[Eu^{+3}] \cdot [H_2A_2]^3} \tag{2}$$

For such a liquid surfactant membrane system, the following assumptions can be made:

1. No convection occurs inside the emulsion globules and the position of internal phase droplets is fixed.
2. The rate of reaction between carrier and europium ion is much faster than the rate of diffusion of the complex in the membrane phase.
3. The diffusion resistance resides primarily in the membrane phase.

Therefore, in the course of mass transfer, the transported component moves from the outside to the inside of emulsion globules layer by layer. The effective membrane thickness increases with the time. Assuming that the transported component can permeate into the next layer only when the internal phase droplets in one layer are saturated and the internal phase droplets are very small and the layers are very thin, then there is a sharp demarcation between the saturated region and the unsaturated region in the internal phase. Decomplexation occurs only at this border, which is known as the internal border. This simplified model is

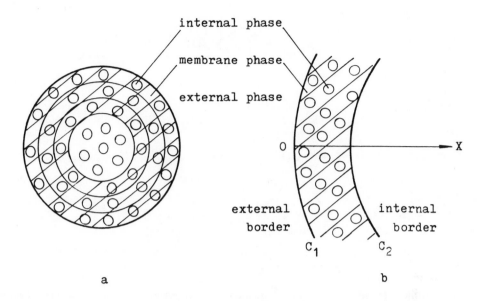

FIGURE 1.  Mass transfer scheme.

shown schematically in Figure 1. Figure 1a shows the movement of the internal border with the time, which makes the effective membrane thickness a function of the time. Figure 1b shows the concentration of complex at the external border, $C_1$, and the concentration of complex at the internal border, $C_2$.

## B. Effective Membrane Thickness

According to this model, the concentration of transported component from the internal phase in the saturated region is a constant value, $C_R$. When the concentration of transported component and pH value in the external phase is high enough, the carrier in the membrane phase is fully loaded, i.e., the concentration of complex in the membrane phase at the external border, $C_1$, is also constant. If the decomplexation power of the internal phase is very strong, the concentration of transported complex in the membrane phase at the internal border is approximately equal to zero. This means that the concentration of transported component of internal phase in the unsaturated region is nearly equal to zero.

When the effective membrane thickness, $\xi$, is much smaller than the diameter of the emulsion globules, Cartesian coordinates may be used. According to Fick's second law:

$$\frac{\partial C}{\partial t} = D \frac{\partial^2 C}{\partial x^2} \qquad (0 \leq x \leq \xi) \tag{3}$$

where $C$ = the concentration of transported complex in the membrane phase, and $D$ = effective diffusivity of complex in the emulsion globule. The boundary conditions for Equation 3 are

$$C_{(x,t)} = C_1 \qquad (x = 0) \tag{4}$$

$$C_{(x,t)} = C_2 \qquad (x = \xi) \tag{5}$$

Thus, solution of Equation 3 yields:

$$C_{(x,t)} = \frac{-C_1}{\phi\left(\frac{\alpha}{\sqrt{D}}\right)} \cdot \phi\left(\frac{Z}{\sqrt{D}}\right) + C_1 \tag{6}$$

where

$$\phi\left(\frac{Z}{\sqrt{D}}\right) = \frac{2}{\sqrt{\pi}} \int_0^{\frac{Z}{\sqrt{D}}} e^{-y^2} dy \tag{7}$$

$$Z = \frac{x}{2\sqrt{t}} \tag{8}$$

$$\alpha = \frac{\xi}{2\sqrt{t}} \; (\text{constant}) \tag{9}$$

Suppose in the interval of time, $\Delta t$, the quantity of transported component passing through the external border ($x = 0$) is

$$\Delta N_1 = -DF \frac{\partial C}{\partial x}\bigg|_{x=0} \cdot \Delta t \tag{10}$$

where F = surface area of emulsion globules. Then, the effective membrane thickness increases by $\Delta \xi$, and in the same time the quantity of transported component accepted in this layer is

$$\Delta N_2 = \frac{1}{2} \cdot \Delta \xi \cdot F \cdot C_R \tag{11}$$

The factor of 1/2 arises since the volume of the internal phase is one half of the total volume of the emulsion. A simple material balance yields:

$$\Delta N_1 = \Delta N_2 \tag{12}$$

As $\Delta t$ approaches zero, then

$$-D \frac{\partial C}{\partial x}\bigg|_{x=0} = \frac{1}{2} \cdot \frac{\partial \xi}{\partial t} \cdot C_R \tag{13}$$

Substituting Equation 6 and 9 into Equation 13:

$$\frac{2C_1\sqrt{D}}{\sqrt{\pi} \, \phi\left(\frac{a}{\sqrt{D}}\right)} = \alpha \cdot C_R \tag{14}$$

When

$$\frac{\alpha}{\sqrt{D}} < 1 \tag{15}$$

$$\phi\left(\frac{\alpha}{\sqrt{D}}\right) \approx \frac{2}{\sqrt{\pi}} \cdot \frac{a}{\sqrt{D}} \tag{16}$$

When

$$\alpha = \sqrt{\frac{DC_1}{C_R}} \tag{17}$$

$$\xi = \sqrt{\frac{4DC_1 t}{C_R}} \tag{18}$$

Equation 18 describes the variation of the effective membrane thickness with t, $C_R$, and $C_1$. It clearly shows that, apart from the $\sqrt{t}$ dependence of $\xi$, the higher the loading of transported complex and the smaller the accepting capacity of the internal phase, the faster is the increasing rate of the effective membrane thickness with the time. This equation may be called the "membrane thickness function".

## C. Mass Transfer Flux through Liquid Membranes

The mass transfer flux, J, may be expressed as

$$J = \frac{D}{\xi} \Delta C \tag{19}$$

where $\Delta C = C_1 - C_2$. When $C_2 \approx 0$, then

$$J = \sqrt{\frac{DC_1 C_R}{4t}} \tag{20}$$

$C_1$ and $C_R$ are functions of the concentration of components taking part in the complexation reactions. In the case of using HDEHP as carrier and europium as the transported component, under certain experimental conditions we can set $C_1 \approx \bar{C}/3$, where $\bar{C}$ is the concentration of carrier in the membrane phase. According to the reaction between europium ion and HDEHP, $C_R = C_H'/3$, where $C_H'$ is the exhausted concentration of $H^+$ in internal phase. Thus, Equation 20 can be transformed into

$$J = \sqrt{\frac{D\bar{C}C_H'}{36t}} \tag{21}$$

When initial acidity in the internal phase, $C_H$, is higher than 1 $N$, the results of the calculations show $C_H' \approx C_H$, so that Equation 21 may be simplified to

$$J = \sqrt{\frac{D\bar{C}C_H}{36t}} \tag{22}$$

since

$$J = \frac{-V}{F} \frac{dC_t}{dt} \tag{23}$$

and

$$-\frac{V}{F}\frac{dC_t}{dt} = \sqrt{\frac{D\overline{C}C_H}{36t}} \tag{24}$$

By integrating Equation 24, Equation 25 can be obtained:

$$1 - \frac{C_t}{C_o} = \frac{FD^{1/2}}{3VC_o}\sqrt{\overline{C}C_H t} \tag{25}$$

where $C_o$ and $C_t$ are the concentration of europium in the external phase at $t = 0$ and $t = t$, respectively, and V is the volume of the external phase.

## III. EXPERIMENTAL

### A. Materials

Europium nitrate solution was prepared by dissolving europium oxide (spectroscopic grade) in 1:1 nitric acid. Europium radioactive tracer was prepared by irradiation of spectroscopic grade europium oxide in the reactor (Institute of Nuclear Energy Technology, Qinghua University) and dissolving in 1:1 nitric acid. HDEHP was a reagent-grade product, and Span-80 was reagent grade. Polybutadiene had a molecular weight of 963 g/g-mol and viscosity of 840 cp at 25°C. Kerosene was 240# kerosene (boiling range 180 to 220°C, purified by distillation).

### B. Preparation of the Emulsion

The membrane phase consisted of Span-80 (2 vol %), polybutadiene (10 vol %), HDEHP (0.5 vol %), and kerosene (87.5 vol %). The internal phase was an aqueous solution of nitric acid. The volume ratio of the membrane phase to the internal phase was 1:1. The emulsion was prepared by stirring the membrane phase and the internal phase at 2000 rpm for 20 to 30 min.

### C. Mass Transfer Experiments

The europium nitrate solution containing europium radioactive tracer was placed in a 500-mℓ beaker and its pH value was adjusted by a sodium acetate-acetic acid buffer solution. The volume ratio of external phase to emulsion was 10:1. The external phase and the emulsion were gently mixed by a stirrer at 150 rpm. Samples of the external phase were taken at definite time intervals and the europium concentration was determined by radioactivity measurements with a type FH-421 single-channel γ-spectrometer.

## IV. RESULTS AND DISCUSSION

### A. Influence of the pH Value of the External Phase on the Mass Transfer Rate

Experimental results are shown in Figure 2. The mass transfer rate increases with increasing pH. An increase in pH shifts the reaction in Equation 1 to the right, thereby increasing $C_1$ and the driving force for mass transfer. The calculation according to the reaction of Equation 1 shows that in the region of pH > 2, $C_1$ increases weakly with pH. This is just the case observed in curve (4) of Figure 2. Also, as shown in Figure 2, under certain conditions the highest percentage of europium removal may be greater than 99%.

### B. Influence of the Acidity of the Internal Phase on the Mass Transfer Rate

The experimental results are shown in Figure 3. The mass transfer rate generally increases

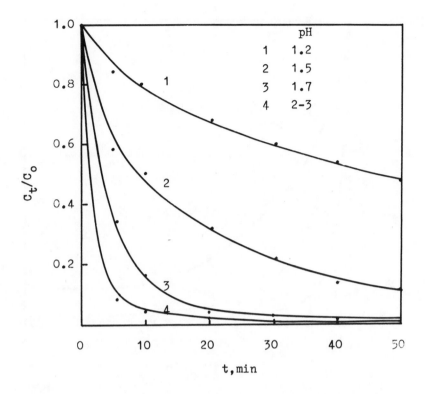

FIGURE 2. The influence of pH in the external phase on the mass transfer rate. ($C_o$ = $1.3 \times 10^{-3}$ $M$; $\bar{C}$ = $7.5 \times 10^{-3}$ $M$; $C_H$ = $4$ $N$)

with an increase in acidity of the internal phase. Curve (4) of Figure 3 shows that in the $C_H > 0.5$ $N$ region, the influence of the acidity of the internal phase on the mass transfer rate is not evident.

## C. Influence of Carrier Concentration on the Mass Transfer Rate

Experimental results are shown in Figure 4. Because the change of carrier concentration gives rise to the change in $C_1$, the mass transfer rate increases with an increase in the carrier concentration.

## D. Verification of Equation 25

In order to verify Equation 25, six experiments were carried out, the conditions for which are given in Table 1. All these experiments fulfill the condition of $C_1 \approx \bar{C}/3$ ($1 - C_t/C_o \leq 80\%$).

If $C_H$ and $\bar{C}$ are constant, then $1 - C_t/C_o \propto \sqrt{t}$. The $1 - C_t/C_o$ vs. $\sqrt{t}$ curves of these experiments are shown in Figure 5, and good linear relationships are obtained. Because carrier concentration $\bar{C}$ is the same and only the internal phase acidity is different, the plot of $1 - C_t/C_o$ vs. $\sqrt{C_H \cdot t}$ should be a straight line in experiments 1, 2, and 3. This plot is shown in Figure 6 and the relationship is basically a straight line. In experiments 4, 5, and 6 the internal phase acidity is the same and only the carrier concentration is different. When $1 - C_t/C_o$ is plotted against $\sqrt{\bar{C} \cdot t}$, the points of these experiments should lie on a straight line if the equation is obeyed. The plot is shown in Figure 7, and a rather good straight line is obtained.

The effective diffusivity of a complex in the emulsion phase may be roughly calculated by using data listed in Table 1. The average diameter of emulsion globules is estimated to

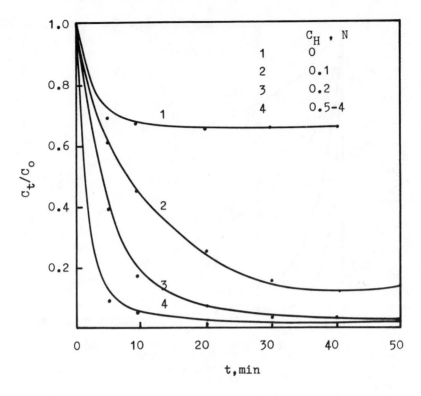

FIGURE 3.   The influence of acidity in the internal phase on the mass transfer rate. ($C_o =$ 1.3 × 10⁻³ $M$; $\bar{C} = 7.5 \times 10^{-3}\ M$; pH = 3)

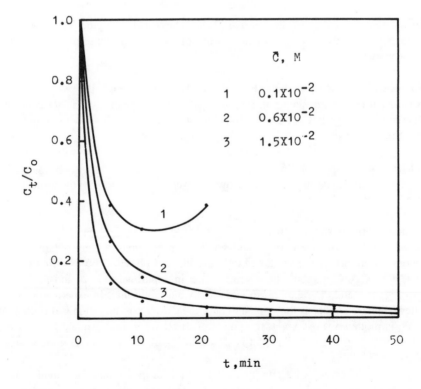

FIGURE 4.   The influence of carrier concentration on the mass transfer rate. ($C_o$ = 1.3 × 10⁻³ $M$; $C_H$ = 4 $N$; pH = 3)

**Table 1**
**EXPERIMENTAL CONDITIONS**

| Run | $C_o$ (M) | pH | $C_H$ (N) | $\bar{C}$ (M) |
|-----|-----------|-----|-----------|---------------|
| 1 | $6.5 \times 10^{-3}$ | 3 | 1 | $7.5 \times 10^{-3}$ |
| 2 | $6.5 \times 10^{-3}$ | 3 | 2 | $7.5 \times 10^{-3}$ |
| 3 | $6.5 \times 10^{-3}$ | 3 | 4 | $7.5 \times 10^{-3}$ |
| 4 | $6.5 \times 10^{-3}$ | 3 | 4 | $1.5 \times 10^{-3}$ |
| 5 | $6.5 \times 10^{-3}$ | 3 | 4 | $3.0 \times 10^{-3}$ |
| 6 | $6.5 \times 10^{-3}$ | 3 | 4 | $7.5 \times 10^{-3}$ |

be 1.2 mm. The average value of effective diffusivity, D, is $3.6 \pm 0.6 \times 10^{-7}$ cm²/sec. The difference between D values from these six experiments is rather small.

The preceding results show that the layered structure model can successfully describe the behavior of a component passing through the emulsion globules. Therefore, this model might be used to describe the mass transfer process of liquid surfactant membranes after considering the distribution ratio of a component between the external phase and the membrane phase.

## NOMENCLATURE

$C_1$  Concentration of transported component at the external border, mol/$\ell$
$C_2$  Concentration of transported component at the internal border, mol/$\ell$
$\bar{C}$  Carrier concentration in the membrane phase, mol/$\ell$
$C_o$  Initial concentration of transported component in the external phase, mol/$\ell$
$C_t$  Concentration of transported component in the external phase, mol/$\ell$
$C_H$  Initial concentration of hydrogen ion in the internal phase, mol/$\ell$
$C_H'$  Exhausted concentration of hydrogen ion in the internal phase, mol/$\ell$
$C_R$  Saturated concentration of transported component in the internal phase, mol/$\ell$
D  Effective diffusivity of complex in the emulsion phase, cm²/sec
F  Total surface area of emulsion globules, cm²
t  Contact time, sec
V  Total volume of the external phase, cm³

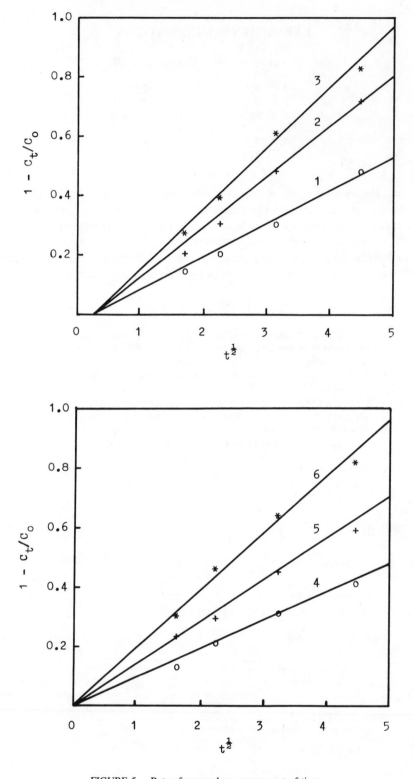

FIGURE 5.   Rate of removal vs. square root of time.

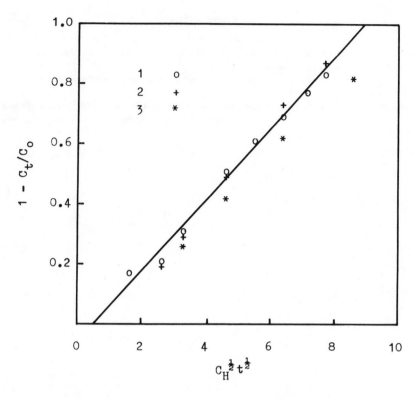

FIGURE 6. Rate of removal vs. square root of time and internal phase acidity.

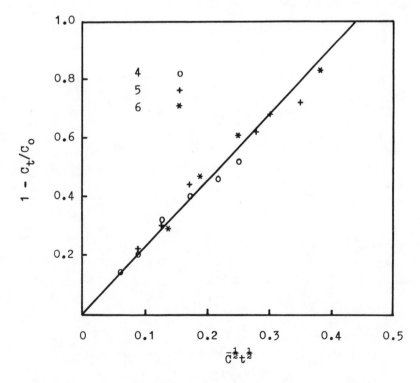

FIGURE 7. Rate of removal vs. square root of time and carrier concentration.

# REFERENCES

1. **Li, N. N.,** Separating Hydrocarbons with Liquid Membranes, U.S. Patent 3,410,794, 1968.
2. **Marr, R., Kopp, A., and Wilhelmer, J.,** Flussig-Membran-Permeation nach der Methode der Multiplen Emulsionnen Ubersicht uber Phanomene, Transportmechanismen und Modellbildungen, *Ber. Bunsenges. Phys. Chem.,* 83, 1097, 1979.
3. **Changyin, J.,** Study on the Mass Transfer Mechanism of Europium Ion through Liquid Surfactant Membranes, M.Sc. thesis, Institute of Nuclear Energy Technology, Qinghua University, Beijing, China, 1981.

Chapter 11

# TRANSFER OF BARBITURATES THROUGH EMULSIFIED LIQUID MEMBRANES

**G. Bouboukas, P. Colinart, H. Renon\*, and G. Trouvé**

## TABLE OF CONTENTS

---

\*   Author to whom correspondence should be addressed.

# I. INTRODUCTION

Li showed about 15 years ago that it is possible to separate the components of a mixture by selective diffusion through a liquid membrane.[1-3] This new separation process was improved by the use of stabilized emulsions. For example, a water-in-oil emulsion is made by a vigorous agitation of an organic phase containing surfactants and an aqueous solution. The resulting emulsion consists of microdroplets of the aqueous phase suspended in a continuous organic medium. This emulsion is mechanically dispersed in an aqueous mixture and its organic continuous phase behaves as a liquid membrane between the two aqueous solutions. Molecules can transfer from one aqueous phase to the other by partition and diffusion through this membrane.

Liquid membrane processes have been applied in various fields such as hydrometallurgy, oil recovery, wastewater treatment, and medical engineering. These applications have been recently reviewed.[4,5]

In medical engineering, liquid membranes can be used either for the in vivo removal of drugs (drinkable emulsions) or for the regeneration of a dialysis liquid in a portable artificial organ. The major interest in liquid membranes for these applications is that droplets in the emulsion can behave as microchemical reactors in which toxic substances are trapped or transformed by chemical or enzymatic reactions. These droplets, isolated from the biological fluids, can offer optimal conditions for reactions which could not take place in the natural gastric or intestinal medium.

Equilibrium concentrations of toxic substances in the solution can be made very low. For example, elimination of phosphate ions by an emulsion containing calcium acetate,[6] and extraction of urea by an emulsion containing urease have been studied.[7] When toxic compounds can be converted to ionic form, they can be trapped in the emulsion as nondiffusive ions.[8,10]

Barbiturates are salts of weak acids which can be trapped by a basic solution in an emulsion. Their physicochemical properties (solubility, diffusivity, acidity, partition, etc.) have been widely studied.[11-14] Transfer of phenobarbital through an emulsified liquid membrane has been studied by Yang and Rhodes.[9] The influence of temperature, membrane viscosity, and volume of dispersed phase in the emulsion on the kinetics of transport have been measured. The results are well represented by an empirical biexponential model. Barbiturates are divided into three classes: slow action, intermediate action, and rapid action barbiturates. Here, the transport through a liquid membrane of three barbiturates belonging to each class — hexobarbital (5-cyclohexenyl-3,5-dimethylbarbituric acid), amobarbital (5-ethyl-5-isoamyl-barbituric acid), phenobarbital (5-ethyl-5-phenylbarbituric acid) — is studied. Partition coefficients are measured and kinetics of transfer are performed in an apparatus with a known interfacial area. A mathematical model is developed to interpret the results, taking into account the composition of the emulsion.

# II. EXPERIMENTAL WORK

## A. Products and Emulsions

Barbituric acids are dissolved in a phosphate buffer at pH 7.3. The inner aqueous phase of the emulsion is either a pH 7.3 buffer or a pH 11.0 (glycin; NaCl; NaOH) buffer.

The organic phase of the emulsion is made of fluid paraffin oil (Fluka®) (viscosity: 21 cP at 25°C) in which a surfactant is dissolved (mannide monooleate or Montanide® 80 from Seppic; density 0.97, viscosity about 300 cP at 25°C, H.L.B. 2.6). The emulsion is made by vigorous agitation of the two phases heated at 80°C with a high speed homogenizer IKA-Ultraturrax, and then it is cooled slowly under mechanical agitation until ambient temperature is reached.

Barbiturate concentrations are determined by UV spectroscopy on 0.25- or 0.5-m$\ell$ samples. A correction for the competitive absorbance of surfactant must be made: the optical density of the sample measured at 280 nm corresponds to the surfactant absorption. From calibration curves, absorption of this surfactant at 210 nm is found and is then subtracted from the sample absorption to obtain the barbituric concentration.

## B. Apparatus

Equilibrium measurements are made in agitated flasks placed in a thermostated bath (37°C). Aqueous solution (2.5 g) and 15 g of organic phase are agitated together with a magnetic stirrer (500 rpm) for 20 min. The phases are then separated by centrifugation (8000 rpm for 30 min).

The kinetics of transfer are measured in the modified Lewis cell shown in Figure 1. The aqueous solution of barbiturate is placed in the lower part of the cell and the emulsion in the upper part. The interface is at the level of the horizontal baffle (2 in Figure 1). Owing to this baffle and to the cylindrical one (3 in Figure 1) placed in the middle of the cell, agitation speeds as high as 250 rpm in the emulsion and 320 rpm in the aqueous phase can be maintained without any movement of the interface. Agitation is provided by a magnetic stirrer in the aqueous phase and a two-bladed impeller in the emulsion phase.

A syringe with a long needle (15 cm) passing through the horizontal baffle allows for sampling of the aqueous solution. The cell contains 70 g of aqueous solution and 70 g of emulsion. The temperature is 37°C.

## C. Measurements of Partition Coefficients

Partition coefficients, m, are defined as the ratio of the barbiturate concentration in the organic phase ($\bar{C}^{eq}$) to the total barbiturate concentration in the aqueous solution ($C_I^{eq}$) at equilibrium:

$$m = \frac{\bar{C}^{eq}}{C_I^{eq}} \tag{1}$$

$\bar{C}^{eq}$ is calculated by a mass balance on the aqueous phase so that:

$$m = \frac{Q_I \, \bar{\rho} \, (C_I^o - C_I^{eq})}{\bar{Q} \, \rho_I \, C_I^{eq}} \tag{2}$$

where $Q_I$ and $\bar{Q}$ are the weights of aqueous and organic phases, $\rho_I$ and $\bar{\rho}$ are their densities, and $C_I^o$ is the initial concentration of the barbiturate in the aqueous phase (50 mg $\cdot$ $\ell^{-1}$). We do not take into account here the dissociation equilibrium of the barbiturates. Measurements are made at 37 $\pm$ 0.5°C.

Table 1 summarizes the results obtained for four different aqueous/organic systems and for the three barbiturates. Values of m are the arithmetic mean values of n repetitive experiments; $\Delta m$ is defined by

$$\Delta m = \frac{t}{\sqrt{n}} \left[ \frac{1}{n-1} \sum_{i=1}^{n} (\bar{m} - m_i)^2 \right]^{1/2}$$

where t is the Student factor.

For all the aqueous/organic systems, the partition coefficients are decreased in the following order: hexobarbital > amobarbital > phenobarbital.

This order is the same as for the solubilities in organic solvent, the action rates in the organism, and the pKa values.

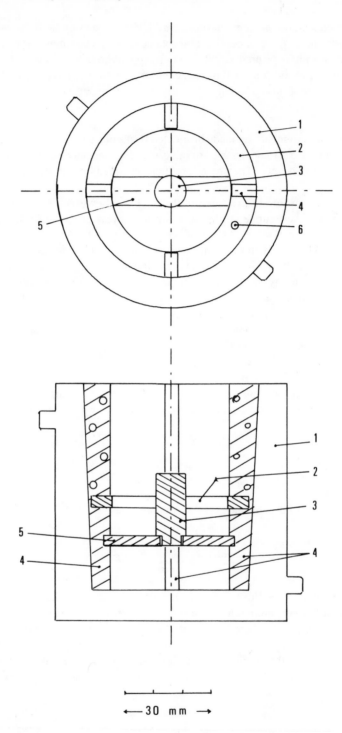

←—— 30 mm ——→

FIGURE 1.   Modified "Lewis cell" for the study of mass transfer. (1) Thermostated jacket; (2) horizontal baffle; (3) central baffle; (4) vertical baffle; (5) support of the central baffle; (6) hole for the sampling needle.

**Table 1**
**PARTITION COEFFICIENTS OF BARBITURATES**

| | | | System no. | | | |
|---|---|---|---|---|---|---|
| | | | 1 | 2 | 3 | 4 |
| $A^a$ | | | Phosphate buffer pH 7.3 | Phosphate buffer pH 7.3 | Phosphate buffer pH 7.3 | Glycin-NaOH pH 11.0 |
| $B^b$ | | | Paraffin oil | Paraffin oil: 95% Montanide® 80: 5% | Paraffin oil: 90% Montanide® 80: 10% | Paraffin oil: 93% Montanide® 80: 7% |
| | | | $\bar{\rho} = 0.844$ kg · m$^{-3}$ | $\bar{\rho} = 0.853$ kg · m$^{-3}$ | $\bar{\rho} = 0.861$ kg · m$^{-3}$ | $\bar{\rho} = 0.858$ kg · m$^{-3}$ |
| Barbiturate | pKa | $C_i^o$ (ppm) | $m^c$ | $m^d$ | $m^e$ | $m^d$ |
| Hexobarbital | 8.27 | 50 | 0.130 ± 0.008 | 0.37 ± 0.02 | 0.35 ± 0.02 | 0.20 ± 0.02 |
| Amobarbital | 7.89 | 50 | 0.017 ± 0.003 | 0.20 ± 0.08 | 0.21 ± 0.05 | 0.11 ± 0.03 |
| Phenobarbital | 7.34 | 50 | 0.011 ± 0.008 | 0.09 ± 0.02 | 0.16 ± 0.02 | 0.09 ± 0.03 |

[a] A: aqueous solution composition.
[b] B: organic solution composition.
[c] Mean value for nine measurements.
[d] Mean value for three measurements.
[e] Mean value for five measurements.

## Table 2
## COMPOSITION (WEIGHT FRACTION) OF THE
## EMULSIONS USED IN THIS STUDY

| | | Emulsion no. | | | |
|---|---|---|---|---|---|
| | | A | B | C | D |
| Organic phase | Paraffin oil Mon- | 0.54 | 0.57 | 0.54 | 0.57 |
| | tanide® 80 | 0.06 | 0.03 | 0.06 | 0.03 |
| Aqueous phase | Weight fraction | 0.40 | 0.40 | 0.40 | 0.40 |
| | pH | 7.3 | 7.3 | 11 | 11 |

The presence of surfactant increases the partition coefficients. But, except for the phenobarbital, an increase from 5 to 10% of the surfactant concentration does not change m. Interactions between barbiturates and surfactant molecules or micelles may explain the leveling of m, as shown by many authors.[16]

In basic medium (system 4) hydrolysis of barbiturates may occur with a nonnegligible rate.[17] We have neglected this phenomenon in our measurements and, thus, the m values in system 4 may be overestimated.

### D. Kinetics of Mass Transfer

The compositions of the four emulsions studied are given in Table 2. Emulsions C and D can trap the ionic barbiturates in the inner dispersed phase where chemical degradation of barbituric may also occur, because of the basic character of the phase. Comparison of emulsions A and B, on one hand, and C and D, on the other hand, indicates the role of surfactant.

The variations of the barbiturate concentrations in the aqueous phase as a function of time are represented in Figures 2 to 4. Several remarks can be made.

- All the curves show an initial period where no transfer seems to take place.
- For each emulsion composition, the rate of transfer follows the same order as the partition coefficient, i.e., hexobarbital > amobarbital > phenobarbital.
- For the three barbiturates, the rates of transfer depend on the emulsion composition in the order D ≥ C > A > B, except for phenobarbital for which this order is C > D > A > B.

## III. MODEL FOR MASS TRANSFER THROUGH LIQUID MEMBRANES

The objective of this model is to calculate the barbiturate concentrations in the aqueous solution as a function of time and emulsion composition.

### A. Hypothesis

1. Two regions are defined in the emulsion phase as shown on Figure 5. A very thin, stagnant film of pure organic phase separates the aqueous barbiturate solution from the core of the emulsion. Mass transfer occurs by molecular diffusion through this film; it is the controlling step in the transport process. The thickness of the film is denoted as $\delta$, and values relative to this film are indicated by the subscript II.
2. The core of the emulsion (subscript III) is well agitated with the mechanical agitator. It consists in an organic phase and droplets of an aqueous solution which are in thermodynamic equilibrium.

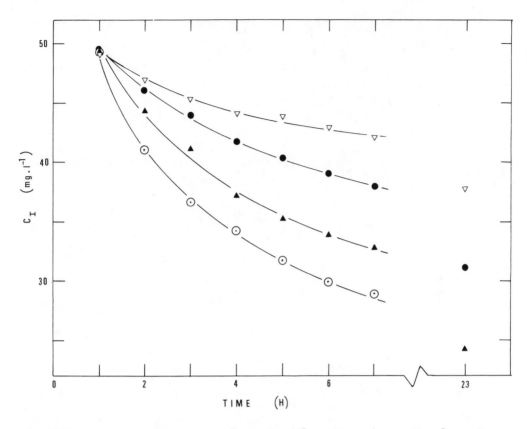

FIGURE 2. Kinetics of hexobarbital transfer. ▽: Emulsion B; ●: emulsion A; ▲: emulsion C; ○: emulsion D.

3. The resistance to mass transfer is negligible in the barbiturate solution (I) and there is equilibrium at the interface (I and II).
4. A steady-state diffusion process takes place in the film of thickness δ.

The hypothesis of the film of pure organic phase has significance both from a physical point of view (because the droplets in the emulsion are really entrapped) and from a hydrodynamical point of view (it is a stagnant film near an interface).

## B. Equations
Due to hypothesis 4, we have:

$$D \frac{\partial^2 C_{II}}{\partial x^2} = 0 \qquad (3)$$

where D is the diffusion coefficient of the barbiturate in the film with the boundary conditions:

$$x = 0; \qquad C_{II,0} = m \, C_I \qquad (4)$$

$$x = \delta; \qquad C_{II,\delta} = m_1 \, C_{III} \qquad (5)$$

m is the partition coefficient of the barbiturate at the interface I and II, and $m_1$ is a pseudo-partition coefficient at the interface II and III. $C_{III}$, the concentration in the core of the emulsion, is defined by Equation 6:

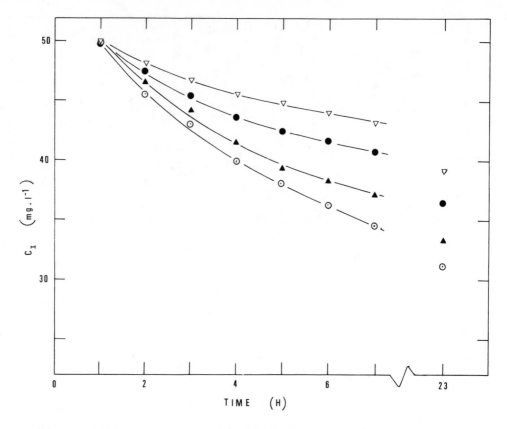

FIGURE 3.   Kinetics of amobarbital transfer. $\nabla$: Emulsion B; $\bullet$: emulsion A; $\blacktriangle$: emulsion C; $\bigcirc$: emulsion D.

$$C_{III} = \phi \, C_i + (1 - \phi) \, C_{II,\delta} \tag{6}$$

where $\phi$ is volumetric fraction of water in the emulsion, and $C_i$ is concentration in the dispersed droplets in the emulsion. Due to assumption 2, we also have:

$$C_i = m' \, C_{II,\delta} \tag{7}$$

where $m'$ is the partition coefficient between organic and aqueous phases in the emulsion. Combining Equations 5 to 7, we get:

$$m_1 = \frac{\phi}{m} + (1 - \phi) \tag{8}$$

By neglecting the barbiturate accumulation in the thin film of thickness $\delta$, a mass balance gives:

$$C_{III} = (C_I^\circ - C_I) \frac{V_I}{V_E} \tag{9}$$

where $V_E$ is the emulsion volume.
   Equation 3 can be easily integrated:

$$C_{II} = (m_1 \, C_{III} - m \, C_I) \frac{x}{\delta} + m \, C_I \tag{10}$$

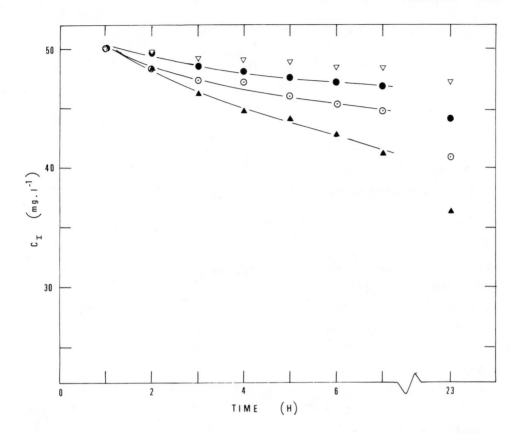

FIGURE 4. Kinetics of phenobarbital transfer. ▽: Emulsion B; ●: emulsion A; ▲: emulsion C; ○: emulsion D.

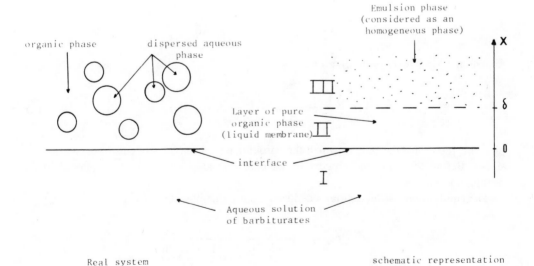

FIGURE 5. Representation of the emulsion for mass transfer modeling.

The diffusion flux is given by Fick's law:

$$N_x = -D \frac{d C_{II}}{dx} \tag{11}$$

or

$$N_x = \frac{D}{\delta} (m_1 C_{III} - m C_I) \tag{12}$$

The mass balance on the aqueous phase yields:

$$N_{x,0} = - \frac{V_I}{S} \frac{d C_I}{dt} \tag{13}$$

where S is the interfacial area between water and emulsion. Because $N_x$ is constant, we can combine Equations 8 to 13 to give:

$$\frac{d C_I}{dt} = -K \frac{S}{V_I} (A C_I + B) \tag{14}$$

with

$$K = \frac{D}{\delta} \tag{15}$$

$$A = m + \frac{m'}{\phi + (1 - \phi) m'} \frac{V_I}{V_E} \tag{16}$$

$$B = - \frac{m' C_I^o}{\phi + (1 - \phi) m'} \frac{V_I}{V_E} \tag{17}$$

Equation 14 is integrated to give

$$\ln \frac{A C_I + B}{A C_I^o + B} = -K \frac{S}{V_I} A (t - t_o) \tag{18}$$

where $t_o$ represents the time at which the barbiturate concentration begins to decrease ($t_o \simeq$ 1 hr). Before this time the process is not operating at steady state, as supposed in the assumption 4.

The equilibrium concentration is deduced from Equations 16 to 18:

$$C_I^{eq} = \frac{m' C_I^o V_I}{m' V_I + m V_E [\phi + (1 - \phi) m']} \tag{19}$$

The variations of $C_i$ as a function of time can also be calculated from Equations 6, 7, 9, and 14:

$$C_i = [\phi + (1 - \phi) m'] \frac{V_I}{V_E} \left( C_I^o + \frac{B}{A} \right) \left[ 1 - \exp\left( -\frac{KSA}{V_I} [t - t_o] \right) \right] \tag{20}$$

**Table 3**
**PARAMETERS OF THE MODEL AND FILM THICKNESS**
**CALCULATED FROM EXPERIMENTAL KINETICS**
**VALUES**

| Emulsion | Parameters | Hexobarbital | Amobarbital | Phenobarbital |
|---|---|---|---|---|
| A | m | 0.35 | 0.21 | 0.16 |
| | m′ | 0.35 | 0.21 | 0.16 |
| | $10^7$ D | 7.24 | 7.63 | 8.88 |
| | A | 0.90 | 0.76 | 0.48 |
| | − B | 27.60 | 27.24 | 16.27 |
| | KSA $V_I^{-1}$ | 0.157 | 0.166 | $3.78 \times 10^{-2}$ |
| | δ | 24.6 | 20.2 | 64.9 |
| B | m | 0.37 | 0.20 | 0.087 |
| | m′ | 0.37 | 0.20 | 0.087 |
| | $10^7$ D | 8.29 | 8.47 | 9.87 |
| | A | 0.87 | 0.55 | 0.26 |
| | − B | 25.06 | 17.46 | 8.90 |
| | KSA $V_I^{-1}$ | $6.76 \times 10^{-2}$ | $7.23 \times 10^{-2}$ | $2.06 \times 10^{-2}$ |
| | δ | 61.9 | 37.4 | 74.0 |
| C | m | 0.35 | 0.21 | 0.16 |
| | m′ | 0.20 | 0.11 | 0.087 |
| | $10^7$ D | 7.24 | 7.63 | 8.88 |
| | A | 0.73 | 0.43 | 0.36 |
| | − B | 18.75 | 10.93 | 10.07 |
| | KSA $V_I^{-1}$ | 0.204 | 0.123 | $8.10 \times 10^{-2}$ |
| | δ | 15.4 | 16.0 | 21.4 |
| D | m | 0.37 | 0.20 | 0.087 |
| | m′ | 0.20 | 0.11 | 0.087 |
| | $10^7$ D | 8.29 | 8.47 | 9.87 |
| | A | 0.75 | 0.42 | 0.27 |
| | − B | 18.89 | 11.09 | 9.41 |
| | KSA $V_I^{-1}$ | 0.297 | 0.174 | $6.33 \times 10^{-2}$ |
| | δ | 11.9 | 11.9 | 24.9 |

*Note:* Units are D (cm² · sec⁻¹); B (mg · ℓ⁻¹); KSA $V_I^{-1}$ (h⁻¹); δ (μm).

## IV. INTERPRETATION OF EXPERIMENTAL RESULTS

### A. Kinetics Curves

Experimental results can be represented as a linear plot in accordance with Equation 18. Values of parameters A and B are calculated from measured values of $V_I$, $V_E$, and $\phi$ and from values of m and m′ given in Table 1. Diffusion coefficients in the organic phase are estimated from the Wilke-Chang equation from experimental values of D in water given by Kakemi et al.[14] The thickness of the diffusion film, δ, is deduced from values of D and K.

Table 3 gives the results for the three barbiturates and the four emulsions studied.

Figures 6 to 9 show that the curves $Y = \ln \dfrac{A\, C_I^o + B}{A\, C_I + B}$ as a function of time are straight lines. This validates the model. The slopes of the curves allow the calculation of the transfer coefficient K.

In Table 3, it can be seen that, in most of the cases, the film thickness is about 10 to 30

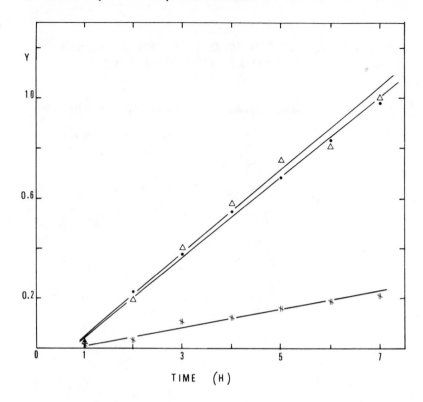

FIGURE 6.   Representation of the kinetics according to Equation 18 for emulsion A. ○: Hexobarbital; △: amobarbital; ≠: phenobarbital.

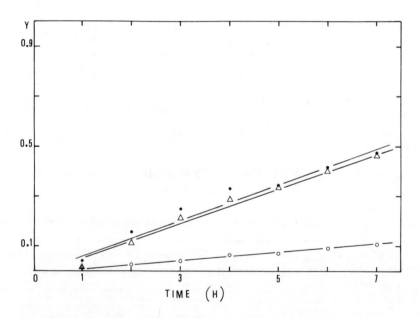

FIGURE 7.   Representation of the kinetics according to Equation 18 for emulsion B. ●: Hexobarbital; △: amobarbital; ○: phenobarbital.

μm, which is in good agreement with values found elsewhere.[4,6] For a given emulsion, the film thickness theoretically should be the same for all the barbiturates. The dispersion of the results could be explained by small variations of the agitator position in the emulsion

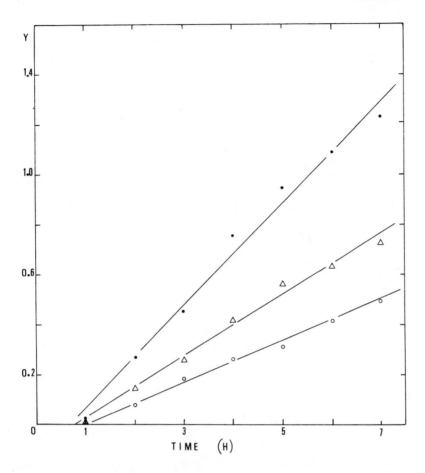

FIGURE 8. Representation of the kinetics according to Equation 18 for emulsion C. ●: Hexobarbital; △: amobarbital; ○: phenobarbital.

($\pm$4 mm) which induce small changes of the hydrodynamics in the emulsion phase. For phenobarbital, however, values of $\delta$ are always greater than those calculated for the other barbiturates. It can be hypothesized that the diffusion coefficients were not well estimated: values of $4.4 \times 10^{-7}$ and $4.9 \times 10^{-7}$ cm$^2 \cdot$ sec$^{-1}$ for D in emulsions A and C and B and D, respectively, would be in better agreement with experimental results.

## B. Rates of Transfer
The experimental rates of transfer are given by the slope of the kinetic curves. These rates can also be calculated from Equations 14 to 17:

$$v_\beta = -\frac{d\,C_I}{dt} = KS\,C_I^o \left[ m\,\frac{\beta}{V_I} + \frac{m'}{\phi + (1 - \phi)m'} \cdot \frac{(\beta - 1)}{V_E} \right] \qquad (21)$$

where

$$\beta = \frac{C_I}{C_I^o} \qquad 0 < \beta \leq 1$$

Table 4 gives experimental and calculated values (Equation 21) of $v_\beta$ for $\beta = 0.9$. These

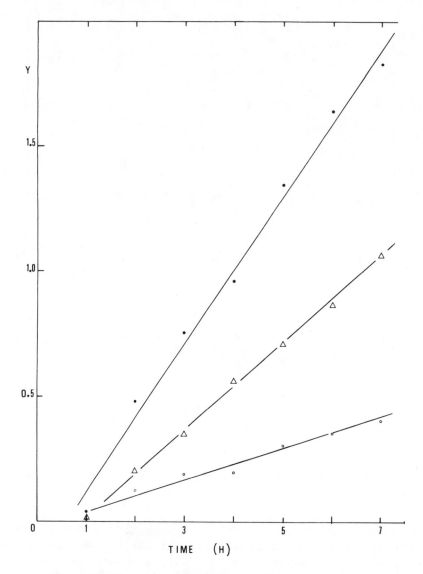

FIGURE 9.   Representation of the kinetics according to Equation 18 for emulsion D. ●: Hexobarbital; △: amobarbital; ○: phenobarbital.

## Table 4
### EXPERIMENTAL AND CALCULATED (*) RATES OF TRANSFER ($mg \cdot \ell^{-1} \cdot hr^{-1}$) FOR $C_1 = 0.9\ C_1^o$

| Emulsion | Rates for $\beta = 0.9$ | Hexobarbital | Amobarbital | Phenobarbital |
|---|---|---|---|---|
| A | $v_{0.9}$ | 2.2 | 1.8 | — |
|   | $v_{0.9}^*$ | 2.25 | 1.53 | — |
| B | $v_{0.9}$ | 1.4 | 0.8 | — |
|   | $v_{0.9}^*$ | 1.10 | 0.96 | — |
| C | $v_{0.9}$ | 4.0 | 2.4 | 1.2 |
|   | $v_{0.9}^*$ | 3.94 | 2.41 | 1.38 |
| D | $v_{0.9}$ | 5.5 | 3.4 | 0.6 |
|   | $v_{0.9}^*$ | 5.88 | 3.24 | 0.64 |

values are in good agreement. Small differences can be observed because of uncertainties in the graphical determination of the slopes, and because the value of K used to calculate $v_\beta$ is a mean value (given in Table 3) and not a local one.

The influence of the emulsion composition (characterized by parameters $\phi$, m, and m') on the rate of transfer can be seen from Equation 21.

The variations of $v_\beta$ with m' (m and $\phi$ being kept constant) are given by Equation 22:

$$\frac{d\ v_\beta}{dm'}\bigg|_{m,\phi} = \frac{(\beta - 1)}{V_E} \frac{\phi}{[\phi + (1 - \phi)m']^2}\ SK\ C_i^o \tag{22}$$

Because $\beta < 1$, this expression is always negative, i.e., $v_\beta$ increases as m' decreases. This traduces the "trapping effect" of the dispersed phase in the emulsion which can be seen by comparing results for emulsions A and C, on one hand, and B and D, on the other hand, for each barbiturate in Table 4.

$v_\beta$ is an increasing function of m (m' and $\phi$ being kept constant) as shown by Equation 23:

$$\frac{d\ v_\beta}{dm}\bigg|_{m',\phi} = \beta\ \frac{SK\ C_i^o}{V_I} \tag{23}$$

This explains the relative efficiencies of emulsions C and D for the three barbiturates. When m and m' are equal, the variations of $v_\beta$ with m (for $\phi$ constant) are given by Equation 24:

$$\frac{d\ v_\beta}{dm}\bigg|_\phi = \frac{SK\ C_i^o}{V_I}\left[\frac{(\beta - 1)\ V_I\ \phi + \beta\ V_E\ [\phi + (1 - \phi)m]^2}{V_E\ [\phi + (1 - \phi)\ m]}\right] \tag{24}$$

Because $\dfrac{C_i^{eq}}{C_i^o} \leq \beta \leq 1$, it can be shown that this expression is always positive and that $v_\beta$ is an increasing function of m.

Values of $v_\beta$ in Table 4 for emulsion A and for hexobarbital (m = m' = 0.35) and amobarbital (m = m' = 0.21) agree with that result. The increase of $v_\beta$ with m is also observed for emulsion B.

If, now, the nature of the two phases of the emulsion is fixed (m' and m fixed), the extraction properties of the emulsion can be modified by varying $\phi$. Equation 25 traduces an intuitive result.

$$\frac{d\ v_\beta}{d\phi}\bigg|_{m,m'} = SK\ C_i^o\ m'\ \frac{(\beta - 1)}{V_E}\ \frac{(m' - 1)}{[m' + (1 - m')\phi]^2} \tag{25}$$

For m' < 1 (trapping emulsion or hydrophilic barbiturate) a high volume fraction of aqueous dispersed phase in the emulsion is needed for a better extraction. For m' > 1 (lipophilic barbiturate and nontrapping emulsion) the presence of water in the extraction phase is unfavorable. These conclusions are confirmed by the analysis of $C_i^{eq}$ variations with $\phi$ (m and m' constant). It must be noted that we assumed constant K, while m, m', and $\phi$ are varied throughout this discussion.

## C. Equilibrium Concentrations

Experimental ($C_i^{eq}$) and calculated values from Equation 19 ($C_i^{eq*}$) are given in Table 5. Because of the errors in partition coefficients (which sometimes are as high as 37%), we

## Table 5
## EXPERIMENTAL AND CALCULATED (*) BARBITURATES EQUILIBRIUM CONCENTRATIONS IN THE AQUEOUS SOLUTION AFTER TRANSFER

| Emulsion | Equilibrium conc | Hexobarbital | Amobarbital | Phenobarbital |
|---|---|---|---|---|
| A | $C_i^{eq}$ | 31.1 | 36.3 | 44.1 |
|   | $C_i^{eq*}$ | 37.9 | 35.8 | 33.9 |
| B | $C_i^{eq}$ | 37.9 | 39.0 | 47.0 |
|   | $C_i^{eq*}$ | 28.8 | 31.7 | 34.2 |
| C | $C_i^{eq}$ | 24.1 | 33.4 | 36.1 |
|   | $C_i^{eq*}$ | 25.7 | 25.4 | 28.0 |
| D | $C_i^{eq}$ | 18.4 | 31.0 | 40.9 |
|   | $C_i^{eq*}$ | 25.2 | 26.4 | 34.9 |
| $\bar{\sigma} = \frac{1}{4} \sum \frac{\left| C_i^{eq} - C_i^{eq*} \right|}{C_i^{eq}}$ | | $16.4 \times 10^{-2}$ | $18.5 \times 10^{-2}$ | $28.4 \times 10^{-2}$ |

can conclude that experimental values are generally in good agreement with calculated values. For phenobarbital, however, they are significantly higher. It is possible that this slowly diffusing species is still not at its equilibrium concentration after 23 hr of contact.

As for the analysis of the rates of transfer, it can be seen that the influence of the variations of m and m' follows the theoretical predictions of Equation 19, i.e., the trapping effect decreases the equilibrium concentration; the increasing m decreases $C_i^{eq}$ (when m = m'), as can be seen for emulsions A and B.

## V. CONCLUSIONS

Partition coefficient measurements have shown the influence of nonionic surfactants on the distribution of three barbiturates between water and a paraffin oil.

Barbiturates can diffuse through the organic phase of an emulsion (liquid membrane). The possibility of trapping barbiturates in a high pH aqueous solution dispersed in the emulsion has been demonstrated.

Kinetics of transfer have been interpreted using a model assuming that the transfer is controlled by diffusion through a thin film of organic phase. The validity of this model should be confirmed by other experiments on other barbiturates or other organic phases (to give values of the partition coefficient over a larger range). Influence of the fraction of dispersed solution in the emulsion should be tested, as well as the influence of the initial concentration of barbiturates in the aqueous solution. However, the good representation of the experimental results by this simple model seems to confirm that the rate of transfer is, indeed, controlled by the diffusion in a thin film.

## NOMENCLATURE

A  Constant defined by Equation 16
B  Constant defined by Equation 17
C  Concentration of barbiturates (mol·dm$^{-3}$)
D  Diffusion coefficient (m$^2$·sec$^{-1}$)

K   Transfer coefficient defined by Equation 15 ($m \cdot sec^{-1}$)
m   Partition coefficient between aqueous solution I and the organic phase
m′  Partition coefficient between organic and aqueous phase in the emulsion
$m_1$  Pseudo-partition coefficient defined by Equation 8
N   Diffusion flux ($mol \cdot m^{-2} \cdot sec^{-1}$)
$Q_1$  Weight of aqueous phase I (g)
$\bar{Q}$  Weight of organic phase.(g)
S   Interfacial area ($m^2$)
V   Volume ($m^3$)
v   Rate of transport ($mol \cdot cm^{-3} \cdot sec^{-1}$)
Y   Function of barbiturates concentration defined by Equation 18

## Greek Symbols

β   Ratio of the concentration at time t to the initial concentration
δ   Thickness of the organic film in the emulsion (m)
φ   Volumetric fraction of water in the emulsion

## Superscripts

eq  At the equilibrium
o   Initial value

## Subscripts

I    Aqueous solution of barbiturates
II   Organic layer in the emulsion
III  Core of the emulsion
E    Emulsion
i    Inner dispersed phase of the emulsion

# REFERENCES

1. **Li, N. N.,** Separating Hydrocarbons with Liquid Membranes U.S. Patent 3,410,794, 1968.
2. **Li, N. N.,** Permeation through liquid surfactant membranes, *AIChE J.,* 17(2), 459, 1971.
3. **Li, N. N.,** Separation of hydrocarbons by liquid membrane permeation, *Ind. Eng. Chem. Process. Des. Dev.,* 10(2), 215, 1971.
4. **Trouvé, G.,** Separation by Mass Transfer through Emulsified Liquid Membrane. Application to the Treatment of Kidney Failure, thesis, Ecole des Mines, Paris, 1982.
5. **Way, J. D., Noble, R. D., and Flynn, T. M.,** Liquid membranes: a survey, *J. Membr. Sci.,* 12(2), 239, 1982.
6. **Trouvé, G., Mahler, E., Colinart, P., and Renon, H.,** Liquid membranes: a moving film apparatus for measurement of mass transfer in emulsions and its application to active transport of phosphate ions, *Chem. Eng. Sci.,* 37(8), 1225, 1982.
7. **Trouvé, G., Jore, D., Duranel, C., and Renon, H.,** Detoxification of biological liquids by mass transfer through emulsions, *Inn. Tech. Biol. Med.,* 3(6), 635, 1982.
8. **Cahn, R. P. and Li, N. N.,** Separation of phenol from waste water by the liquid membrane technique, *Sep. Sci.,* 9(6), 505, 1974.
9. **Yang, T. T. and Rhodes, C. T.,** Transport across liquid membranes: effect of formulation variables, *J. Appl. Biochem.,* 2, 17, 1980.
10. **Chilamkurti, R. N. and Rhodes, C. T.,** Transport across liquid membranes: effect of molecular structure, *J. Appl. Biochem.,* 2, 17, 1980.
11. **Otagari, M., Miyaji, T., Vekama, K., and Ikeda, K.,** Inclusion complexation of barbiturates with β-cyclodextrin in aqueous solution, *Chem. Pharm. Bull.,* 24, 1146, 1976.

12. **Khall, S. A.,** The use of the solubility parameters as an index of drug activity, *Can. J. Pharm. Sci.,* 11(4), 121, 1976.
13. **Doyle, T. D. and Proctor, J.,** Drugs: effect of solvent composition on the partition of barbiturates, *J. Assoc. Off. Anal. Chem.,* 56(4), 864, 1973.
14. **Kakemi, K., Takaichi, A., Ryohei, H., and Ryoji, K.,** Absorption and excretion of drugs: absorption of barbituric acid derivatives from rat stomac, *Chem. Pharm. Bull.,* 15(10), 1534, 1967.
15. **Koprivc, L., Hranilovic, J., and Zlata, S.,** Transfer of some barbituric acid derivatives across liquid barrier as model membrane, *Acta Pharm. Jugosl.,* 27, 147, 1977.
16. **Treiner, C.,** *Galenica: Agents de Surface et Émulsions,* Puisieux, F. and Seiller, M., Eds., Technique et Documentation Lavoisier, Paris, 1983, chap. 6.
17. **Garret, R. E., Bojarski, J. T., and Yakatan, G. T.,** Kinetics of hydrolysis of barbituric acids derivatives, *J. Pharm. Sci.,* 60(8), 1145, 1971.

Chapter 12

# GAS SEPARATION DESIGN WITH MEMBRANES

**D. L. MacLean, W. A. Bollinger, D. E. King, and R. S. Narayan**

## TABLE OF CONTENTS

# I. INTRODUCTION

Membranes are as old as life itself. Many of the body functions include separations using membranes. Industrial applications originally centered around liquid-liquid or liquid-solid separations such as reverse osmosis, ultrafiltration, and microfiltration. An interesting personalized history of membrane technology has recently been published by Lonsdale[1] in which he traces the development from pre-1950 to the present. The first successful gas separation system was marketed in 1979 by Monsanto after 3 years of successful commercial demonstration within the company. The membrane is in the form of a hollow fiber. Recently, several other companies, such as Separex and Envirogenics, have begun marketing spiral wound separators using flat films of the cellulose acetate family. The current applications of membrane separators are in the recovery or removal of hydrogen, helium, or carbon dioxide from methane, heavier hydrocarbons, carbon monoxide, and nitrogen. Gas separation with synthetic membranes has been reviewed by Matson et al.,[2] while MacLean et al.[3] have discussed fundamentals of gas permeation. Several of the key steps in the development, design, and use of membrane separators for gas separation are discussed below.

# II. SELECTION OF THE POLYMER

The first step in the development of a separation system is the selection of the polymer for the membrane. One wants a polymer which has high transport rates for the gases to be recovered or, better yet, high rates of transport for the gases to be rejected at low pressures, and low rates for the gases to be recovered at high pressure. In either case, a reasonable ratio of fast gas rates to slower gas rates (separation factor) is required. The rates and ratio of rates are functions of the particular process stream to be separated, and thus, process design requires some field test data on similar streams. The polymer chosen should have a high glass transition temperature such as the 190°C $T_G$ for the polysulfone polymer used in Prism® separators. This provides good temperature and chemical resistance. High modulus and high collapse pressures are required for applications at high differential pressures. The membrane should also be easily fabricated and resistant to common contaminants such as water and alkanes.

# III. MECHANISM OF TRANSPORT

In order to understand membrane processes, some analysis of the fundamental mechanism of transport is beneficial. The basic theory involves solution of the gas in the membrane followed by diffusion across a thickness $\ell$. The flow of component i across the membrane at any given point is

$$q_i = \frac{P_i}{\ell} A \, \Delta p_i \tag{1}$$

and the separation factor is

$$\alpha_{ij} = P_i/P_j \tag{2}$$

where $P_i$ is the permeability of component i, $\Delta p_i$ is the partial pressure differential for i, and A is the effective area available for transport.

There are at least two fundamental theories for the actual solution-diffusion mechanism. One of the most widely held views of gas transport is the dual-mode theory. According to this theory there is a linear sorption isotherm, similar to Henry's law, coupled with sorption in microvoids created in the glassy polymer during formation in quenching or coagulation. This is followed by molecular diffusion across the membrane according to Fick's law. The basic equation for the sorbed concentration for a single component is

$$C = k_D p + \frac{C'_H bp}{1 + bp} \tag{3}$$

and for the permeability of the gas

$$P = k_D D_D + \frac{D_H C'_H b}{1 + bp} \tag{4}$$

where $k_D$ is a Henry's law coefficient, $D_D$ and $D_H$ are diffusion coefficients, and $C'_H$ and b are parameters.[4] Analogous equations can be obtained for a blend of gases.

Another molecular theory, the matrix model,[5] leads to similar results except where there is strong interaction between the vapor and polymer. The microvoid mechanism has been replaced with an increased diffusion coefficient due to increased chain mobility from the absorbed gases. The permeability coefficient is given by:

$$P = \frac{D_o C'}{p} \exp(\beta\kappa) \tag{5}$$

Where $D_o$ is the diffusion coefficient with no vapor/polymer interaction, $\beta$ is a constant, and $\kappa$ (C) is a variable related to the free energy of mixing or the change in glass transition with concentration.

## IV. SEPARATOR DESIGN

In order to fully utilize the intrinsic transport properties of the membrane, the separator should have ideal flow patterns and low pressure drops. The engineer can choose from several designs. Two of the separator designs are the hollow fiber separator (see Figure 1) and the flat film, spirally wound separators. In the hollow fiber separator the shell flow is parallel to the fibers. The internal bore flow is countercurrent to the shell flow. In the spiral wound separator a complex multidirection flow pattern is obtained. A hollow fiber separator design was developed for Prism® separators.

To aid in the design of the separator and the prediction of performance, a mathematical analysis of the system was developed for Prism® separators. For countercurrent plug flow with perfect radial mixing and constant permeabilities, a system of nonlinear ordinary differential equations has been developed and solved numerically. For countercurrent flow this involves split boundary conditions. Accurate mathematical models for spirally wound separators are extremely difficult to develop because of the complex multidirectional flow patterns (see Figure 2). For hollow fiber separators, various flow pattern models can be developed and compared to experimental data. Some of these are the plug flow model, the perfect backmixing model, the dispersion model, the mixing cell model, and the by-passing model. Under design operating conditions, the plug flow model is quite accurate.

In order to visualize the flow patterns, some of the earliest separator development work in Monsanto was done with separators for liquid separations where shell-side flow patterns were investigated by both dye injection and by salt injection with electrical conductivity measurement. One system investigated was a reactive dialysis separation where one species would pass through a polymer membrane and react with a sweep fluid on the bore side.

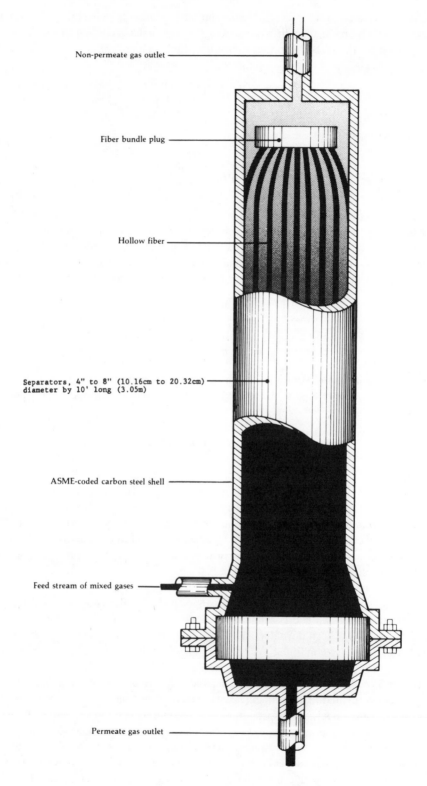

Non-permeate gas outlet

Fiber bundle plug

Hollow fiber

Separators, 4" to 8" (10.16cm to 20.32cm)
diameter by 10' long (3.05m)

ASME-coded carbon steel shell

Feed stream of mixed gases

Permeate gas outlet

FIGURE 1.    Prism® separator.

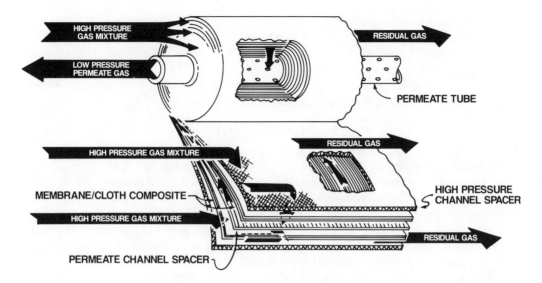

FIGURE 2.    Spiral-wound element construction.[6]

The bore side was open on both ends and the sweep fluid was pumped down the inside bore of the fiber. This system was simple to analyze analytically. For plug flow:

$$P/\ell = \frac{Q_F \ln C_i/C_o}{A} \tag{6}$$

while for a dispersion model

$$C_i/C_o = \frac{4 \, b \, \exp\left(\frac{1}{2}\frac{vL}{D}\right)}{(1 + b)^2 \exp\left(\frac{bvL}{2D}\right) - (1 - b)^2 \exp\left(-\frac{bvL}{2D}\right)} \tag{7}$$

where $b = 1 + 4 \, (P/\ell) \dfrac{A}{Q_F} (D/vL)$, $C_i$ = the inlet concentration, $C_o$ = outlet concentration, $v$ = velocity, and $D$ = dispersion coefficient. The dispersion model is analogous to the Wehner and Wilhelm model for chemical reactors and allows calculation of dispersion coefficients. Considerable understanding was gained for this system which translated to the gas permeation separator design.

Another useful analysis is the effect of boundary layers on the mass transfer and, ultimately, the efficiency of the separator. For thin membranes it was derived that

$$\frac{1}{Eff_{H_2}} = \frac{1}{k_G} (P/\ell)_{H_2} \, p_s + 1 \tag{8}$$

where the

$$H_2 \text{ efficiency } (EFF_{H_2}) = \frac{(P/\ell)_{H_2} \text{ actual}}{(P/\ell)_{H_2} \text{ pure gas}} \tag{9}$$

$k_G$ is the mass transfer coefficient of hydrogen, and $p_S$ is the shell pressure. This assumes efficiency loss is only a result of boundary layer effects.

Under normal conditions, boundary layer efficiency loss is not a factor. This situation can be compared to heat exchangers where boundary layers are important, since fluid side heat transfer is much slower than conduction through the metal surface. In contrast, diffusion in polymers is usually slower than gas phase diffusion.

## V. PILOT PLANTS

Commercial separation systems are typically shop fabricated and skid mounted, thus minimizing project time, field installation time and cost, and space requirements. In order to develop the complete system, including any required pretreatment, portable pilot plant systems ("miniskids") were developed. These have been operated within Monsanto with approximately 20 applications outside Monsanto. The basic miniskid is shown in Figure 3. This contains two small-scale separators (for which the scale-up factors for commercial size separators are well known) and various pressure, flow, and temperature measurement instruments and controls. The separators can be operated either in parallel or series, at two different feed or product pressures, and at two different feed temperatures. A variety of pressures up to 2050 psi (14,135 kPa) and temperatures up to 90°C have been utilized. Various pretreatment steps are available if required.

## VI. APPLICATION DEVELOPMENT

Within Monsanto, demonstration tests have been conducted to develop the oxo-syn gas ratio adjustment application plus hydrogenation, ammonia, and methanol purge recovery applications. These applications have now been successfully commercialized and proven.

The oxo-syn gas application (Figure 4) is a ratio adjustment of $H_2/CO$. Prism® separators are able to produce a variety of $H_2/CO$ ratios by varying operating parameters. This, coupled with the low shell-side ($H_2/CO$ stream) pressure drop, makes the system quite versatile and economically attractive.

Hydrogenation purges, such as the one at Monsanto's plant in Decatur, Ala., are difficult to treat with conventional technology because of extreme fluctuations in feed rates (a fivefold variation in rate is not uncommon). Hollow fiber separators can handle these effectively, yet simply. One method is to control the feed pressure or temperature. Thus, by changing either the driving force for the separation ($\Delta P$) or the intrinsic permeability properties of the separators, hydrogen recovery or purity can be controlled. Alternatively, the recovery and purity may be allowed to vary.

The ammonia purge gas recovery system was developed at the Monsanto ammonia plants in Luling, La. Figure 5 shows a two-stage pressure operation which maximizes recovery while minimizing recompression costs. In some cases, recompression costs are small, and a single-stage system returning the hydrogen to one of the available suction pressures of the syn-gas compressor may be used. Systems can now be designed with hydrogen recoveries as high as 95%. Also, differential pressures up to 1650 psi (11,377 kPa) have now been demonstrated. Figure 6 shows an installed skid-mounted system for this nominal 600 TPD plant. In Figure 6 the large vessel is an ammonia absorber (water scrubber). This serves a dual purpose of protecting the fibers and recovering the ammonia. This water scrubber, coupled with an ammonia distillation system already at the plant or supplied by Monsanto, pays for itself in product ammonia. For example, with 5-TPD recovered ammonia at $200/T, this is $329,000 for 330 day/year, minus minimal operating costs. This savings is in addition to the principal savings resulting from recovered purge hydrogen. Monsanto has designed and delivered numerous systems of Prism® separators to other operators of nominal 1000-TPD ammonia plants. The systems are quite similar to the one above in concept.

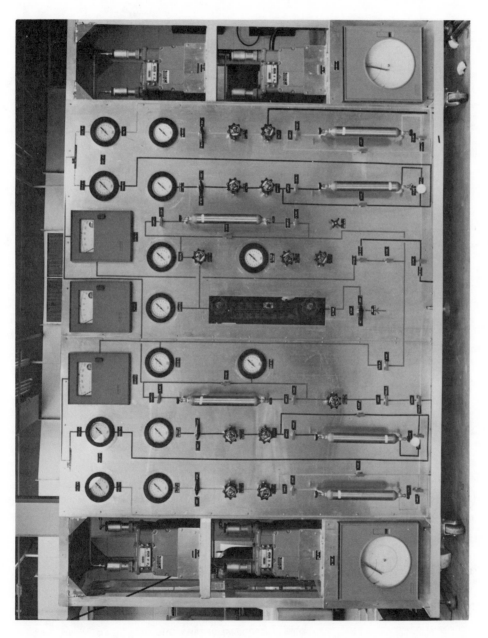

FIGURE 3.    Prism® separator demonstration unit.

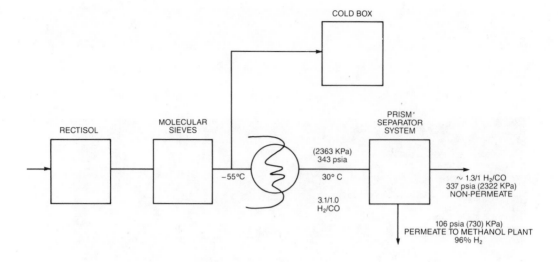

## PROCESS FLOW DIAGRAM FOR H₂/CO ADJUSTMENT

FIGURE 4.   Oxo-syn gas application.

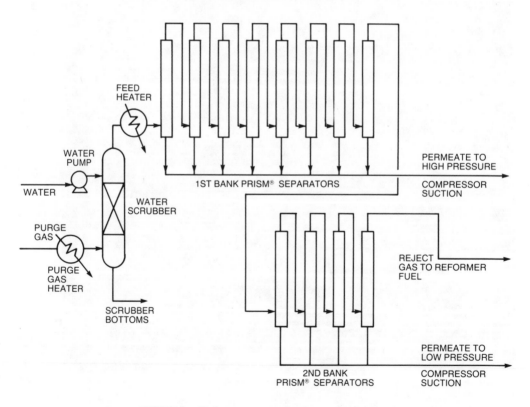

FIGURE 5.   Hydrogen recovery from ammonia plant purge.

Our methanol purge recovery application (see Figure 7) also uses a water scrubber for a similar purpose, and it, too, pays for itself in recovered methanol. The methanol/water mixture is simply sent to the existing crude methanol distillation column. Prism® separators

FIGURE 6. Prism® separator system — 600 TPD ammonia plant (Luling, La.).

have operated on stoichiometric as well as nonstoichiometric $H_2/CO_x$ ratio methanol plants at 750 (5171 kPa) and 1500 (10,343 kPa) psi at differential pressures up to 1000 psi (6895 kPa).

Outside Monsanto, in addition to ammonia and methanol tests, application development has been accomplished on numerous refinery processes (see Table 1). Commercial systems have also been developed for $CO_2$ removal from casing head gas for enhanced oil recovery and from landfill gas.

Many of the refinery hydrogen recovery systems, as illustrated by the ARCO Petroleum Products Company commercial unit in Figure 8, are characterized by their ease of operation and minimal pretreatment. The ARCO commercial unit, operating on a naphtha hydrotreater

FIGURE 7.    Methanol purge application.

unit for about 3 years, uses only temperature control and liquid knockout as pretreatment. The hydrotreater application is shown in Figure 9. Hydrotreater process gas, typically around 600 psi (4137 kPa), can be sent directly to the Prism® separator system and the hydrogen product, usually 250 psi (1724 kPa) and >95 mol % $H_2$, can be recycled to the hydrotreater or sent to other hydroprocesses, such as a hydrocracker.

The flexibility of membrane separations is evident by the wide range of operating conditions that can be handled. Operating conditions can vary from 2000 psi feed pressure,

## Table 1
## PRISM® SEPARATOR REFINERY PILOT TESTS

| Location | Refinery process application | Pressure (psig) | | H₂ purity (mol %) | | Operating dates |
|---|---|---|---|---|---|---|
| | | Feed | Product | Feed | Product | |
| Texas | Heavy gas oil HDS | 610 | 50/150 | 60 | 95 | 4—7/79 |
| Oklahoma | Naphtha HDS | 480 | 150 | 80 | 97 | 7—12/79 |
| California | Naphtha HDS | 600 | 250 | 85 | 98 | 8—12/79 |
| Canada | Diesel HT | 530 | 170 | 70 | 95 | 8—10/80 |
| Louisiana | Lt. cycle oil HT | 700 | 100 | 61 | 94 | 6—8/80 |
| Texas | FCCU | 850 | 100 | 12 | 60 | 3—5/80 |
| Louisiana | FCCU | 680 | 85 | 26 | 84 | 8/80—1/81 |
| | FCCU | 650 | 50 | 17 | 73 | 10—11/80 |
| Texas | THDA | 430 | 50 | 51 | 91 | 4—5/80 |
| | CRU | 575 | 175 | 80 | 97 | 10/80—6/81 |
| | THDA | 400 | 100 | 75 | 95 | 1—9/81 |
| Louisiana | Gas oil HT | 950 | 600 | 88 | 98 | 5—7/81 |
| Texas | HCU | 1050 | 250 | 65 | 94 | 5—11/81 |
| | Lt. gas oil HT | 850 | 250 | 68 | 92 | 7—9/81 |
| Mississippi | CRU | 720 | 220 | 79 | 98 | 6—12/81 |
| Texas | THDA | 450 | 50 | 70 | 96 | 5—6/81 |
| | HCU | 1520/1200 | 880/400 | 67 | 94 | 9—12/81 |
| California | HCU | 50 | 50 | 50 | 83 | 5—8/82 |
| | VGO HT | 860 | 520/260 | 89 | 98 | 10/82—1/83 |

1000 psi differential pressure, and 80 mol % $H_2$ feed for high pressure units, such as hydrocrackers, to low pressure applications with 250 psi feed pressure, 200 psi differential pressure, and <50 mol % $H_2$ feed.

Operation at high temperatures, 150 to 180°F (65 to 82°C), allows membrane systems to handle high levels of aromatics which are common in toluene hydrodealkylation and catalytic reformer applications. High operating temperature is also used for dewpoint control and to maximize the recovery rate of hydrogen.

Stripping hydrogen from fuel or FCCU gas is a good example of membranes producing a synthetic natural gas (SNG) stream. SNG can be used directly as a high-BTU fuel or as feed for LPG recovery. These streams are characterized by a wide range of components, including olefins and aromatics, and low purity hydrogen (10 to 30%). These low pressure streams, 100 to 200 psi (689 to 1379 kPa), require compression to 500 to 600 psi (3450 to 4137 kPa), but can recover over 80% of the hydrogen. The hydrocarbons are concentrated and are recovered at near-feed pressure.

The performance of Prism® separators for a $CO_2$ separation application is governed by the following system parameters:

1. Feed gas composition, operating pressure, and temperature
2. Nonpermeate $CO_2$ content requirement
3. Permeate $CO_2$ purity requirement
4. Hydrocarbon recovery requirement
5. $CO_2$ recovery requirement

For a given set of system conditions (e.g., composition, flow rate, pressure, and temperature), there exists a definite relationship between the amount of $CO_2$ recovered and the purity of the recovered $CO_2$ depending upon the processing scheme. Figure 10 illustrates a

FIGURE 8.   ARCO commercial unit.

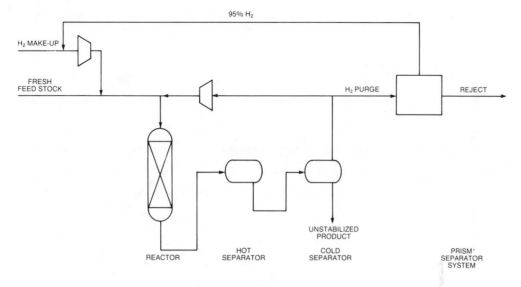

FIGURE 9.   Hydrotreater application.

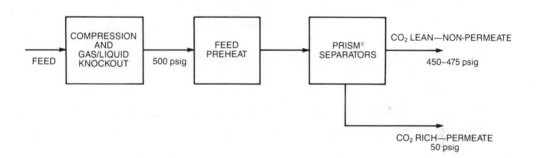

FIGURE 10.   Block flow diagram for single-stage system.

simplified block flow diagram for a single-stage system. Figure 11 shows the relationship between $CO_2$ purity and the amount of $CO_2$ recovered for a feed gas containing 80% $CO_2$. Figure 12 shows the relationship between $CO_2$ purity and the percent $CO_2$ remaining in the nonpermeate.

As can be seen, a significant reduction in $CO_2$ levels can be accomplished while simultaneously recovering high purity $CO_2$. Based on the system parameters described above, a $CO_2$ purity of 95% can be achieved in most cases. In this respect, the single-stage membrane gas separator system is recommended as a topping or bulk $CO_2$ removal process. However, at higher operating pressures (above 500 psi) the single-stage system is capable of removing most of the $CO_2$ present in the hydrocarbon stream to produce pipeline quality gas.

The $CO_2$ in the nonpermeate hydrocarbon stream can always be reduced to 5% or less and the recovered $CO_2$ can be maintained at purities of 95% if a two-stage system is employed. This is illustrated in Figure 13. As can be seen, additional compression is required. Such a two-stage scheme can essentially accomplish the following:

1.   Reduce $CO_2$ in the hydrocarbon containing nonpermeate gas (residue) to pipeline quality
2.   Recover $CO_2$ at 95% or higher purity
3.   Achieve hydrocarbon recoveries of 90% or higher

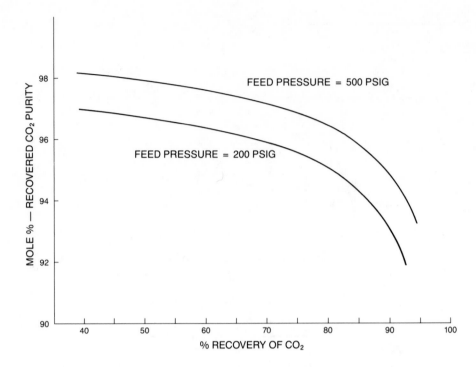

FIGURE 11.   Relationship of recovered $CO_2$ purity vs. % recovery of $CO_2$.

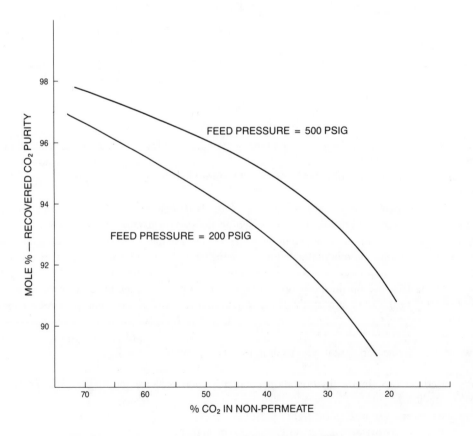

FIGURE 12.   Relationship of recovered $CO_2$ purity vs. % $CO_2$ in nonpermeate.

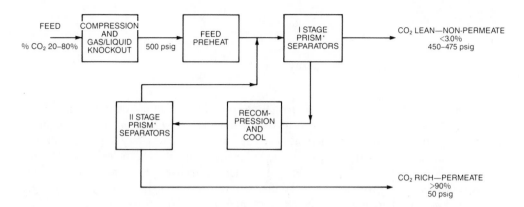

FIGURE 13.   Block flow diagram for two-stage system.

In this processing scheme, therefore, the membrane gas separator would essentially replace a conventional $CO_2$ removal process. Further processing of the gas residue to meet pipeline sulfur requirements of typically 4 ppm may be necessary. This, of course, would depend on the sulfur content of the feed gas.

A processing scheme, such as illustrated in Figure 13, can be practically employed for feed gas with a wide range of $CO_2$ levels. The additional capital and incremental energy requirements are directly related to the five parameters previously mentioned.

Membrane gas separation by Prism® separators offers a low cost processing alternative to conventional $CO_2$ removal processes. Once the needed partial pressure driving force is achieved by compressing the feed gas, no further expenditure of energy is necessary to separate the $CO_2$. This is especially true when the process is used for topping or bulk removal of $CO_2$. Goddin[7] has presented an independent detailed cost analysis for recovering $CO_2$ from casing head gas where bulk removal is effected by membrane separation followed by a conventional gas treating process. Goddin's analysis shows that for feed gas containing more than 3% $CO_2$ and higher, the process heat requirements for a membrane separation process are approximately 30 to 40% lower than for a cryogenic process. When compared to an acid gas removal process using DEA as a solvent, the membrane separation process requires only 20 to 25% of the process heat.

## VII. PROCESS CONTROL

Membrane separators are easily controlled, which is particularly attractive to the manufacturing engineer who wants minimum disruption to his plant and minimum operator attention. Possible manipulated parameters are feed rate, feed pressure, product pressure, membrane area, and/or feed temperature. Controlled variables include purities (or purities above a certain minimum value) and recoveries (or recoveries within a certain range). Designs have been implemented where different hydrogen product rates are furnished on demand by automatically closing the bore flow from certain separators upon receiving a signal from the downstream process.

## VIII. PROCESS DESIGN

One simple overview of process design using hollow fiber separators can be obtained from the following equations:

$$\text{recovery of } i = \frac{(P/\ell)_i \, A \, \Delta p_i}{Q_{Fi}} \tag{10}$$

where $Q_{Fi}$ is the feed flow rate of i, and

$$\text{product purity of i} = \frac{\alpha \, \Delta p_i / \Delta p_j}{\alpha \, \Delta p_i / \Delta p_j + 1} \tag{11}$$

for two components where $\Delta p_i$ is the partial pressure differential of component i.

These are idealized cases for zero recovery, but they show the impact of the major parameters and are good approximations at low recoveries. The recovery (e.g., of hydrogen or carbon dioxide) is directly proportional to the permeability rate coefficient ($P/\ell$, i.e., the higher the permeability rate coefficient, the greater the recovery for a given area, or the fewer separators for a given recovery. To maintain recovery at higher feed rates, more area is required or lower bore pressures are required (higher $\Delta p_i$). Higher feed pressures and feed purities require less area for a given recovery, or allow more recovery for a given area. The product purity of $H_2$ is related to the separation factor and the ratio of partial pressure driving forces. Another way to write that equation is to solve for the mole fraction of hydrogen in the product:

$$X_i = \frac{-B + \sqrt{B^2 + 4(1 - \alpha)\alpha \frac{p_S}{p_B} Y_i}}{2(1 - \alpha)} \tag{12}$$

where $B = (\alpha - 1) \frac{p_S}{p_B} Y_i + \frac{p_S}{p_B} + \alpha - 1$, $p_S$ = shell total pressure, $p_B$ = bore total pressure, and $Y_i$ = shell or feed composition (assumed nearly equal for low recovery).

Figures 14 and 15 show how, for a given feed composition, a certain $\alpha$ and $p_S/p_B$ are required for a given purity. For example, with a separation factor of 40, product purities of 96.8 to 98.5% would be predicted for pressure ratios of 2 to 4 with 70% feed. With a 40% feed and a pressure ratio of 4, a 92% product purity is obtained for the same separation factor of 40. Not shown are techniques such as cascading separators, such that either the shell exit gas or bore product gas is sent to another separator system operating at a different $p_S/p_B$ to improve purity and/or recovery. Other techniques, such as recycling the shell exit gas from the second stage back to the feed gas, can be used to increase purity and recovery.

Exact purities and recoveries differ from the simple calculation because purity decreases at high recoveries and many options using staging or recycle are available to increase purity, if so desired.

## IX. SUMMARY

In conclusion, we have presented an analysis of some of the key parameters in the commercially demonstrated technology of gas separation with membranes. The analysis began with a selective and chemically resistant polymer, continued to a separator with nearly ideal flow patterns and low pressure drops, and finished with a high recovery, easily controlled process for recovering gases such as hydrogen or carbon dioxide from such diverse processes as ammonia, methanol, hydrotreaters, hydrocrackers, enhanced oil recovery, as well as natural gas streams.

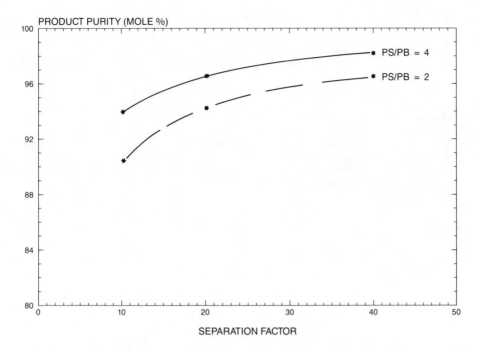

FIGURE 14.   Product purity vs. separation factor at different shell/bore pressures and at 70% H$_2$ feed composition.

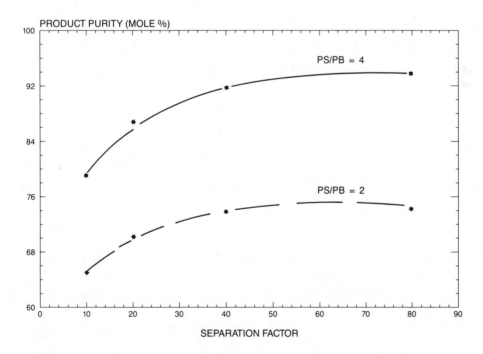

FIGURE 15.   Product purity vs. separation factor at different shell/bore pressures and at 40% H$_2$ feed composition.

# REFERENCES

1. **Lonsdale, H. K.,** The growth of membrane technology, *J. Membr. Sci.,* 10, 81, 1981.
2. **Matson, S. L., Lopez, J., and Quinn, J. A.,** Separation of gases with synthetic membranes, *Chem. Eng. Sci.,* 38, 503, 1983.
3. **MacLean, D. L., Stookey, D. J., and Metzger, T. R.,** Fundamentals of gas permeation, *Hydrocarbon Process.,* 8, 47, 1983.
4. **Paul, D. R. and Koros, W. J.,** *J. Polym. Sci. Polym. Phys. Ed.,* 14, 675, 1976.
5. **Sefcik, M. O. and Rauchel, D.,** Polymer preprints, *Am. Chem. Soc.,* preprint.
6. **Schell, W. J. and Hoernschemeyer, D. L.,** Principles of Membrane Gas Separation, paper presented at AIChE Symp., Anaheim, Calif., 1982.
7. **Goddin, C. S.,** Comparison of Processes for Treating Gases with High $CO_2$ Content, paper presented at the 61st Annu. GPA Convention, Dallas, Tex., 1982.

Chapter 13

# A COMPARATIVE ANALYSIS OF THE ROLE OF RECYCLE OR REFLUX IN PERMEATORS SEPARATING A BINARY GAS MIXTURE

**Steven Teslik and K. K. Sirkar**

## TABLE OF CONTENTS

# I. INTRODUCTION

Separation of gas mixtures by selective permeation through a nonporous polymeric membrane is of increasing importance due to the recent commercial introduction of a number of processes. Low fluxes and/or poor separation factors have been limiting their wider commercial adoption. The traditional solution for a low separation factor, i.e., multistaging, is not particularly welcome for membrane separation processes. Hwang and Thorman[1] and Hwang et al.[2] have therefore recently proposed the continuous membrane column configuration which uses reflux at the enricher top. An alternative configuration would recycle a part of the enriched permeated gas stream to the feed stream to increase the enrichment level of the permeated stream (Kimura and Walmet,[3] Teslik and Sirkar[4]). Consequently, comparative evaluation of the performance of the two different permeation gas enrichment schemes is necessary for selecting the optimum separation scheme.

Teslik and Sirkar[4] have analyzed the partial light fraction recycle permeator under the condition of a zero pressure ratio. They have used a flat plate permeator configuration with no axial pressure drop on either side of the membrane. They have also compared the performance of such a system with that of the flat plate membrane enricher of a continuous membrane column obtained by Hwang et al.[2] for zero pressure ratio and an optimized compressor load. The binary gas mixture chosen was air (i.e., $O_2$-$N_2$) and silicone membrane material was selected. The results of Teslik and Sirkar[4] show that for any permeate $O_2$ mole fraction up to 0.44, the recycle permeator has a higher productivity and requires a lower compressor load per unit net permeated product than those for the optimized enricher of a continuous membrane column.[2]

Practical membrane separation processes operate at finite values of pressure ratio. In addition, there is often a significant pressure drop on at least one side of the membrane. We have, therefore, continued our comparative analysis of the recycle system against the refluxed enricher here with two values of pressure ratio for a silicone capillary permeator system with axial pressure drop on the higher pressure feed side. The separation system continues to be air, considered as a binary mixture of 20.9 mol % oxygen and 79.1 mol % nitrogen. Further, the same feed flow rate and operating conditions are selected for each configuration with a given permeator of fixed membrane area. One of the pressure ratios is high, while the other one is quite low to facilitate comparison with the behavior obtained earlier with zero pressure ratio.[4] The separation behavior of the competitive schemes have been compared in a number of ways. The extent of separation approach used by Teslik and Sirkar[4] has also been used here for the higher pressure ratio to illustrate its usefulness.

The permeator models used here incorporate the refined pressure drop expressions and capillary deformations suggested by Thorman and Hwang[5] and used by Hwang and Thorman.[1] An additional objective of the present modeling effort was to predict the experimentally observed behavior of a silicone capillary permeator with partial permeate recycle for oxygen enrichment. This comparison is being treated elsewhere (Heit et al.[6]). Further, no effort has been made here to compare other variations of recycle scheme or reflux scheme, even if some of them may be better than the separation schemes considered in this chapter.

# II. DESCRIPTIONS OF SEPARATION SCHEMES

Consider Figure 1A which shows a compressor feeding atmospheric air into the permeator. A fraction, $\eta$, of the permeated stream at atmospheric pressure is recycled back to the fresh feed air to be compressed and sent into the permeator. The fraction, $(1 - \eta)$, of the permeated stream is the net permeated product. Figure 1B is a variation on the separation scheme of Figure 1A in that the compressor on the feed side is replaced by a vacuum pump on the permeate side. For given pressure levels on the two sides, the separation calculations by the

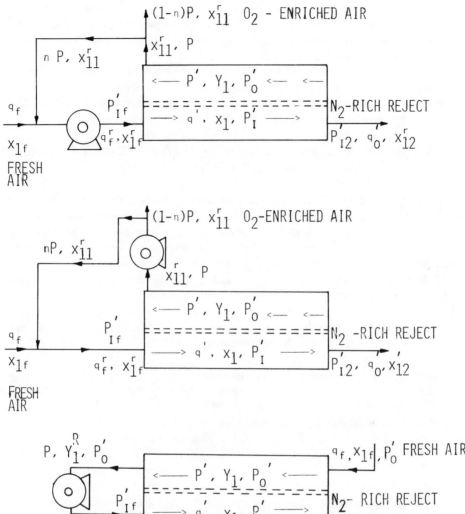

FIGURE 1. Three separation schemes for $O_2$-enriched air by a silicone capillary membrane permeator. (A) Partial permeate recycle membrane permeator; (B) partial permeate vacuum recycle membrane permeator; (C) enricher of a continuous membrane column.

two schemes would yield identical results. However, the compressor load required in each case for the same fresh feed flow rate, permeator, and operating conditions will be different, since in Figure 1B only the total permeate is being compressed, whereas in Figure 1A, fresh feed as well as fraction of the permeate are to be compressed (Kimura and Walmet[3]). The latter sum is usually larger than the permeate flow rate.

The third scheme shown in Figure 1C represents the "enricher alone" configuration of the continuous membrane column (Hwang et al.[2]). This usually operates with the permeate side at atmospheric pressure with the feed introduced into the permeate side. The combined feed-permeate is compressed and introduced on the other side of the membrane. If the permeate side is somehow maintained under vacuum, the top compressor may be replaced by a vacuum pump. However, the type of compressor load advantages enjoyed by the vacuum pump recycle in Figure 1B does not exist here.

For given pressure levels on either side of the membrane, feed flow rate, and a fixed capillary silicone permeator, the equations necessary for describing the two systems will be presented now. For such purposes, Figures 1A and 1B and essentially identical. Further, countercurrent flow exists in both the configurations, e.g., recycle or reflux.

## III. A THEORETICAL MODEL FOR A COUNTERCURRENT CAPILLARY PERMEATOR WITH RECYCLE OR REFLUX

A capillary permeator with permeate flowing countercurrent to the high pressure feed has been chosen for this study, since this produces the highest separation. The equations and the boundary conditions governing the capillary permeator with partial permeate recycle (Figure 1A or 1B) will be considered first. The system consists of N capillaries of length Z and inside and outside radii $r_i$ and $r_o$, respectively. High pressure gas flows through the inside of the capillaries and the permeated gas flows on the shell side of the permeator.

The following assumptions were utilized:

1.  The gas mixture contains only two species, 1 and 2, both being permeable. Species 1 is more permeable through the membrane.
2.  The permeability coefficients $Q_1$ and $Q_2$ of species 1 and 2, respectively, are independent of gas concentration or gas pressure.
3.  Both the tube side and shell side streams are perfectly mixed in the radial direction.
4.  Shell side axial pressure drop is negligible.
5.  Plug flow model can be used for both the streams.
6.  Axial diffusion is negligible compared to convective terms.

Since silicone capillaries are being used as membranes, the refined pressure drop expressions of Thorman and Hwang[5] as well as their expressions for axial and radial capillary deformations are incorporated into the governing equations. The gas pressure at capillary inlet is $P'_{1f}$ and at outlet is $P'_{12}$. The local gas pressure at any axial coordinate z (measured from the high pressure gas inlet) is $P'_1$ (see Figure 1). The gas pressure on the shell side is assumed constant at $P'_0$. The two governing equations for high pressure side gas flow rate $q'$ and the more permeable gas mole fractions $x_1$ are (Hwang and Thorman[1])

$$\frac{dq'}{dz} = - \frac{2\pi N}{\ln(r_o/r_i)} \left[ (Q_1 - Q_2)(x_1 P'_1 - y_1 P'_0) + Q_2 (P'_1 - P'_0) \right] \tag{1}$$

$$\frac{dx_1}{dz} = \left[ \frac{-2\pi N\, Q_1(x_1 P'_1 - y_1 P'_0)}{\ln(r_o/r_i)} - x_1 \frac{dq'}{dz} \right] \Big/ q' \tag{2}$$

with $y_1$ being the permeate mole fraction of the more permeable component (i.e., 1) at any given location. If $P'$ is the permeated product flow rate at any axial location in countercurrent flow and $q'_o$ is the high pressure reject flow rate at the permeator end with a species 1 mol fraction of $x'_{12}$, then the following material balances are valid at any location of the countercurrent permeator:

$$q' = P' + q'_o \tag{3}$$

$$x_1 q' = P' y_1 + q_0' x_{12}^r \tag{4}$$

The value of $y_1$ at the reject end, i.e., $y_{10}$, is to be evaluated from the quadratic representing pure cross flow

$$\frac{y_{10}}{(1 - y_{10})} \frac{Q_1}{Q_2} \left[ \frac{x_{12}^r P_{12}' - y_{10} P_0'}{(1 - x_{12}^r) P_{12}' - (1 - y_{10}) P_0'} \right] \tag{5}$$

The inside axial pressure drop equation is[5]

$$\frac{dP_1'}{dz} = \frac{K_1 \mu q' RT}{\pi N r_i^4 P_1'} \left[ \frac{1}{K_2} + \frac{4 Re_w}{Re_z} \frac{z}{r_i} \right]$$

$$+ \frac{4/r_i}{\dfrac{Re_z}{3P_1'} - \dfrac{P_1' N \pi r_i^3}{2 \mu q' RT}} + \left[ \frac{8 \pi q' RT}{N \pi r_i^4 P_1'} \right] \tag{6}$$

where

$$K_1 = 8(-1 - 0.75\, Re_w + 0.0407\, Re_w^2 - \dots\dots) \tag{6a}$$

$$K_2 = (1 - 0.056\, Re_w + 0.0153\, Re_w^2 \dots\dots) \tag{6b}$$

$$Re_w = \frac{r_i V_{rw}\, \rho}{\mu} \quad ; \quad Re_z = \frac{r_i V_z \rho}{\mu} \tag{6c}$$

The radial and axial deformations of the capillary are taken into account in the following way.[5] Replace $r_i$ in Equations 1 to 6c by $r_i^*$ defined by

$$r_i^* = r_i + u_r/r = r_i \tag{6d}$$

where

$$u_r = \frac{(P_i' - P_o')^{r_i^2/r_o^2}}{E \left( 1 - \left[ \dfrac{r_i}{r_o} \right]^2 \right)} [(r/2) + (3\, r_o^2/2\, r)], \quad r_i \leq r \leq r_o \tag{6e}$$

Similarly, replace $r_o$ in Equations 1 to 6c by $r_o^*$ defined by

$$r_o^* = r_o + u_r/r = r_o \tag{6f}$$

Here E is Young's modulus for the capillary material. The axial displacement at location z is given by

$$u_z = - \frac{[P_i' - P_o'](r_i^2/r_o^2)}{E \left( 1 - \left[ \dfrac{r_i}{r_o} \right]^2 \right)} z \tag{6g}$$

The capillary length of every small increment in z direction during numerical integration should be multiplied by $[1 + (u_z/z)]$ to accommodate axial deformation. Note that $P'_o$ is constant by assumption along z, whereas $P'_1$ varies from $P'_{1f}$ at feed location ($z = o$) to $P'_{12}$ at the reject stream location on the high pressure side.

Since the simulation is for a partial permeate recycle permeator with a fresh feed gas flow rate of $q_f$ having a more permeable (species 1) mole fraction $x_{1f}$, the high pressure feed flow rate actually entering the permeator is $q_f^r$ with a mole fraction of $x_{1f}^r$. If P is the total molar flow rate of permeated stream from the permeator and a fraction $\eta(<1)$ of this permeated stream is recycled to the fresh feed, then the following total and component material balances are valid:

$$q_f^r = q_f + \eta\,P \tag{7}$$

$$q_f^r\,x_{1f}^r = q_f\,x_{1f} + \eta\,P\,x_{11}^r \tag{7a}$$

The recycle permeator operates at a cut $\theta^r$ defined by

$$\theta^r = \frac{P}{q_f^r} \tag{7b}$$

whereas the net cut $\theta_{net}^r$ defined with respect to the net permeated product and the fresh air feed is

$$\theta_{net}^r = \frac{(1 - \eta)\,P}{q_f} \tag{7c}$$

Note that Equations 7 and 7b lead to

$$q_f^r = \frac{q_f}{(1 - \eta\theta^r)} \tag{8}$$

and Equation 7a may be written as

$$x_{1f}^r = (1 - \eta\theta^r)\,x_{1f} + \eta\theta^r\,x_{11}^r \tag{9}$$

The numerical simulation is carried out for specific values of $q_f$, $x_{1f}$, $P'_{1f}$, $\eta$, $r_i$, $r_o$, $P'_o$, N, and capillary length. The integration begins at the reject end and proceeds to the feed inlet location where the computed values of q', P', $x_1$, and $y_1$ are estimates of $q_f^r$, P ($= \theta^r q_f^r$), $x_{1f}^r$, and $x_{11}^r$. The integration is started with guesses of $q'_o$, $x_{12}^r$, and $P'_{12}$ at the high pressure reject stream location. To determine whether these guesses were right, one compares at the feed end simultaneously the computed values of q' with $q_f^r$ determined from Equation 8, $x_1$ with $x_{1f}^r$ determined from Equation 9, and $P'_1$ with the specified $P'_{1f}$. Note that $\theta^r$ is known from the calculated values of q' and P', since at the feed end $\theta^r = P'/q'$) for the countercurrent arrangement. If the computed values of q',$x_1$ and $P'_1$ at the feed end are different from $q_f^r$, $x_{1f}^r$, and $P_{1f}$ by more than a predetermined amount, an iterative Runge-Kutta technique is adopted for generating the next guesses of $q'_o$, $x_{12}^r$, and $P'_{12}$. This method is described in a thesis by Teslik[7] and is adopted directly from a detailed description given by Fan.[8] This method was also utilized by Antonson et al.[9] For backward integration replace z by $-z$ in the derivatives of Equations 1, 2, and 6.

Basically, nine additional equations are formed by differentiating Equations 1, 2, and 6 with respect to each of the variables $q'_o$, $x_{12}^r$, and $P'_{12}$. These equations upon integration over the length of the permeator generate constants that prescribe the changes in each of these variables for generating the next guesses. For example, the new guess of the exit pressure $P'_{12}$, $P'_{12}{}^{(N)}$, is given by

$$P_{12}^{\prime(N)} = \frac{P_{12}^{\prime(N-1)} G_{P_1}(F) - P_{1f}^{\prime(N-1)} + P_{1f}'}{G_{P_1}(F)} \tag{10}$$

where $P'_{12}{}^{(N-1)}$ is the last guess used to obtain a value of $P'_{1f}{}^{(N-1)}$ for the inlet pressure whose exact value for the problem is $P'_{1f}$ and $G_{P_1}(F)$ is the constant generated by the solution of these additional equations (incorporated in the iterative method) that are unique to the pressure variable. The argument F indicates that $G_{P_1}(F)$ is obtained from the solutions of the additional equations at the feed end of the permeator. The additional equations are given in Teslik,[7]

In exactly the same manner, the new guess for the reject end mole fraction $x_{12}^r$, $x_{12}^{r(N)}$, is given by

$$x_{12}^{r(N)} = \frac{x_{12}^{r(N-1)} G_{x_{12}}r(F) - x_{1f}^{r(N-1)} + x_{1f}^{r*}}{G_{x_{12}}r(F)} \tag{11}$$

where $x_{12}{}^{r(N-1)}$ is the last guess of $x_{12}^r$ used to obtain $x_{1f}{}^{r(N-1)}$ at the feed end, $G_{x_{12}}r(F)$ is the constant (generated by the iterative integration of the additional equations) unique to the composition variable, and $x_{1f}^{r*}$ is the value of $x_{1f}^r$ obtained from Equation 9 for given $x_{1f}$, $\eta$, and the values of $x_{11}^r$ and $\theta^r$ obtained after integration with the last guess $x_{12}{}^{r(N-1)}$. Similarly, a new guess for $q'_o$ is generated. Iterative integration was carried out until the calculated values of all three variables at the feed end converged to specified values. Such calculations were carried out for a large number of $\eta$ values varying between 0 (conventional permeator) and close to 1 (corresponding to total recycle) for any given value of fresh feed flow rate. This fresh feed flow rate was varied over a wide range next with all other conditions remaining constant for a given permeator and feed gas composition. This is tantamount to varying the nondimensional membrane area over a wide range, although the membrane area remains fixed.

The results obtained from such computations could be represented in a number of ways, amongst them being the extent of separation representation. The extent of separation, defined by Rony,[10,11] has been applied by Sirkar[12] to single entry separators. The extent of separation for a recycle permeator, $\xi^r$, may be represented following Teslik and Sirkar[4] as

$$\xi^r = |\dot{Y}_{11}^r - \dot{Y}_{21}^r| = \left| \frac{\theta^r q_f^r (1 - \eta) x_{11}^r}{q_f x_{1f}} - \frac{\theta^r q_f^r (1 - \eta) x_{21}^r}{q_f x_{2f}} \right| \tag{12}$$

$$\xi^r = \frac{(1 - \eta) \theta^r x_{11}^r}{(1 - \eta \theta^r) x_{1f}} \left| 1 - \frac{x_{21}^r x_{1f}}{x_{2f} x_{11}^r} \right| \tag{12a}$$

The computational result obtained could be utilized to calculate the extent of separation $\xi^r$ for the recycle permeator and plot $\xi^r$ against either $x_{11}^r$, permeate mole fraction of more permeable species 1, or any other suitable quantity.

The comparative power requirement may be estimated by comparing the suction flow rate to the compressor-per-unit net permeated product for a given permeate composition. This assumes isothermal operation of the compressor for schemes being compared with fixed pressure levels at the compressor inlet and exit. For the separation scheme of Figure 1A, the required quantity is then $[1/\theta^r (1 - \eta)]$. For the vacuum recycle scheme of Figure 1B, the required quantity is $[1/(1 - \eta)]$.

We next consider the enricher of a continuous membrane column (Figure 1C) and present a brief overview of the calculation procedure employed by us. Only then the results from the two schemes will be compared. First, the permeator in both schemes remains unchanged. Secondly, the fresh feed flow rate $q_f$ and fresh feed composition $x_{1f}$ are held constant so that the two schemes to be compared have the same input. Third, the pressure of the fresh feed gas entering the CMC enricher in Figure 1C is $P'_o$, while the pressure of the compressed gas leaving the compressor at column top is $P'_{1f}$ (where $P'_o$ and $P'_{1f}$ values are those adopted for the recycle permeator study). The notation for the enricher of the CMC is given in Figure 1C. The high pressure side flow rate and light component mole fraction (inside the capillary) at any location is indicated by $q'$ and $x_1$, respectively. The corresponding values on the permeate side are $P'$ and $y_1$. Thus, the notation used for the variables on both sides of the membrane are the same for the refluxed enricher as well as the recycle permeator. The reject conditions are $q'_o$, $x^R_{12}$, $P'_{12}$ with $x^R_{12}$ instead of $x^r_{12}$ because of the reflux scheme instead of the recycle scheme. The total high pressure feed flow rate entering the capillary is $q^R_f$ (instead of $q^r_f$ in the recycle scheme). The net product flow rate is $P^R$ of composition $x^R_{11}$. The two mass balance Equations 3 and 4 for the recycle permeator are to be replaced by

$$q' - q'_o = P' - q_f \tag{13}$$

$$q' x_1 - q'_o x^R_{12} = P' y_1 - q_f x_{1f} \tag{14}$$

For the reflux scheme, $q'$ varies between $q^R_f$ to $q'_o$. The permeate side flow rate varies between P (at the top) to $q_f$ at the bottom. The reflux ratio for the scheme is defined by

$$RFL = q^R_f / P \tag{15}$$

such that $O < RFL < 1$ (Hwang and Yuen[13]) with $P^R$, the net product flow rate, being equal to $P - q^R_f$. Note that in this refluxed enricher scheme, the value of $x_1$ at the top (compressor outlet), $x^R_{11}$, is such that $y_1$ at the top is equal to $x^R_{11} = y^R_1$.

For a given calculation, $P_{1f}$, $q_f$, $x_{1f}$, $P'_o$, as well as the permeator dimensions are known. For any set of $q'_o$, $x^R_{12}$, and $P'_{12}$, the conditions at the top should be such that $x^R_{11} = y^R_1$ and the column top pressure on the high pressure side becomes $P'_{1f}$. One first chooses a value of $q'_o < q_f$ (normally it is taken as some fraction of $q_f$) and then guesses $x_{12}$ and $P'_{12}$ at the bottom such that column top conditions are satisfied. Equations 1, 2, 6, 6a, 6b, 6c, 6d, 6e, 6f, and 6g are used with proper boundary conditions and the two new material balance Equations 13 and 14. The iterative Runge-Kutta procedure with backward integration from column bottom to column top is utilized by forming nine additional equations exactly as in the recycle permeator. New guesses for $x^R_{12}$ and $P'_{12}$ are generated by using equations very similar to 11 and 10, respectively, since one of the three initial conditions $q'_o$ on the high pressure side is known in this case by specification. Consequently, convergence is achieved somewhat more easily than with a recycle permeator. Calculations for a given $P'_{1f}$, $q_f$, $x_{1f}$, $P'_o$, are carried out over a range of values of $P^R$ or $q'_o$ to achieve a range of reflux ratio RFL. Then, $q_f$ is changed to explore the effect of feed flow rate for the given conditions and the given permeator.

One could calculate the value of the extent of separation, $\xi^R$ for the CMC enricher for comparing it with $\xi^r$ for the same permeator and operating conditions. This quantity can be calculated from[4]

$$\xi^R = \left(\frac{P^R}{q_f}\right) \frac{y^R_1}{x_{1f}} \left|1 - \frac{y^R_2 x_{1f}}{y^R_1 x_{2f}}\right| \tag{16}$$

where $y_2^R = (1 - y_1^R)$. Note that the $\theta_{net}^r$ for the CMC enricher is $(P^R/q_f)$ and $y_1^R$ is equivalent to the permeate composition $x_{11}^r$ of the recycle permeator.

## IV. NUMERICAL RESULTS

Extensive computer calculations have been carried out for both types of separation schemes with an $O_2$-$N_2$ system and $x_{O_{2f}} = 0.209$ where species 1 is $O_2$. The objective has been to study the comparative performance of the two schemes for producing oxygen-enriched air (and not for producing tonnage oxygen!). All calculations are based on silicone capillary permeators containing capillaries of length 3.11 m and underformed outside and inside diameters of 0.61 and 0.238 mm, respectively. The permeabilities of $O_2$ and $N_2$ through the silicone capillary material were chosen to be, respectively, $2 \times 10^{-3}$ and $0.935 \times 10^{-3}$ (mol) (m)/(Pa) (sec) (m$^2$). The temperature is 25°C. The viscosities of the gas mixtures were estimated from Wilke.[14] The Young's modulus for the silicone material was taken to be $30.8 \times 10^5$ Pa.

Two sets of calculations were carried out. *In the first set*, the value of the highest pressure $P'_{1f}$ was taken to be $2.265 \times 10^5$ Pa, whereas the value of $P'_o$ was $1.025 \times 10^5$ Pa. The fresh air feed flow rate was varied between $q_f = 1.34 \times 10^{-4}$ gmol/sec (i.e., $3 \times 10^{-6}$ std m$^3$/sec) and $3.12 \times 10^{-6}$ gmol/sec (i.e., $0.07 \times 10^{-6}$ std m$^3$/sec) for the recycle scheme, and between $q_f = 0.446 \times 10^{-4}$ gmol/sec (i.e., $1.0 \times 10^{-6}$ std m$^3$/sec) and $3.12 \times 10^{-6}$ gmol/sec (i.e., $0.07 \times 10^{-6}$ std m$^3$/sec) for the refluxed enricher scheme. The total number of capillaries in the permeator was 35. In this set of calculations, then, the inlet pressure ratio $\gamma$ defined by $(P'_o/P'_{1f})$ is high, i.e., 0.4525.

In the *next set of calculations*, the highest pressure $P'_{1f}$ was taken to be $1.132 \times 10^5$ Pa. The shell side constant low pressure $P'_o$ was taken to be $0.0932 \times 10^5$ Pa. The inlet pressure ratio is, thus, quite low, 0.0823, due to a very high vacuum on the permeate side. Under these conditions, convergence in the recycle permeator becomes difficult with very low feed flow rates. So, the total number of capillaries in this set of calculations was increased to 300. Correspondingly, the feed flow rate was increased by an order of magnitude. Since under this pressure ratio one approaches the ideal permeator condition[4] and the comparative behavior of the recycle permeator with the CMC enricher is expected to be close to that observed by Teslik and Sirkar,[4] the number of fresh flow rates investigated was limited. Specifically, the following fresh air feed flow rates $0.8 \times 10^{-6}$ and $0.4 \times 10^{-6}$ std m$^3$/sec were selected. Note that the nondimensional membrane area in both sets of calculations are similar in the higher ranges, since it is proportional to the product of high pressure times membrane area divided by feed flow rate. Further, only somewhat lower feed flow rates were considered, since these were more important from a comparison point of view as will become clear soon.

The second set of calculations with a very low inlet pressure ratio for a recycle permeator represents more realistically the schematic of Figure 1B with a vacuum pump. Consequently, the compressor load requirement per unit net permeated product is proportional to $(1/1 - \eta)$. No such reduction in compressor load is possible in the CMC enricher scheme under the second set of pressure conditions, even if one uses a vacuum pump instead of a compressor. Notice, however, the nondimensional separation behavior is unaffected if schemes of Figure 1A or 1B are interchanged.

We first present the results of the first set of computer calculations at a high pressure ratio of 0.4525 for the partial light fraction recycle permeator. In Figure 2, the permeated net product flow rate in standard m$^3$/sec are plotted against the permeate oxygen mole fraction for six different fresh feed flow rates varying between $3 \times 10^{-6}$ and $0.07 \times 10^{-6}$ std m$^3$/sec. The curve for each fresh feed flow rate incorporates a wide range of recycle ratios, with the highest value of $x_{11}^r$ corresponding to the highest recycle ratio and the lowest $x_{11}^r$

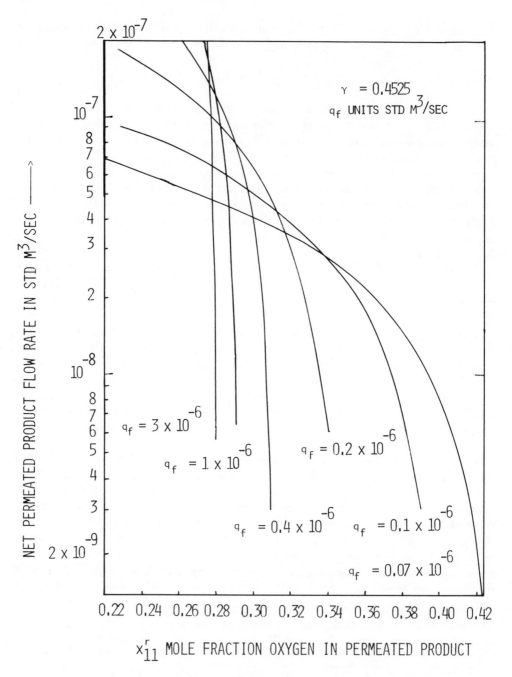

FIGURE 2.    Productivity of a partial permeate recycle silicone capillary permeator for $O_2$ enrichment with various fresh air feed flow rates.

value corresponding to the lowest recycle ratio and highest cut. For a given fresh feed flow rate through the given permeator under specified values of $P'_{If}$ and $P'_o$, the permeated net product flow rate decreases as the recycle ratio increases. Simultaneously, the oxygen enrichment increases significantly. Obviously, a low fresh feed flow rate combined with a high recycle would lead to oxygen compositions substantially greater than the theoretical maximum attainable in a nonrecycle permeator. One notices, also, that for a given permeated

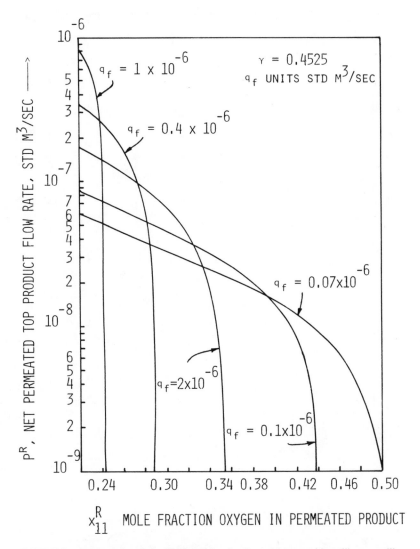

FIGURE 3. Productivity of the CMC enricher for $O_2$ enrichment with a silicone capillary permeator for various fresh feed flow rates.

net product flow rate, oxygen enrichment increases as the fresh feed flow rate decreases. We should point out here that for these high enrichment regions, although the $\theta^r_{net}$ is somewhat low, the permeator is operating at substantial values of $\theta^r$ exactly as was observed for an ideal permeator by Teslik and Sirkar.[4] We note further that at high fresh feed flow rates, recycling is not very fruitful and observed enrichments are minimal due to very little recycle-induced changes in the feed composition $x^r_{1f}$ entering the permeator.

A similar behavioral pattern is observed if we consider the results of computer calculations for the enricher of a continuous membrane column. In Figure 3, we present these results as the net top product flow rate $P^R$ against its oxygen composition $x^r_{11} = y^R_1$ for various fresh feed flow rates. For a given fresh feed flow rate, higher oxygen enrichment is obtained at a higher reflux ratio with a consequent reduction in the rate at which enriched product is withdrawn. One notices further that significant oxygen enrichment is achieved in the CMC enricher.[13]

It is of much more interest to compare the performance of the recycle scheme against the

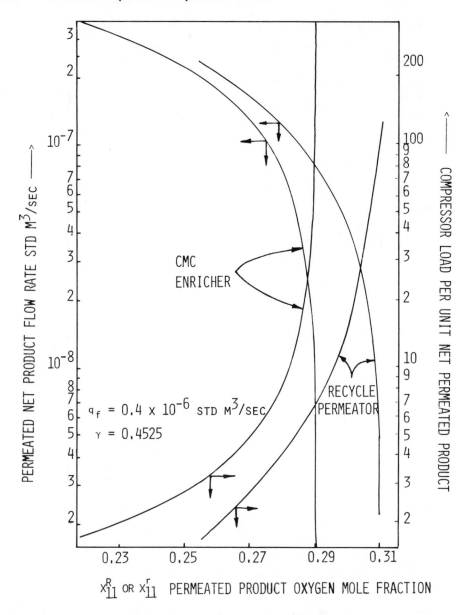

FIGURE 4.   Comparative performance of a recycle permeator and a CMC enricher at a high $\gamma$ = 0.4525 and a high fresh feed flow rate.

reflux scheme under identical inputs and operating conditions at a high pressure ratio. Such a comparison is partially provided in Figures 4 and 5. Figure 4 represents the comparative separation behavior as well as the power requirement for both schemes using the same permeator, same operating conditions, and a fresh feed flow rate of $q_f = 0.4 \times 10^{-6}$ std m$^3$/sec. The recycle permeator has a higher net permeated product flow rate at a given permeate composition compared to that from a CMC enricher. If one uses the ratio of the compressor suction load per unit net permeated product as an estimate of the power requirement (as has been done with an ideal permeator[4]), we see, from Figure 4, that the power requirement at a given permeate composition is considerably lower for the recycle permeator at all compositions. Of course, for this particular fresh feed flow rate, the enrichments are not very high.

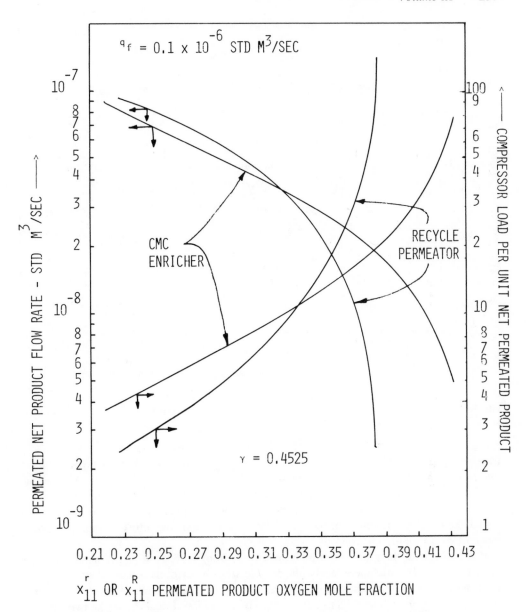

FIGURE 5. Comparative performance of a recycle permeator and a CMC enricher at a high $\gamma = 0.4525$ and a lower fresh feed flow rate.

The enrichments are considerably higher if we consider Figure 5 where the fresh feed flow rate is much smaller, i.e., $0.1 \times 10^{-6}$ std m³/sec. We notice in Figure 5 a lower net permeated product rate from the CMC enricher compared to a recycle permeator at all product compositions up to 0.327 mol fraction oxygen, beyond which the CMC enricher has a higher net permeated product rate at a given product composition. Thus, for the lower values of permeate compositions, the recycle scheme has a higher production rate, whereas at higher permeate compositions, the reflux scheme has a higher production rate. On the other hand, the compressor load for the recycle scheme is lower than that of CMC enricher until a product composition of 0.34 mol fraction oxygen is reached. The region of better performance by the refluxed enricher scheme then certainly exists at permeate oxygen mole

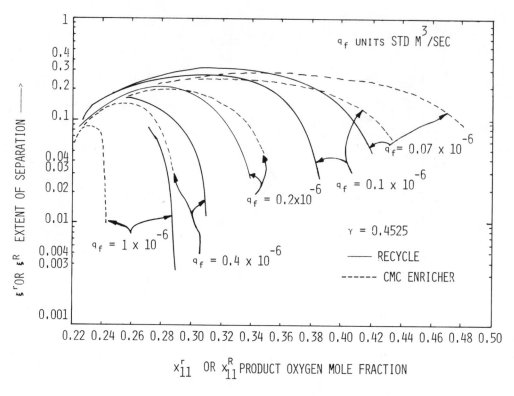

FIGURE 6.    Comparative performance of a recycle permeator and a CMC enricher at a high $\gamma = 0.4525$ using $\xi$.

fractions greater than 0.34 at this low feed flow rate of $0.1 \times 10^{-6}$ std m³/sec. In the intermediate composition range between 0.327 and 0.34 mol fraction oxygen, an economic analysis can only resolve which one is the better scheme. Since energy requirement often constitutes more than 50% of the cost of gas permeation plants (Ohno et al.[15]), one can appreciate the possible utility of the recycle scheme vis-à-vis the reflux scheme for a wide range of oxygen enrichment levels.

It would appear then that, for a high pressure ratio of 0.4525, at higher fresh feed flow rates and at lower enrichment levels, the recycle scheme performs better, whereas at lower fresh feed flow rates and higher enrichment levels, the reflux scheme is better. A better perspective is gained by considering the extents of separation, $\xi^r$ and $\xi^R$, for the two schemes for various fresh feed flow rates against the permeated product composition. Figure 6 shows the results of such a plot for feed flow rates varying between $1 \times 10^{-6}$ and $0.07 \times 10^{-6}$ std m³/sec. For lower values of fresh feed flow rates, it is observed that the recycle scheme has higher extents of separation at given lower oxygen enrichment levels than the CMC enricher. Further, the recycle permeator is performing better than the CMC enricher in these regions. The index $\xi$ clearly demonstrated this aspect. However, when considerable $O_2$ enrichment is required at a high pressure ratio, it appears that the CMC enricher has higher extents of separation exactly as observed before. This shows again how useful $\xi$ is. It ought to be recalled that at these high levels of $O_2$ enrichment, an extremely high compressor load is required. An estimate of such power requirements is available in Teslik[7] for the recycle and refluxed enricher schemes for a wide range of oxygen enrichments at the given values of fresh feed flow rates.

We will now briefly present the results of the *second set of computer calculations* at a very low pressure ratio of 0.0823. First, we note that Heit et al.[6] carried out their experiments

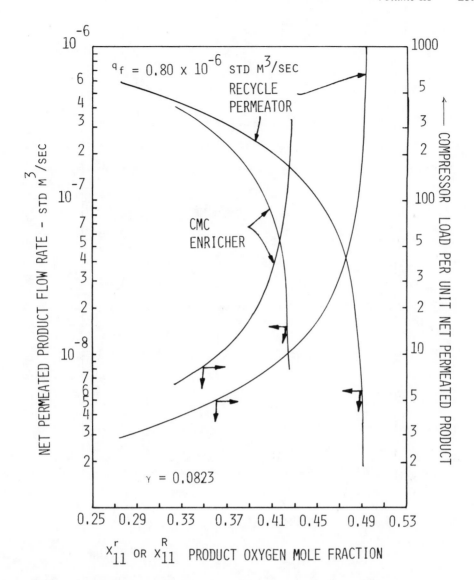

FIGURE 7.   Comparative performance of a recycle permeator and a CMC enricher at a low γ = 0.0823 and a higher fresh feed flow rate.

under similar conditions. Figure 7 shows the permeated net product flow rate in std m³/sec against the mole fraction of oxygen in permeate for both the recycle as well as the CMC enricher scheme for a fresh feed air flow rate of $0.80 \times 10^{-6}$ std m³/sec. The nondimensional membrane area, in this case, would be similar to that in the first set of calculations (at a higher pressure ratio with less membrane area) for a fresh air feed flow rate of $0.195 \times 10^{-6}$ std m³/sec. We notice from Figure 7 that the recycle permeator has a significantly higher net rate of permeate production than the CMC enricher everywhere, and especially so in the higher ranges of oxygen mole fractions in the permeate, i.e., around $x^r_{11} \sim 0.40$ and beyond. If one goes to a value of nondimensional area twice that in Figure 7 by decreasing the fresh feed flow rate to half the value, the relative performance does not change. In Figure 8, the fresh air feed flow rate is only $0.40 \times 10^{-6}$ std m³/sec. One notices that due to the low pressure ratio, the recycle permeator has a higher net permeate production rate at any given $x^r_{11}$ when compared with the CMC enricher. Further, the values of $x^r_{11}$ attained are

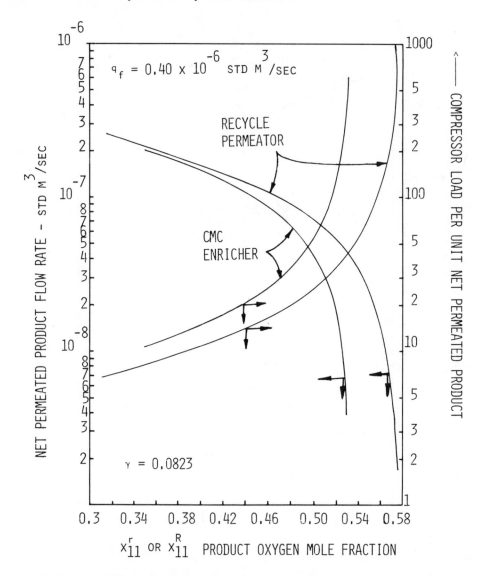

FIGURE 8.    Comparative performance of a recycle permeator and a CMC enricher at a low $\gamma$ = 0.0823 and a lower fresh feed flow rate.

very high, i.e., 0.5 to 0.57. The comparison has not been carried out at lower fresh feed flow rates since silicone membrane is hardly the right candidate for producing a highly $O_2$-enriched stream. However, notice that CMC enricher performance vis-à-vis the recycle permeator at this low pressure ratio is improving as the fresh feed flow rate is reduced. In addition, the compressor loads required for achieving these high compositions are extraordinarily high.

Figures 7 and 8 also indicate the net compressor load per unit net permeated product stream against $x^r_{11}$ for both the schemes. The only difference between the compressor loads for the recycle scheme here and in Figures 4 and 5 is that we have utilized $[1/(1 - \eta)]$ here instead of $[1/\theta^r (1 - \eta)]$ used in Figures 4 and 5. This is possible because of the vacuum pump in Figure 1b. Even if one were to use $[1/\theta^r (1 - \eta)]$ in Figures 7 and 8, the compressor load requirements for the recycle scheme would still be lower than those for the CMC enricher at all $x^r_{11}$, since $\theta^r$ has large values for these calculations. We thus observe that, at very low pressure ratios approaching ideal permeator, the recycle permeator has a higher rate of net permeate production and a lower compressor load requirement than the CMC enricher when the two separation schemes are used with the same permeator, same fresh feed flow rate, and same operating conditions. The numerical calculations performed by Teslik and Sirkar[4] on an ideal permeator operating at zero pressure ratio would naturally suggest such a behavior.

## V. DISCUSSION AND CONCLUSIONS

An interpretation of the observed comparative separation behavior between the recycle scheme and the CMC enricher scheme will be offered here. Teslik and Sirkar[4] had observed for flat permeators that, at zero pressure ratio, for any required permeate compositions up to $x_{l1}^r = 0.44$, the recycle permeator yielded a higher net rate of permeate production than the optimized CMC enricher (Hwang et al.[2]). Conversely, at a given rate of net permeate production, the enrichment from a recycle system is higher. Such a behavior could come about due to the substantial dilution of the permeate in the CMC enricher by the fresh feed as well as due to the requirement of a substantial reflux for the permeator to operate. Since all of the fresh feed mixes with all of the permeate in the CMC enricher entering the compressor, the power requirements are, in general, likely to be higher than that of the recycle scheme (Figure 1A or 1B.)

Obviously, there are competing effects of a number of phenomena occurring here. In a recycle permeator, there is enrichment of fresh feed by partial permeate recycle and then the mixed feed concentration is amplified by the membrane. In the CMC enricher, the high pressure feed concentration is increased by mixing fresh feed with all of the permeate in a distributed way, and then this mixed feed concentration is amplified by the membrane. One must superimpose on this the effect of pressure ratio. At a very low pressure ratio, the permeate from the recycle permeator is highly enriched. There is no dilution of this permeate quality. On the other hand, the permeate from the CMC enricher is diluted heavily by the feed stream resulting in poorer permeate quality. Further, no driving force advantage is gained by mixing the permeate with feed continuously on the low pressure side of the CMC enricher. At a *high pressure ratio,* however, mixing the permeate with feed on the low pressure side of a CMC enricher is tantamount to increasing the driving force compared to that in a recycle permeator. The observed behavior is a complex summation of these various factors.

The effect of a high fresh feed flow rate or low dimensionless membrane area is considerably clearer. Regardless of pressure ratio, if introduced on the permeate side as in a CMC enricher, it depresses CMC enricher performance compared to that of a recycle permeator due to considerable permeate dilution. This effect is also seen in the case of a very low pressure ratio. In another sense, however, the CMC enricher scheme may be better. The mixing here involves the fresh feed composition with the lowest permeate composition, whereas in the countercurrent recycle permeator, mixing involves the highest permeate composition with fresh feed. The CMC enricher then accommodates "no-mixing" criteria somewhat more than the recycle scheme.

Obviously, shifting the location of the feed introduction around the countercurrent permeator could provide significant changes in the performance of a recycle or reflux scheme. Further, splitting the permeate stream in the recycle scheme to prevent mixing of poorer permeate with the richest permeate would be desirable. Simultaneous achieving of both would, perhaps, lead to the optimum permeation separation scheme with a single membrane permeator. Such a scheme is practically feasible only with flat permeators and cannot be physically implemented with a *simple* hollow fiber or capillary permeator.

The separation index $\xi$, the extent of separation, appears to be quite useful in delineating the regions of optimum performance. This was earlier found to be true for the ideal permeator.[4] We substantiate it for real permeators, also.

The comparisons carried out here are limited to gas separation with a low separation factor membrane. For such a process, the CMC enricher will be more efficient in providing high enrichments at high pressure ratios. For lower pressure ratios, the recycle permeator will be more efficient in producing enriched permeate streams. A recycle permeator with silicone membranes would operate quite efficiently for oxygen enrichment under vacuum recycle conditions. The role of a recycle or reflux in permeators with membranes having a high separation factor, however, needs to be investigated. A similar behavioral scenario is most likely.

## NOTATION

| | |
|---|---|
| $E$ | Young's modulus for silicone rubber, Pa |
| $K_1$ | Function defined by Equation 6a |
| $K_2$ | Function defined by Equation 6b |
| $N$ | Number of capillaries in the silicone capillary permeator |
| $P$ | Total permeate side gas flow rate at exit, mol/sec |
| $P^R$ | Net high pressure permeated product flow rate from CMC enricher, mol/sec |
| $P'$ | Permeate side gas flow rate at any distance from high pressure gas inlet, mol/sec |
| $P'_1$ | Gas pressure on the high pressure side of permeator at any axial location, Pa |
| $P'_{1f}$ | Value of $P'_1$ at high pressure gas inlet, Pa |
| $P'_{12}$ | Value of $P'_1$ at high pressure gas exit, Pa |
| $P'_o$ | Pressure of gas on the low pressure side of permeator, Pa |
| $q'$ | High pressure gas flow rate in permeator at any distance from inlet, mol/sec |
| $q_o$ | Gas flow rate out of the permeator at high pressure gas exit, mol/sec |
| $q_f$ | Fresh air feed flow rate, mol/sec |
| $q_f^r$ | High pressure feed gas flow rate actually entering recycle permeator, mol/sec |
| $q_f^R$ | High pressure gas flow rate actually entering the CMC enricher at top, mol/sec |
| $Q_1, Q_2$ | Permeability of gas species 1 ($O_2$) and 2 ($N_2$), respectively, (mol) (m)/(sec) (m²)(Pa) |
| $r_i$ | Inside radius of capillary, m |
| $r_o$ | Outside radius of capillary, m |
| $r_i^*, r_o^*$ | Deformed internal and external capillary radius, Equations 6d and 6f, respectively, m |
| $Re_w$ | Permeation Reynolds number in capillaries, Equation 6c |
| $Re_z$ | Axial Reynolds number in capillaries, Equation 6c |
| $T$ | Absolute temperature, °K |
| $v_{rw}$ | Radial permeation velocity at capillary wall (is negative), m/sec |
| $v_z$ | Axial average velocity of gas in a capillary, m/sec |
| $x_{if}$ | Mole fraction of species i in fresh feed stream to permeator |
| $x_{if}^r$ | Mole fraction of species i in feed stream actually entering the recycle permeator high pressure side |
| $x_{ij}$ | Mole fraction of species i in stream j leaving permeator |
| $x_1$ | Mole fraction species 1 ($O_2$) on high pressure side of permeator at any location |
| $x_{11}^R$ | Mole fraction species 1 ($O_2$) entering high pressure top section of CMC enricher |
| $x_{12}^R$ | Mole fraction species 1 ($O_2$) leaving high pressure bottom section of CMC enricher |
| $y_1$ | Mole fraction of species 1 (oxygen) at any location on the permeate side |
| $y_1^R$ | Mole fraction of oxygen in stream leaving CMC enricher at the top on permeate side |
| $y_{10}$ | Value of $y_1$ at recycle permeator reject end, Equation 5 |
| $z$ | Capillary length coordinate from high pressure feed entry, m |
| $Z$ | Total length of capillaries, m |

### Greek Letters

| | |
|---|---|
| $\rho$ | Density of high pressure gas in capillary, Equation 6c |
| $\theta$ | Permeator cut |
| $\theta^r$ | Cut of a recycle permeator, defined by Equation 7b |
| $\theta_{net}^r$ | Net cut in a recycle permeator based on fresh feed, defined by Equation 7c |
| $\gamma$ | Pressure ratio, defined by ($P'_o/P'_1$) |

| | |
|---|---|
| μ | Viscosity of high pressure gas mixture in capillary, Pa-s |
| η | Recycle ratio in a partial permeate recycle permeator |
| $\xi^r$ | Extent of separation in a recycle separator, defined by Equation 12 |
| $\xi^R$ | Extent of separation in a CMC enricher, defined by Equation 16 |

## Superscripts

| | |
|---|---|
| (N) | Nth iteration in capillary recycle permeator calculation |
| r | Recycle separator |
| R | CMC enricher |

## Subscripts

| | |
|---|---|
| f | Refers to feed stream entering permeator |
| i | Subscript with i = 1, refers to $O_2$, i = 2 to $N_2$ |
| j | Subscript with j = 1 indicating permeate, j = 2 indicating reject |

# REFERENCES

1. **Hwang, S. T. and Thorman, J. M.,** The continuous membrane column, *AIChE J.*, 26, 558, 1980.
2. **Hwang, S. T., Thorman, J. M., and Yuen, K. H.,** Gas separation by a continuous membrane column, *Sep. Sci. Technol.*, 15, 1069, 1980.
3. **Kimura, S. G. and Walmet, G. E.,** Fuel gas purification with perm-selective membranes, *Sep. Sci. Technol.*, 15, 1115, 1980.
4. **Teslik, S. and Sirkar, K. K.,** Ideal Silicone Flat Plate Permeator with Partial Permeate Recycle for $O_2$-enriched Air, unpublished.
5. **Thorman, J. M. and Hwang, S. T.,** Compressible flow in permeable capillaries under deformation, *Chem. Eng. Sci.*, 33, 15, 1978.
6. **Heit, L. B., Majumdar, S., and Sirkar, K. K.,** Oxygen enrichment by a recycle permeator, in Symp. on Gas Separations, Paper No. 46b, AIChE Meet., Anaheim, Calif., June 8, 1982.
7. **Teslik, S.,** Analysis of a Light Fraction Recycle Permeator and Oxygen Enrichment Separation Schemes, M. Eng. thesis, Stevens Institute of Technology, Hoboken, 1983.
8. **Fan, L. T.,** *The Continuous Maximum Principle,* John Wiley & Sons, New York, 1966.
9. **Antonson, C. R., Gardner, R. J., King, C. F., and Ko, D. Y.,** Analysis of gas separation by permeation in hollow fibers, *Ind. Eng. Chem. Proc. Des. Dev.*, 16, 473, 1977.
10. **Rony, P. R.,** The extent of separation: a universal separation index, *Sep. Sci.*, 3, 239, 1968.
11. **Rony, P. R.,** The extent of separation: on the unification of the field of chemical separations, *AIChE Symp. Ser.*, 68(120), 89, 1972.
12. **Sirkar, K. K.,** On the composite nature of the extent of separation, *Sep. Sci.*, 12, 211, 1977.
13. **Hwang, S. T. and Yuen, K. H.,** Optimization of continuous membrane column, in 73rd AIChE Annu. Meet., Paper No. 22a, Chicago, November 19, 1980.
14. **Wilke, C. R.,** A viscosity equation for gas mixtures, *J. Chem. Phys.*, 18, 517, 1950.
15. **Ohno, M., Morisue, T., Ozaki, O., and Miyauchi, T.,** Comparison of gas membrane separation cascades using conventional separation cell and two-unit separation cell, *J. Nucl. Sci. Technol.*, 15(5), 376, 1978.

Chapter 14

# LASER-INDUCED SEPARATIONS (LIS)

**N. M. Lawandy and J. M. Calo**

## TABLE OF CONTENTS

## I. INTRODUCTION

The fact that separation processes are notoriously inefficient (in a thermodynamic sense) provides a continual driving force to improve established separation schemes and to develop new ones. However, in order for a particular separation scheme to become adopted, its net impact on the specific application is even more important than its relative efficiency. In order to illustrate this point, consider the minimum theoretical work required to perform a specified separation. This is the work required to reverse the entropy increase on mixing; viz.,

$$W_{min} = T\Delta S = RT \sum_i x_i \ln x_i \tag{1}$$

where T is the absolute temperature, R the gas constant, and the $x_i$ are the mole fractions of all species in the mixture. Thus, the minimum work required to separate an arbitrary 50 mol % binary mixture is

$$W_{min} = RT \sum_{i=1}^{2} x_i \ln x_i \cong 1.7 \text{ kJ/mol} \tag{2}$$

(or 0.018 eV per molecule). In comparison, consider the energy (or work) required to volatilize a unit of product in a distillation system. The latent heat of vaporization of *n*-butane (taken as typical of light hydrocarbons) is 22.3 kJ/mol (at its normal boiling point of 272.66 K). The relatively large difference between 22.3 and 1.7 kJ/mol represents a substantial potential improvement that can be made in separating *n*-butane from a binary mixture. However, it is important to note that the separation cost of *n*-butane via distillation is still only a small fraction of its product value (~0.5% for *n*-butane; e.g., see Shinskey[1]). Thus, the impetus to develop better separation techniques for *n*-butane is not very great.

From the preceding, it is obvious that the development, acceptance, and adoption of new, more efficient separation schemes will be most propitious for situations where existing separation techniques are grossly inefficient and/or where the separation costs represent a significant fraction of the product value or process cost.

The current chapter is concerned with a review of the various laser-induced separation (LIS) schemes that have been proposed and proven experimentally. Of these, the laser isotope separation (also denoted as LIS in the literature) schemes for uranium enrichment are the most notable and the best developed (i.e., to the scale-up stage[2]). This situation exists today precisely because this application fulfills the criteria discussed above, i.e., the gross inefficiencies of the major current processes. The energy required to enrich $^{235}U$ from its 0.7% natural abundance to a 3% product is about $2.4 \times 10^{11}$ J/mol (2.5 MeV per molecule) for gaseous diffusion and $2.4 \times 10^{10}$ J/mol (250 keV per molecule) for gas centrifugation; while the minimum work required for separation is only about 2.4 kJ/mol (25.2 meV per molecule).[3] The basic energy requirement for a uranium LIS enrichment process is the ionization energy of the species (6.2 eV for uranium) or the energy required to rupture a bond (e.g., 5 eV for $UF_6$). If the photon energy is deposited in $^{235}U$ only, then the total energy efficiency depends on the "wall socket" efficiency (i.e., the efficiency with which the laser converts electrical to optical energy), which is typically on the order of 0.1 to 10%. Thus, the energy consumption of an ideal uranium LIS process would be between 50 eV and 5 keV per molecule.[3] For a more accurate estimate, additional losses should also be included to take into account factors such as the fractional photon utilization, optical element losses, and undesired absorptions. In any case, the resultant value will still be orders-of-magnitude closer to the ideal minimum value than the current major schemes. However, in spite of this large incentive for process improvement, the fact that there has been and continues to be significant activity in developing gas centrifugation (e.g., see Wilkie[4]) and LIS (e.g., see Rhodes[2]) processes, it is instructive to note that uranium is still being enriched primarily via the gaseous diffusion process of WW II vintage. This underscores the overwhelming influence of development and construction costs on process economics.

Another important consideration in evaluating and comparing separation schemes is the separation factor that can be achieved. One of the major sources contributing to the inefficiency of many separation processes is the energy expended in transporting fluids through the many stages or cascades required to achieve high separation factors (even disregarding the large incremental capital costs involved). The staging requirement increases rapidly to infinity for high separation factors, approaching prohibitively large numbers of stages quite readily. This is also a distinct characteristic of the gaseous diffusion process. It has been estimated[3] that it requires about the same amount of energy to reduce the tailings concentration of $^{235}U$ from 0.35 to 0.17% as it does to reduce it from 0.7 to 0.3%, with only half as much enriched material recovered. A tailings concentration of 0.35% (common in a conventional gaseous diffusion plant) results in a loss of about 43% of the fissionable material.

Theoretically, a laser isotope separation scheme is capable of practically complete separation in a single stage. In practice, of course, this idealization would probably not be attainable, but, in any case, a separation factor much higher than that achievable by current processes is quite probable.

The preceding discussion on the efficiency of laser-induced separation methods has concentrated exclusively on the specific subclass of laser isotope separation which is the most well developed. This is precisely because of the point emphasized above that new separation processes are most likely to be developed for applications where the separation step is grossly inefficient and/or represents a relatively large fraction of the product value. This is indeed the case for obtaining highly pure isotopes. Apart from perhaps deuterium and the isotopes of carbon and oxygen, isotopes are generally quite expensive as a result of highly inefficient separation schemes. For these applications, energy requirements on the order 50 eV to 5 keV per molecule, as for uranium, may be quite attractive. However, in order for laser-induced separation schemes to be competitive with other traditional nonisotopic separation schemes, the energy requirements must be much less than this; e.g., approximately 0.23 eV per molecule for the *n*-butane distillation example. Energy requirements on this order, however, can be realized with infrared photons (0.12 eV for 10.6 $\mu$m $CO_2$ radiation). When characteristically high separation factors are taken into account, even multiphoton LIS schemes may eventually become competitive for certain applications.

## II. LASER-INDUCED SEPARATION (LIS) METHODS

In general, all separation processes can be conceptualized as schematized in Figure 1. The mixture to be separated is brought into contact with a separating agent which causes the system to form one or more distinct regions or phases, differing in composition from one another. These regions or phases are then physically separated, allowing the recovery of an enriched or product fraction and a depleted or residue fraction. This conceptualization also applies to LIS techniques. The distinct characteristic which categorizes a separation process as laser-induced is that the separating agent is the *selective* excitation of a species in the mixture due to the absorption of one or more photons from a laser source. The primary incentive for using a laser source is the ability of the laser to *selectively* interrogate the spectroscopic signature of a specific species at sufficiently high intensity. Laser sources typically exhibit linewidth-to-frequency ratios less than $10^{-5}$ and are, therefore, capable of interacting with a single pair of molecular or atomic energy levels. The laser photon energy

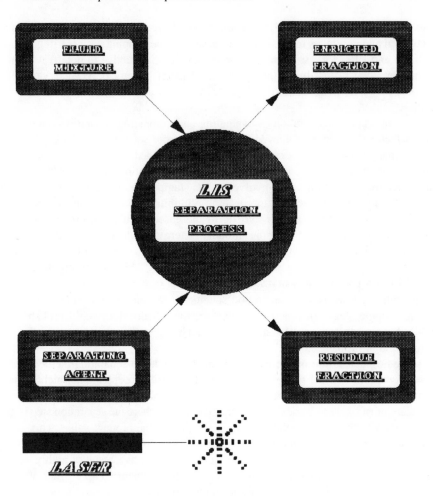

FIGURE 1.    Schematic of general laser-induced separation scheme.

is thus deposited *selectively,* thereby resulting in a low *optical* energy per separated atom or molecule. This ratio is far better than any achievable by depositing the same energy into the entire mixture ensemble. This characteristic selective excitation excludes, therefore, any process that uses a laser simply as a thermal heat source (which would tend to be quite inefficient in any event). Due to this same selective nature, laser separation schemes also usually suffer from interatomic/molecular interactions which tend to distribute the energy deposited to all the constituents in the mixture.

In the following sections we identify and outline six distinct classes of LIS methods. The emphasis in this discussion is on providing clear explanations of the physical mechanisms involved in the selectivity of the radiative interaction and the physical separation steps. The first three classes are of a type wherein the laser radiation produces a selected state of the desired species, followed by a separate removal step. The principal difference among these is the mechanism of excitation of the atom or molecule. Laser-induced formation of particulates is included together with selective photodissociation of molecules. The fourth class, the mechanism of photodeflection of atomic and molecular beams, differs from these in that it relies on combined direct laser-induced selection/removal of one species. The last two classes are laser-induced selective transport and adsorption. These methods rely on the selective adsorption of laser radiation to produce states that possess different transport and adsorption properties than their parents. These latter techniques are relatively more recent than the others and thus, for the most part, are still in the research stage with less well-defined applications.

## A. Selective Photoionization of Atoms

### 1. Overview

The photoionization method of gas mixture separation is based on the selective production of atomic ions, which are then either collected by electric and magnetic fields or scavenged by plasma chemical reactions. The photoionization process for nonisotopic separations exhibits a relatively large selectivity and can often be employed using single photon excitation. The single photon process is often not practical, however, for isotopic separations due to the overlap of ionizing transitions which are commonly encountered in the Doppler regime. This overlap is caused by Doppler broadening, whereby the absorption line of the transition is broadened by the distribution of thermal velocities along the direction of the incident laser beam. The degree of broadening is usually greater than the isotopic difference in the transition frequency for most practical conditions, thus destroying the intrinsic selectively. Two techniques used to overcome this problem are (1) confinement of the atoms to a narrow beam traveling perpendicular to the laser radiation and (2) cooling by adiabatic expansion. In this manner, the velocities of the atoms in the direction of irradiation may be kept low enough so that the resultant Doppler broadening is less than the isotope shift. The primary problem with the first technique is the relatively low density to which the beam is restricted.

Another way to produce high selectivity in isotopic separations using atomic photoionization is via a multistep process in which the ionization is produced by a sequence of excitations using a highly monochromatic laser source. The final ionizing step is then accomplished with either a sufficiently intense incoherent source or a second laser. In this manner, only the single isotope desired will acquire the correct energy required for ionization in the multistep absorption process.

The first proposal for using selective photoionization in which the final step results in an electron in the continuum was due to Letokhov.[5] This process, which in principle is also applicable to molecular species, has been primarily restricted to atoms. This is due to the additional spectral congestion caused by the rotational and vibrational modes of molecular species. Since this first work, the method of two-step photoionization has been demonstrated by many groups. The first such experiments in uranium enrichment were initiated as early as 1971 by the Avco-Everett Research Laboratories and resulted in several patents.[6] The first report of the selective two-step process to appear in the open literature was by the Lawrence Livermore Laboratory group in 1974.[7] This work utilized a C.W. dye laser at 591.54 nm which could be tuned to pump the $5L_6 \rightarrow 7M_7$ ground to excited state transition in either $^{238}U$ or $^{235}U$. Photoionization then takes place from the $7M_7$ state to the continuum by optical excitation with the output from a mercury arc lamp with a wavelength longer than 210 nm. The lamp output was filtered to prevent direct single photon ionization of both isotopes of uranium.

Since this initial work there has been an explosion in the volume of experimental studies relating to the two-step photoionization technique. This work has shown that the method is best suited for isotopic separation of the alkali, alkaline-earth, rare earth, and actinide groups, which are not easily incorporated into molecular species suitable for photodissociation methods.[8] In particular, the technique is ideal when atomic beam conditions are required and is, therefore, of use in solid-state deposition technology. The availability of pulsed dye and excimer lasers with high powers and high repetition rates has had a pronounced impact on the number of atomic species that can be separated using this approach. The tunability of dye lasers and their associated second and third harmonics have resulted in coverage of the photon energy range from 1 to 6 eV. Within this broad range, 53 elements out of 95 (from H to Am) can be accessed by dye laser fundamentals, while an additional 20 elements can

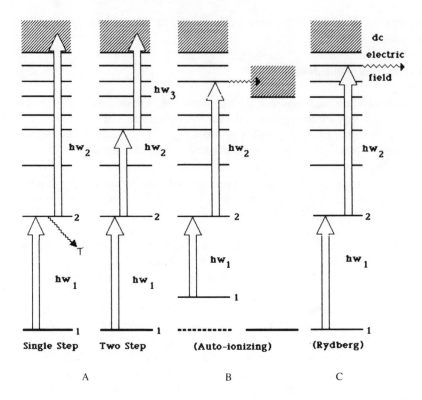

FIGURE 2.    Energy level schematics for : (A) single and two step; (B) auto-ionizing; and (C) Rydberg schemes for photoionization.

be accessed using second-harmonic generation methods. The remaining elements are the lighter species which can usually be chemically bonded and separated as molecular compounds.

The physical mechanisms involved in the two-step ionization process are presented in the following section. The general trends in atomic structure leading to separability, as well as some of the fundamental kinetics and loss mechanisms, are also discussed.

## 2. Selective Photoionization Methods

The selective photoionization of atoms has been accomplished using a variety of multistep processes. All these methods share the common feature of initial photoexcitation followed by ionization. The basic schemes can be classified into three types. The first consists of a single or two-step, highly selective photoexcitation of a given species, followed by an optical ionization step in which the bound electronic state is pumped directly to the continuum with a photon. This latter photon can have energies in the ultraviolet, visible, or infrared, depending on the degree of excitation achieved by the selective preparatory step or steps. This mechanism is schematized in Figure 2A.

The second scheme is based on an initial single or multistep, highly selective excitation followed by an excitation to a bound auto-ionizing state. Auto-ionizing states usually arise due to excitations of inner shell electrons and result in shifted terms from simultaneous excitation of two or more valence electrons. These bound levels lie above the ionization limit for single valence electron excitations and can decay to ionized continuum states of the atom. The auto-ionization can have a high probability of decay to the continuum because of the large density of final states. Unfortunately, auto-ionizing states can also decay radiatively without expulsion of electrons. This latter process results in a loss of ion production and is clearly detrimental to the yield of this separation method. This method is schematized in Figure 2B.

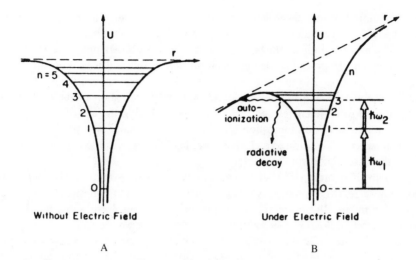

FIGURE 3.   The effect of an applied uniform electric field on the Coulomb potential of a hydrogenic system.

The third type of photoionization method is based on the multistep selective pumping of a given species to highly excited Rydberg states (large principal quantum numbers). Once the selected atoms are in a Rydberg state, electric field-affected ionization takes place. This scheme is shown in Figure 2C. Static electric field-affected ionization is a tunneling process which arises from the bending of the Coulomb potential due to a perturbation of the form:

$$H_{pert} = -eE_o z \tag{3}$$

where $E_o$ is the magnitude of an applied static field in the z direction. The effect of such a field on the screened Coulomb potential results in a penetrable barrier as shown in Figures 3A and 3B. Simple Wentzel, Kramers, and Brillouin (WKB) methods are sufficient to show that the DC field dependence is highly nonlinear and a strong function of the barrier area and, consequently, the quantum number of the Rydberg state.

Each of the above methods of selectively producing ions of one species in a mixture has particular facets which merit additional consideration. These include minimum excitation thresholds and loss mechanisms. It can generally be said that the most efficient coupling of photons to an atom is via transitions between bound states. This observation implies that most often the auto-ionizing state and Rydberg state techniques are more efficient. However, the auto-ionizing levels, it must be pointed out, have comparatively broad absorptions due to the short lifetime of the shifted configuration. These states are as much as $10^6$ times broader than typical atomic transitions ($10^{-5}$ nm).

All of the above methods suffer from interatomic interaction and radiative loss processes. In the case of isotopic mixtures, resonant energy transfer between isotopes is possible and can readily take place when collisions occur. Also, fluorescence from levels involved in the selective excitation stage is a common loss mechanism for all these methods. The Rydberg and auto-ionization schemes can suffer fluorescent losses from the final photoexcitation steps as well. The first method discussed of direct photoionization followed by multistep preparation does not suffer from this latter loss mechanism.

### 3. Selective Photoexcitation

In mixtures of different *molecular* species, selective photoexcitation steps are often easily accomplished, provided that tunable laser sources are available. The excitations take place on $\Delta S = 0$ rovibronic transitions in one of the species. The question of selectivity becomes considerably more difficult when isotopic molecular species such as 2CsCl: $^{238}UO_2Cl_2$ and 2CsCl: $^{235}UO_2Cl_2$ are considered.[9] The gas phase visible and ultraviolet absorptions of most molecular isotopes are indistinguishable, primarily due to broadening mechanisms arising from collisions and Doppler broadening. Careful high resolution measurements of the $UF_6$ molecule in the 360- to 480-nm region have revealed no usable features. The problem is, however, tractable utilizing low temperatures and matrix isolation methods. This complication results in low efficiencies and yield, thereby excluding the use of molecular isotope precursors for efficient photoionization schemes. The molecular isotope shifts are comparatively much larger in the infrared and, therefore, lend themselves more to photodissociation techniques discussed later in this chapter.

The problem of photoionizing dissimilar atoms to separate mixtures of atomic species does not require selective preparative steps, since ionization energies are strong functions of atomic Z number. Furthermore, different atomic species have substantially different chemistries, thereby allowing the possibility of chemical separation.

The problem of selective photoionization of atomic isotopes is considerably more complicated. Although the effects of the physical environment (e.g., temperature, pressure) under which a separation is conducted will be discussed at the end of this section in the context of uranium enrichment, these matters must be considered briefly at this point. The reason for this is that the widths of atomic lines relative to the frequency difference between the same transitions in two isotopes is of paramount importance for achieving any selective preparation using laser sources. The formation of neutral gaseous atomic species from batch mixtures of isotopes is often accomplished via evaporation. This results in relatively high temperatures (over 1000°C) which, in turn, can cause significant Doppler broadening for the lighter elements. This problem is not alleviated for heavier elements since the atomic line-shifts between isotopes also vary inversely with mass.

Considering Doppler broadening to be the largest contribution to the width of the narrow *bound state* transitions used to reach the lower level of the ionizing step, it is required that

$$\delta\nu > \Delta\nu_D \tag{4}$$

where $\delta\nu$ is the difference in frequency between the same transition in two isotopic species and $\Delta\nu_D$ is the Doppler width of the transition, as given by:

$$\Delta\nu_D = 2\nu_o \left(\frac{2kT\ln2}{Mc^2}\right)^{1/2} \tag{5}$$

$\nu_o$ is the transition frequency, M is the atomic mass, and c is the speed of light. With this condition (generalizable to any broadening mechanism characterized by a width $\Delta\nu$), the physical origins and magnitude of $\delta\nu$ for isotopic atomic transitions used in preparative steps can be examined.

The isotopic change in an atomic nucleus results in transition frequency shifts due primarily to three associated nuclear effects: (1) mass shift, (2) shape and size effects, and (3) nuclear magnetic moments. These changes in the nucleus alter the reduced mass, electronic wavefunction, and the fine structure of electronic levels.

The nuclear mass effect causes a shift in the spectra of isotopes due to an alteration of the kinetic energy portion of the atomic Hamiltonian. The kinetic energy in the center of mass frame is composed of the two terms given below:

$$T = \frac{1}{2\mu} \sum_i P_i \cdot P_j + \frac{1}{M} \sum_{i \neq j} P_i \cdot P_j \tag{6}$$

where M is the nuclear mass, $\mu$ is the reduced mass of the electron and the nucleus given by:

$$\mu = \frac{Mm_e}{M + m_e} \tag{7}$$

and $P_j$ is the momentum of the $j^{th}$ electron. This expression shows that the mass effect between isotopes is greatest for the lightest atoms. The shift in the spectra of light atoms can be estimated using hydrogen-like systems whose frequencies are given by the Rydberg formula:

$$\nu_{12} = \frac{R}{1 + \dfrac{m_e}{M}} \left( \frac{1}{n_1^2} - \frac{1}{n_2^2} \right) \tag{8}$$

where R is the Rydberg constant, and $n_1$ and $n_2$ are the lower and upper principal quantum numbers of the transition. This expression for the frequency, $\nu_{12}$, shows that for isotopes varying by a mass $\delta M$, the shift of a spectral line is given by:

$$\delta\nu = \nu_{12} \frac{m_e(\delta M)}{M^2} \tag{9}$$

This result shows that the largest shift occurs for the lightest isotopes, falling off as $M^2$ with increasing nuclear mass. The shifts are about 0.02% for the H isotope lines and a factor of $10^3$ less for K isotope transitions. Thus, the effect is negligible for atomic numbers above 20 and is completely masked for uranium.

For heavier nuclei, the dominant isotope effect in bound-state spectra is primarily due to nuclear size and shape changes. The larger nuclei possess radii which scale as $A^{1/3}$, where A is the total number of nucleons. This results in a relative radius change. $\delta R/R$, due to a nucleon number increase $\delta N$ given approximately by:

$$\frac{\delta R}{R} \cong \frac{1}{3} \frac{\delta N}{A^{1/3}} \tag{10}$$

This shape and volume change causes the intranuclear field to differ from the Coulomb field, thereby altering the electronic wavefunction and state energy. The shift may be approximated for S states from the Racah formula:[10]

$$\delta\nu \cong \frac{2}{3} \frac{\pi}{h} |\psi(0)|^2 Z\, e^2 <r^2> \tag{11}$$

where $\psi(0)$ is the electron wavefunction at the nuclear center, and $<r^2>$ is the expectation value of $r^2$ over the electron wavefunction space. $|\psi(0)|^2$ scales as Z and, therefore, leads to a volume shift proportional to $Z^2$. It should be noted that the isotope volume shift exhibits the opposite sign of the mass shift. This effect dominates all other effects for $A > 100$ and even A nuclei. The even isotopes have no hyperfine structure due to their net nuclear moments. The shape change in nuclei also contributes to the volume-related effects and accounts for large shifts for atoms with $85 < A < 115$.[11]

Nuclei which are large, deformed, and/or have an odd number of nucleons have strong effects on the hyperfine structure and thus exhibit isotope shifts. Magnetic hyperfine splittings arise from the interaction of the nuclear magnetic dipole moment with the net magnetic field of the electrons. This effect scales linearly with Z. The change in the hyperfine effects between isotopes can change the number of levels, as well as the level spacings, since the nuclear angular momentum is altered by the addition of nucleons. In addition, this effect can alter selection rules between transitions in different isotopes. An example of this behavior occurs in the odd isotopes of Hg where forbidden lines of the even isotopes were observed as early as 1928 by R. W. Wood. This type of selective excitation is very useful when the desired isotopes fall in this category. Furthermore, the shift due to hyperfine effects are often larger than those previously discussed.

We conclude this section on the physical basis for selectivity by pointing out that the multistep photoionization approach possesses advantages from the viewpoint of selectivity or enrichment, and drawbacks due to the increase in the number of interaction steps. If each sequential step $i \rightarrow j$ has a photon conversion efficiency (proportional to the absorption cross section) $\epsilon_{ij}$, a loss (spontaneous emission) rate $\gamma_i$, and a selectivity ratio $S_{ij}$ (ratio of cross section at a given frequency between the two species or isotopes), then the following scaling is expected:

$$\text{total selectivity} = \Pi S_{ij} \tag{12}$$

$$\text{total photon efficiency} = \Pi(q_{ij}\epsilon_{ij} - \gamma_i) \tag{12a}$$

where $q_{ij}$ is the photon density at the $i \rightarrow j$ transition. It becomes clear that the more steps one has, the more selective the net process becomes, but at a possible sacrifice in photon energy conversion.

### 4. Ion Separation

Physical separation of the selectively photoionized species is most often accomplished either by electric fields or by plasma chemical reactions. The choice of method depends on the density regime where the system is operated and on the availability of suitable molecules to participate in reactions specific to the ionized species.

Although the detailed kinetics of a particular mixture dictate the laser power required to saturate the selective photoexcitation processes, typical cross sections result in values of a few watts. The photoionization steps have by far the lowest cross section (about $10^{-5}$ less) and require the most energy. This argues for the use of auto-ionizing states if their typically broader lines can be tolerated relative to the shifts required.

Assuming that the optical pumping situation has been optimized, one can then consider the atomic densities involved. When the atomic densities are low, electric field separation is most useful, since this mechanism can be applied to gas mixtures and does not rely on the availability of molecular plasma scavengers. The use of a charged collector plate configuration for uranium isotope separation is shown in Figure 4. The uranium is evaporated in an oven operated at 2000 K. This procedure must be carefully controlled since uranium changes phase at temperatures not far above this point at these low densities. The atomic vapor is then photoionized using a two-step process. If the gas density is low enough, the ions have ballistic trajectories and the transit time to the cathode is independent of density. At higher densities, the ion yield increases linearly with density and the separation coefficient falls off inversely proportional to the density. This regime is followed by one in which collisions between the photoexcited atoms and the isotope become important. In this regime resonant energy transfer between isotopes can occur, resulting in a substantial loss in selectivity. Since the auto-ionizing states are wide and the continuum is broad, the second ionizing step is not selective and can produce ions of both isotopes. In addition to this complication, the ion motion becomes a driven diffusive drift in this regime. These considerations clearly show that the yield of ions decreases with increasing density in this regime.

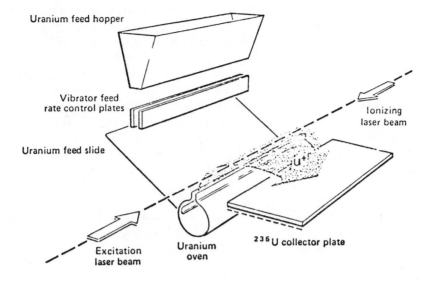

FIGURE 4. Uranium evaporator and electrostatic removal arrangement for photoionization isotope enrichment process. (From Rhodes, G. W., *Quantum Electronics,* Vol. 4, Jacobs, S. F., Sargent, M., III, Scully, M. O., and Walker, C. T., Eds., Addison-Wesley, Reading, Mass., 1976. With permission.)

For typical resonant energy transfer rates and a time of interaction with the radiation of $10^{-5}$ sec, the optimum vapor density is on the order of $10^{14}$ cm$^{-3}$. This is many orders of magnitude higher than the densities required for collisionless molecular beams.

The use of plasma reaction chemistry to isolate the selectively photoionized ions is based on finding suitably reactive species. These species must exhibit reaction cross sections which are considerably larger than the gas kinetic cross sections. Common problems encountered in such removal schemes are (1) the requirement of a different scavenger molecule for each of the species to be separated, and (2) loss of selectivity due to pressure broadening effects and limited reactivity with excited states of the scavenger species. If these difficulties can be effectively eliminated, plasma chemistry removal would be most promising from the viewpoint of scalability and ease of separation.

## B. Selective Photodissociation of Molecules

The process of selective photodissociation of molecules to effect a separation is somewhat similar to that of photoionization. The primary differences are in the molecule-laser interaction, the causes of molecular isotope shifts, and the removal steps. In this section three mechanisms of selective photodissociation and the associated criteria on molecular parameters are discussed; these are (1) single photon processes, (2) two-step photodissociation, and (3) infrared multiphoton photodissociation. Since these techniques usually result in neutral fragments, the removal step is generally chemical. The related mechanism of laser-induced particulation (LIP) is also briefly discussed here, because these schemes are usually initiated by selective dissociation of a single species.

### 1. Single Photon Bond Rupture

The use of photo-predissociation as a mechanism for selective formation of stable dissociation products is extremely efficient, but, unfortunately, is not universal. The primary difficulty occurs when isotopic separations are desired. The process is one in which an ultraviolet photon puts a molecule in a predissociative state. Predissociative states are often short-lived and, consequently, have very broad absorptions, often masking isotope shifts in electronic molecular spectra. Moreover, the spectroscopic data available are often not comprehensive enough to lead to an *a priori* determination of separation feasibility.

One of the most detailed studies employing photo-predissociation is one on formaldehyde.[12] The absorption at the edge of the first singlet state in $H_2CO$, leading to $H_2$ and CO products, has been shown to have a high quantum efficiency.[13] Furthermore, the lifetime of the predissociative state is long enough to allow for the separation of D and H in $H_2CO$ and HDCO mixtures. These experiments yielded separation factors K(D/H) = 9. Separation factors of as high as K = 14 were achieved using a C.W. He-Cd laser at 325.03 nm. From an initial mixture ratio of $H_2{}^{12}CO:H_2{}^{13}CO$ = 1:10, an 80-fold enrichment of $^{12}CO$ was produced.[13] Other work on $Br_2$ has shown that halogen isotopes are also amenable to single-step photo-predissociation. Reactions of photoexcited $Br_2$ with HCl resulted in $K(^{81}Br/^{79}Br)$ = 5.[14,15]

Selective photo-predissociation has also been demonstrated in condensed phases. Separation of nitrogen and carbon isotopes by laser irradiation of s-tetrazine of natural isotopic abundance has been demonstrated in both the condensed phase at low temperature[17] and in the gaseous state.[18]

Single-step selective photo-predissociation promises to be a useful high efficiency separation method for C, N, and O isotopes. Although the method is highly efficient, it suffers from a lack of general applicability. However, recent increases in narrow band, intense ultraviolet sources, such as excimer lasers, promises to increase the general applicability of the method.

### 2. Multistep Photodissociation

Photodissociation of a molecule can also be accomplished using multiple photoexcitation steps. The effect of multiple steps is to greatly increase selectivity over single photon processes due to the rules for combining conditional probabilities. In such a scheme, selectivity is most often achieved by use of IR and UV fields. The IR pumping of vibrational levels in the ground-state electronic well is used to shift the continuous photoabsorption band, which results in dissociation. The mechanisms of two-step IR and UV selective photodissociation and multistep photodissociation are schematized in Figures 5A and 5B.

The process of two-step photodissociation using a resonant selective IR field and subsequent UV dissociation utilizes the vibrational isotopic shifts in vibration-rotation spectra. The increase in nuclear mass of a given atom in a molecule shifts the energy level spacing of the vibrational ladder of all modes as well as the rotational moment of inertia. As an example, we may consider the stretch mode ($\nu_3$) of $^{12}CH_3F$, where the methyl group is vibrating to the first order against the F atom. The $Q_{12,2}$ transition in $\nu_3$ (v = 0 → 1) is coincident with the 9P(20) line of the $CO_2$ laser; the $Q_{12,1}$ transition is 70 MHz lower in frequency. If one considers the same transition in $^{13}CH_3F$ and treats the motion in $\nu_3$ as a diatomic stretch, the ratio of absorption frequencies is

$$\nu^{13}/\nu^{12} = (\mu^{12}/\mu^{13})^{1/2} \tag{13}$$

where

$$\mu^i = \frac{(M_{C^iH_3})M_F}{M_{C^iH_3} + M_F} \tag{13a}$$

and $M_{CH_3}$ and $M_F$ are the methyl group and F atom masses, respectively. This simple calculation, neglecting rotation, yields a shift of 550 GHz (the vibrational frequency is $\sim 3 \times 10^{13}$ sec$^{-1}$). Even more dramatic is a substitution in the hydrogens for the $\nu_4$ mode. The

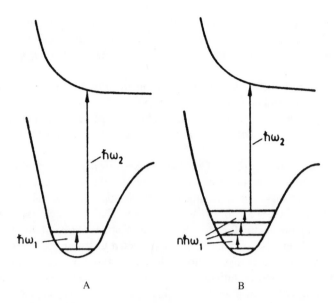

FIGURE 5. Potential energy schematic of (A) single step and (B) multistep, infrared optical photodissociation of molecules.

$\nu_4$ mode absorption lies at 3.3 $\mu$m for $CH_3F$ and at 9 $\mu$m for $CD_3F$. These examples serve to illustrate the sensitivity of IR absorption spectra to molecular structure and mass. In view of the narrow bandwidth of the most often utilized IR pumping source, the $CO_2$ laser, it is clear that IR pumping can be extremely selective. Typical C.W. $CO_2$ lasers have linewidths of 10 kHz and are tunable within a bandwidth of 100 MHz (single line), while the pulsed TEA $CO_2$ lasers have linewidths of 100 MHz (Fourier transform limit of 40 to 100 nsec pulses) and are continuously tunable from 9 to 11 $\mu$m.

The process of selective IR and UV photodissociation has most often been applied in the limit of two steps. The major drawbacks are primarily related to kinetics, which influence the efficiency and selectivity of the mechanism; viz.: (1) thermal populations of vibrational states in the absence of IR pumping: (2) broad electronic photoabsorption tails; (3) resonant energy transfer between isotopes; and (4) rotational "bottlenecking". The first three effects influence selectivity, while the last one limits the rate of energy uptake and affects the overall efficiency in bulk conversion. These effects are examined in detail in review articles specific to this process.[19] The problems associated with selectivity can often be overcome by pumping high-lying modes, cooling the gas, or by overtone excitation. Another alternative is that of multiple step IR preparation as was mentioned earlier and shown in Figure 5B.

The earliest studies of two-step, selective photodissociation were performed on $^{14}NH_3$/$^{15}NH_3$ mixtures.[20,21] This molecule was an obvious target since the $NH_3$ UV spectrum was well studied, and $^{14}NH_3$ has several coincidences with $CO_2$ laser lines. This early work resulted in separation factors as high as 6. Other isotopes such as $^{10}B$ and $^{11}B$ have been separated using $BCl_3$ gas phase precursors.[22]

Because of its wide applicability, this method is one of the most actively pursued schemes. The increase in UV spectral data and the development of excimer lasers for the 200- to 300-nm range will greatly increase the number of species accessible to a single IR and UV laser system.

## 3. Selective IR Multiphoton Dissociation

This section is concerned with the process of selective dissociation of polyatomic molecules into stable separable fragments using intense IR fields. This process is one of the most global, since almost all isotopic molecular species possess the large relative shifts in the IR region mentioned in the previous section. Multiphoton dissociation involves the absorption of as many as 40 photons in a collisionless process and can result in separation factors as high as 3000.

The first reported isotopically selective process induced by the absorption of high power $CO_2$ lasers was due to Ambartzumyan.[23] Since the early work on $BCl_3$,[24-26] $SF_6$,[27-29] and $OsO_4$,[30] the multiphoton dissociation and associated isotope separation feasibility of other molecules have been studied. These include $S_2F_{10}$,[31] $OsO_4$, $CF_3I$, $CF_3COCF_3$, $SF_5NF_2$, $UF_6$, $C_2H_4$, OCS, and $D_2O$.[32]

The first experiments using intense $CO_2$ laser pulses resonant with the (v = 0) → (v = 1) vibration-rotation transition of a given isotopic vibrational mode resulted in the irreversible generation of molecular dissociation fragments of the isotopic species. This process required a threshold intensity in the $10^7$- to $10^9$-$W/cm^2$ range. The first such experiments on $SF_6$ resulted in an enrichment factor for the $^{34}S$ isotope of about 3000. Many gases, however, show no irreversible composition change when subjected to strong IR fields. In such cases (e.g., $BCl_3$, $OsO_4$) the result is dissociation followed by rapid recombination. This view is supported by the observation of luminescence from fragment species. Separation can still be achieved for these molecules if active scavengers are present to compete with the recombination chemistry. In the case of $BCl_3$ and $OsO_4$, the use of $O_2$ and $C_2H_2$ as scavengers (respectively) has been successful.

The basic physics of the *collisionless* absorption of tens of IR photons by a polyatomic molecule, and its subsequent dissociation, has only reached maturity in recent years. The experimental "workhorse" for these theories has been the $SF_6$ molecule.[33] This is due to its relatively uncomplicated rotational structure (i.e, the spherical top which requires only the total angular momentum quantum number, J, to describe its gross rotational structure), and its strong and well-studied absorption coincidences with $CO_2$ laser lines.

The dissociation of $SF_6$ to form $SF_5$ and F requires 41, 10-μm photons. Bloembergen has hypothesized a detailed mechanism based on a six-photon absorption process in the $\nu_3$ band, followed by the absorption of the remaining 35 photons in the so-called quasi-continuum of all the molecular modes.[34] This mechanism is depicted in Figure 6. The molecular quasi-continuum of modes arises from the strong nonlinear coupling between the excited states of all the modes and is believed to be fairly adequately represented by a nondeterministic evolution leading to ergodic behavior. The two most arbitrary assumptions in dealing with a theoretical model which employs this picture are (1) the inception point of the quasi-continuum, and (2) the magnitude of the matrix elements involved in the absorption of the remaining 35 photons.

The n-photon process, followed by subsequent coupling to the continuum, leaves the molecule in a state of superexcitation above the dissociation limit of several bonds. At this stage the molecule has access to several energetically favorable decay channels. The bulk of the experimental evidence indicates that the *weakest* bond is most often broken. This type of behavior is, in fact, predicted by the RRKM theory of unimolecular reactions.[35-37] The detailed application of this theory to $SF_6$ accurately predicts the shift of the optimum dissociation wavelength to the red, as well as the threshold energy fluence of about 1 MW/$cm^2$. The theory also predicts the rapid increase in the dissociation probability with energy fluence. The results of some experiments with $SF_6$ are presented in Figure 7.

In addition to direct, selective, multiphoton dissociation, other schemes utilizing a preparative excitation step have also been proven successful. One such approach uses a C.W. or pulsed $CO_2$ laser to pump a (v = 0) → (v = 1) transition or a strong overtone. This step is then followed by another intense $CO_2$ laser pulse which causes the dissociation.

The process of IR multiphoton dissociation suffers from collisional energy randomization if pulse widths are not kept well below the inverse V-V (vibrational-vibrational energy)

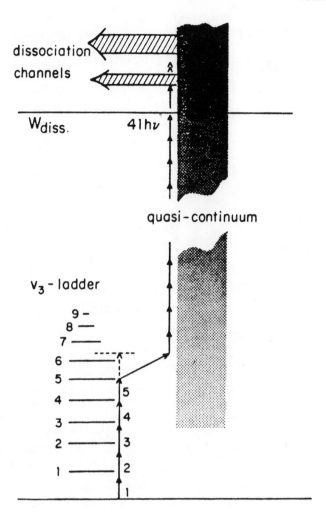

FIGURE 6. Proposed excitation path for the infrared multiphoton dissociation of $SF_6$. (From Bloembergen, N., *Opt. Comm.*, 15, 416, 1975. With permission.)

transfer rate. In addition, the selectivity is often poor when heavy molecules such as $UF_6$ are utilized due to the high rotational state density. In such cases, cooling by adiabatic expansion or other methods must be utilized in order to limit the vibrational thermal population to low temperatures, and to insure that Doppler broadening does not cause rotational states to strongly overlap.

This method is, perhaps, the most promising scheme for commercial isotope separation because of its global applicability and its high efficiency. The latter is due to the near-unity dissociation probabilities per molecule that can be achieved, and the high "wall socket" efficiency of the $CO_2$ TEA laser.

### 4. Laser-Induced Particulation (LIP)

Apparently, the first reported light-induced particle formation phenomenon was by Tyndall in 1869 concerning the illumination of "nitrite of butyl". Since then, numerous observations of light-induced particulation have been reported in the literature (e.g., see Luria et al.[38]). Thus, in order to limit the current discussion, here we concentrate solely on studies employing laser sources. This category of LIS schemes (i.e., LIP) is included at this point in the chapter,

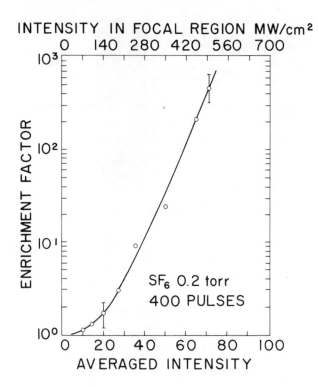

FIGURE 7.   Enrichment factor vs. average intensity of $CO_2$ laser pulses for $SF_6$ sulfur isotope separation using multiphoton excitation. (From Ambartzumyan, R. V., Gorokhov, Yu., A., Letokhov, U. S., and Makarov, G. N., *JETP Lett.*, 21, 171, 1975. With permission.)

because, in most cases, these processes are initiated by selective dissociation of a certain species by laser radiation, followed by *spontaneous* formation of solid particulates, which, of course, comprise a phase separate from the original gas phase, differing in composition, which can then be easily removed by conventional means. However, it must be clearly understood that the detailed mechanisms of the various LIP processes vary depending upon the particular chemistry that is operative in a specific system. Some of these LIP schemes can also be categorized or viewed as laser-assisted or induced chemistry (LIC), which is rapidly becoming quite a large field in and of itself (e.g., see Knudtson and Eyring[39]). Thus, once again, in order to limit the scope of the current chapter, here we concentrate exclusively on those LIP schemes that produce a second solid phase upon irradiation of a fluid phase and seem to have some clear separations connotation.

The work of Ronn et al.[40-42] falls into this category. Upon irradiation of various species in the vapor phase with a $CO_2$ TEA laser (P[20] line of the 10.6-$\mu$m branch), Ronn[40] reported substantial particulate formation. More specifically, irradiation yielded solid particulates of iron from ferrocene [$Fe(C_5H_5)_2$], molybdenum metal from molybdenum hexacarbonyl [$Mo(CO)_6$], and sulfur from OCS. The mechanism for these processes was nonresonant laser-induced breakdown (LIDB) produced by the very high ac electric field generated by the intense IR radiation, which breaks the weakest bond available in the system to initiate the nucleation/particulation process.

In subsequent work, Ronn and Earl[41] used a somewhat similar LIDB method to produce $UF_5$ particulates from $UF_6/H_2$ mixtures. Unlike the previous results, when $UF_6$ was present alone in the irradiation cell, no particulate formation occurred. However, in the presence of

$H_2$, the introduction of radiation caused a dramatic pressure drop in the cell, the appearance of particulates, identified as $UF_5$, the disappearance of major $UF_6$ bands in IR absorption spectra, and the appearance of HF bands. Thus, the mechanism was postulated to be

$$2UF_6 + H_2 \rightarrow 2UF_5 + 2HF$$

Apparently, hydrogen acts as a scavenger for F, which would otherwise rapidly recombine with $UF_5$. The efficiency of the process was characterized by the energy deposited by the laser per $UF_5$ molecule formed, which was about 612 to 750 eV per molecule.

Since LIDB seeks the weakest bond in the system for dissociation, it is not a very selective process in excitation, but is quite selective thermodynamically; i.e., the subsequent particulation reactions produce species that are more thermodynamically stable and do not back-react to any appreciable extent. For this reason, LIDB seems to have some potential for bulk chemical processing operations. However, due to the low selectivity of the excitation step, it would not be suitable for isotope separation. In order to investigate this point further, Lin and Ronn[42] examined particulate formation in $SF_6$ and $SF_6/H_2$ mixtures, using both resonance multiphoton absorption and the off-resonance LIDB process. In the resonance mode, pulsed $CO_2$ laser emission at 944.15 cm$^{-1}$ was strongly absorbed by the $v_3$ vibrational mode of $SF_6$. In pure $SF_6$, certain products of the subsequent fluorine atom chemistry with contaminants and cell materials were observed (by IR and mass spectrometric analyses), as well as some fine particles (via scattering of He-Ne laser light). However, although particles were observed during the laser pulse, no particulates were observed after the experiment. This was attributed to the fast recombination rate of fluorine and sulfur atoms, allowing little sulfur particulate matter to be produced. In 1:1 $SF_6$:$H_2$ mixtures at total pressures below 100 torr, however, in addition to similar species detected in the vapor phase, particulate matter was formed and precipitated isotropically throughout the cell. At total pressures less than 100 torr, these aggregates were so small that they remained suspended for seconds, At pressures exceeding 100 torr, very "thick" particulates were deposited on the wall. The size of the large particles was measured (by rate of fall and electron micrography) to be in the 1- to 3-μm-size range for 100 torr and was observed to increase with pressure. The smaller nucleation embryo particulates were estimated at several tens of sulfur atoms. The threshold pressure for LIDB of these same mixtures using the nonresonant P(20) 9.6-μm laser line was found to about 20 torr. Similar large particulate matter was formed, but deposited solely around the focal point. Product analysis showed identical species to those found in the resonance work. It was concluded that the reaction of $SF_6$ with or without $H_2$, initiated by laser radiation, occurred in two different stages: (1) inside the radiation volume, $SF_6$ molecules are completely dissociated; and (2) subsequent recombination processes produce highly excited species, followed by thermodynamically controlled production of final vapor phase products and particulates.

Another series of reports that has appeared in the literature is concerned with the production of particulates upon irradiation of alkali metal vapors in a hydrogen (or deuterium) buffer gas. Tam et al.[43] observed particulate formation upon dissociative excitation of $Cs_2$ to Cs atoms in the 7P state with an $Ar^+$ laser at 454.5 or 457.9 nm, resulting in the production of large densities of CsH molecules in the vapor phase. Due to the very low vapor pressure of CsH, this creates a supersaturated condition with respect to CsH crystals, leading to nucleation and the formation of dense clouds of particulates. A similar effect was observed for rubidium mixed with $H_2$ (150 torr or greater) at 350°C when irradiated with 476.5 or 488 nm line of an $Ar^+$ laser.

In a similar fashion, Tam et al.[44] reported on the production of dense clouds of CsD particulates in periodic pulses when mixtures of $D_2$ gas and Cs vapor were irradiated with a single mode C.W. dye laser (operated with R6G) at 601.05 nm. The initiation step in this case was traced to the dissociative excitation of $Cs_2$ molecules to Cs atoms in the $8D_{3/2}$ state. The oscillatory production rate was attributed to a growth/depletion feedback mechanism, whereby the production of CsD in the laser beam results in the production of CsD nuclei which, in turn, act as very efficient sinks for additional CsD molecules. This then depresses the concentration of CsD within the laser beam enough to slow down additional particle growth. The larger particulates then fall away from the beam due to the influence of gravity, thereby depleting the number density of nucleation sinks for the CsD produced within the laser beam, and initiating a new cycle in the particulate formation process. The particulate intensity decreased (as determined by simultaneous sideward Mie scattering), the oscillations weakened, and the period increased as the laser was detuned. Due to the narrow bandwidth of the effectiveness of this process, it was suggested that this scheme might be developed into a method for isotope enrichment for the alkali metal or hydrogen isotope.

Even more recently, Allegrini et al.[45] reported the formation of ultrafine particles by resonant laser excitation of potassium vapor in a closed cell, water-cooled at the ends, or, more accurately, a heat pipe oven (HPO). Potassium metal was heated in the center section of the pipe to form potassium vapor at about 480 to 650°C. The gas in the two ends of the cell consisted mostly of the buffer gas, since the potassium vapor was concentrated at the midsection of the HPO. Upon irradiation with the 647.1- or 676.4-nm line of a $Kr^+$ laser, a fog or snow was formed, and then a macroscopic deposit of microcrystals was observed at the laser entrance end of the HPO, while the other end remained practically free from the deposit. The deposition rate was estimated at about $10^{-6}$ g/W sec. It was noted that particle formation was observed only under certain operating conditions, and only at the buffer gas/potassium vapor boundary located near the laser entrance window end of the HPO. It was also noted that the particulate fog produced was fairly independent of the buffer gas used. Fine particle formation was simultaneously observed by scattering of the 514.5-nm line of an $Ar^+$ laser. These latter observations revealed oscillatory behavior of the fine particulate concentration, which was explained in terms of a feedback mechanism whereby the $Kr^+$ laser radiation, which was responsible for the particulation, began to be scattered at high-enough particulate concentration, thereby reducing its effectiveness. Once the particulate formation rate was decreased, scattering of the laser radiation also decreased, which caused the particulate formation rate to increase once again, thereby initiating the next cycle.

The general mechanism for particulate formation was hypothesized as follows. It is believed that resonant laser excitation of $K_2$ molecules in the vapor phase by the $Kr^+$ laser radiation produces intense laser heating of the gas at the boundary between the buffer gas and potassium vapor caused by collisions of the excited species with the rest of the gas. This process essentially converts internal energy provided by the laser to translational energy of the local gas phase. This induces a transient temperature/pressure disequilibrium which tends to cause the potassium vapor to expand into the buffer gas, and the concomitant supersaturation results in particle formation. The fact that no chemical reaction occurred was substantiated by X-ray spectroscopy that clearly proved the macroscopic particles to be pure potassium. In a sense, this process is similar to seeded supersonic free jet techniques for producing clusters and agglomerates as a result of the supersaturation caused by the adiabatic expansion (e.g., see References 45 and 46) except for the fact that there are no constricting orifices for maintaining the pressure difference required for forming the expansion. Thus, transient laser heating in this case takes the place of the constricting orifice to establish the requisite temperature/pressure difference between the source region and the lower pressure expansion region consisting of the buffer gas.

Obviously, the preceding LIP schemes are relatively recent and still quite limited in application. However, the distinct feature of producing a separate solid phase upon laser irradiation, which is easy to remove, makes LIP schemes likely candidates for future separation-oriented research.

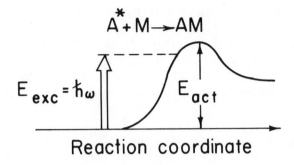

FIGURE 8. Potential energy schematic of the effective lowering of the activation energy of a chemical reaction by internal vibrational or electronic excitation.

## C. Photochemical Separations

Photochemical separation of isotopes is based on the selective optical enhancement of molecular or atomic reaction rates with a third species. This process has been demonstrated using electronic excitation as well as vibrational state pumping. In its most elementary form, this process may be viewed as lowering the effective activation energy along the reaction coordinate, as shown in Figure 8.

### 1. Vibrational Excitation

The first proposal for using vibrational excitation of reacting molecules to separate gases was due to Gilbert in 1963.[48] However, the first experimental vibrational photochemical effect was shown by Hall and Pimentel in the same year.[49] The effects of vibrational excitation on reaction rates can be dramatic. Examples of vibrational state reaction rate enhancements are

$$N_2 \ (v = 4) + O^+ \rightarrow NO^+ + N$$

and

$$HCl \ (v = 1) + K \rightarrow KCl + H$$

The rate of the first reaction is enhanced by a factor of $10^3$, while that of the second reaction by a factor of $10^2$ above the room temperature thermal rates of these reactions. The more remarkable fact is that if the same energy per molecule is put into translation, the rate of the HCl reaction only increases by about a factor of 5. These reactions belong to the class with energy barriers located relatively late in the reaction coordinate and thus are very sensitive to vibrational excitation.

The selective excitation of vibrational levels may be accomplished either singly or in multiples of vibrational quanta. A single photon may be used to pump a $\Delta v = +1$ transition or a $\Delta v = +2$ overtone. Multiphoton absorption using intensities on the order of $MW/cm^2$ can produce molecules possessing many quanta of vibration. In addition, a Raman-type process can be utilized to excite vibrations in modes with zero dipole matrix elements. This method must always be employed when homonuclear diatomics, such as $N_2$, are to be excited. This process however, is not isotope-selective unless a second Stokes field is tuned to generate a coherent two-photon resonant process.

Successful isotope enrichments have been reported for Cl, C, N, K, B, and H species. By making use of the rate enhancement of the HCl + Br reaction with the two-step excitation of HCl to v = 2, $^{35}$Cl was separated.[50,51] HCl was excited using a pulsed HCl chemical laser. The separation indicated that the reaction rate for HCl (v = 2) is a factor of $10^{11}$ larger than that for v = 0 at room temperature*. Intense $CO_2$ laser excitation of $BCl_3$ + $H_2S$ mixtures resulted in high vibrational excitation of $BCl_3$ and photochemical separation of B isotopes.[53] Finally, Raman-excited $N_2$ in air yielded NO molecules enriched in $^{15}$N by a factor of 100. The interpretation of these results, however, is not clear.[54,55]

### 2. Electronic Excitation

Excited electronic state chemistry has been the subject of study for many years. Electronic state excitation energies are of the same order of magnitude as chemical bond energies and, therefore, can often dramatically alter reaction rates. Using orbital correlation diagrams, one can reasonably predict the relative reactivity of excited and ground electronic states. In order to make use of this effect the remaining task is to selectively excite isotopes or components of a gas mixture.

The selective electronic excitation of molecules most often poses no problems when chemically different mixtures are considered. Unfortunately, however, when isotopically selective electronic excitations are required, problems do occur. Most polyatomic molecules do not have well-resolved rotational structure with isotopic shifts in the visible and UV regions. However, some molecules do possess shifts which may be used for photochemical separations.

In order to produce a successful selective photochemical separation, a suitable reactant must be found, and the following requirements must be considered:

1.  The reactant and the molecule must not react under ambient conditions or with the container walls.
2.  The excited state should have the following properties: (1) long fluorescence lifetime, (2) small predissociation cross section, and (3) higher reactivity cross section with the reactant than that for quenching.
3.  Energy transfer to the isotope not optically excited must be avoided.
4.  Pressures must be kept low enough to insure that the homogeneous linewidth does not exceed the isotope shift.
5.  The reaction products must be stable and should not react with either isotope precursor.
6.  If unstable species or free radicals are generated, they must be chemically blocked by appropriate scavengers.

Apparently, the first attempt at electronic photochemical isotope separation was by Hartley et al.[56] Pertel and Gunning enriched $^{202}$Hg from the 30% natural abundance to 85% in an Hg, $H_2O$, and butadiene mixture. The complete reaction mechanism for this separation is not completely understood.

Isotopic enrichment of diatomic species has been accomplished with both atomic resonance lamps and lasers. Photolysis of CO at 206 nm resulted in a 30:1 excitation ratio between $^{13}$CO and $^{12}$CO, or $C^{18}O$ and $C^{16}O$. It is believed that selectivity was partially lost due to resonant isotopic energy transfer.[58] Similar photolysis experiments were performed on $N_2$ and resulted in a photochemical enrichment factor of 4 for $^{15}N^{14}N$.[59] The 514.5-nm line of the argon ion laser has been used to enrich ortho-$I_2$.[60] The reactive partner for the ortho-$I_2$ separation was 2-hexene. The halogenation reaction of the olefin is believed to occur via a radical chain mechanism. The experiment indicated that C.W. irradiation could be used to produce photochemical separation, while simultaneously serving as a pump for fluorescence monitoring.

Dye laser excitation of the $I^{37}$Cl molecule at 605.3 nm resulted in photochemical separation

---

*   The reaction of $CH_3OH$-$CD_3OD$ mixtures in $Br_2$ induced by an HF laser has also been reported to be isotopically selective.[52]

of $^{35}$Cl and $^{37}$Cl.[61] Because of the potential well crossing at 18000 cm$^{-1}$ ICl can readily predissociate. However, in the experiment cited, the photon energy was kept lower than the crossing of the A and X states. The conclusion is that the pumping selectively produced A-state species with enhanced activity. The reacting partners were *cis-* and *trans-*ClHC=CHCl. Other enrichments of Cl have been performed using Ar$^+$ laser radiation (465.78 nm) and Cl$_2$CS as a precursor.[62] The reaction partner selected for these separations was diethoxyethylene.

## D. Photodeflection

The first proposal for using photodeflection of atomic beams was due to Pressman.[63] Following this original suggestion, several other groups studied the possibilities for using this scheme as a general technique for isotope separation.[64]

The physical mechanism upon which photodeflection is based is the net transfer of momentum from photons to atoms in a cycle of induced absorption and spontaneous fluorescence. When an atom absorbs a resonant photon of frequency $v_0$, the atom gains a momentum $hv_0/c$ in the photon direction. If the atom decays back down via one or a series of *spontaneous isotropic* decays, the effect is a net transfer of momentum. Using this mechanism, an atom in a beam can be deflected tangentially to the direction of motion by tuning a laser to the stationary atom's transition frequency (neglecting second-order Doppler shifts). Therefore, if a beam of several isotopes is sent past a region where a laser beam tuned to one of the isotope's shifted frequencies propagates tangentially, a selective deflection and a separation can take place. Since atomic isotope shifts in molecular species are often present and large in comparison to the linewidth and fluctuation width of tunable dye lasers, this method potentially offers wide applicability.

In the presence of a resonant field, the atoms may undergo absorption, stimulated emission, and isotropic (classical) fluorescence. The difference in the number of absorption events minus the number of stimulated emission events during traversal of the laser beam determines the net amount of transverse momentum than can be transferred to the isotopic atom. It is, therefore, important that the rate of stimulated events is kept below the spontaneous transition rate for atoms bathed in a resonant field. If the atomic beam has an average velocity V and angular beam spread of $\Delta\theta$, separation may be achieved when

$$N > 2MVC(\Delta\theta)/hv_0 \tag{14}$$

where N is the difference in the number of absorption events and stimulated emission events per atom, and M is the atomic or molecular mass. Using well-collimated beams and a combination of optimum intensities and interaction lengths, a separation can be achieved by employing appropriate skimmer configurations for physical separation.

The first experimental demonstration of photodeflection for separation of heavy isotopes was due to Bernhardt et al. in 1974.[65] The experiment used the isotopic hyperfine structure of the Ba(I) 553.57-nm resonance line. The pumped transition in Ba(I) is the $6s^{2\,1}S_0 \rightarrow 6s6pP_1^o$ with a spontaneous lifetime of 8.37 nsec. In barium, metastable states limit the relaxation back to the ground state and the efficiency of the photodeflection mechanism. Resonant scattering terminates when an atom reaches the metastable $6s5d^1D_2$ state whose dipole transition to the ground state is parity forbidden (see Figure 9). The limitation on the number of absorption events due to metastable beam trapping results in a minimum required collimation for the atomic beam. The atomic energy levels relevant to the Ba(I) separation are shown in Figure 9.

The Ba(I) experiment utilized a rhodamine 6G dye laser with an output linewidth of 10 MHz. The laser was introduced at right angles using a cylindrical collimation system. The apparatus produced barium atoms with an average velocity of $4 \times 10^4$ cm/sec and a beam divergence of $2 \times 10^{-3}$ rad. Under these experimental conditions the average number of photons absorbed per barium atom was 25, resulting in a tangential velocity of 20 cm/sec.

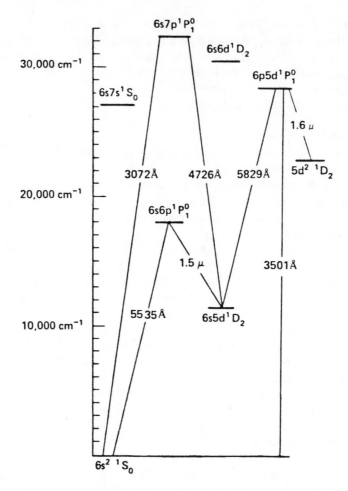

FIGURE 9.    Relevant energy levels for Ba photodeflection scheme. (From Bernhardt, A. F., Duerra, D. E., Simpson, J. R., and Wood, L. L., *Appl. Phys. Lett.*, 25, 617, 1974. With permission.)

Beam separation was observed visually as well as with a mass spectrometer. The most abundant isotope, $^{138}$Ba, could be optically monitored by a decrease in fluorescence at the end of the beam due to metastable state trapping. This interpretation was confirmed using a second dye laser tuned to the 582.9-nm $6s5d^1P_1$ transition. Furthermore, when a mass spectrometer sampling slit was placed at the end of the beam farthest from the oven, the $^{138}$Ba peak could be altered by a factor of 2 to 3 by tuning the laser to the $^{138}$Ba absorption. This method exhibited a 70% efficiency for separation of the $^{138}$Ba isotope. The experimental apparatus is shown in Figure 10.

Subsequent work aimed at increasing the efficiency of this separation was performed using multifrequency pumping of Ba.[66] The multifrequency approach was aimed at circumventing the previously mentioned fluorescence trapping in the $6s5d^1D_2$ state. This scheme utilized a second rhodamine laser at 582.9 nm to pump atoms to the $6p5d^1P_1^o$ state from which a strongly allowed electric dipole transition to the ground state (350.1 nm) can take place. Unfortunately, another metastable state, the $5d^{2\,1}D_2$ state, is also available to the $6p5d^1P_1^o$ state via a fluorescent transition at 1.5 μm. The two-laser system yields an efficiency of 83%, a 13% increase over the single laser system. In summary, this method can approach a 100% efficiency for separation of heavy isotopes. Unfortunately, the low density conditions

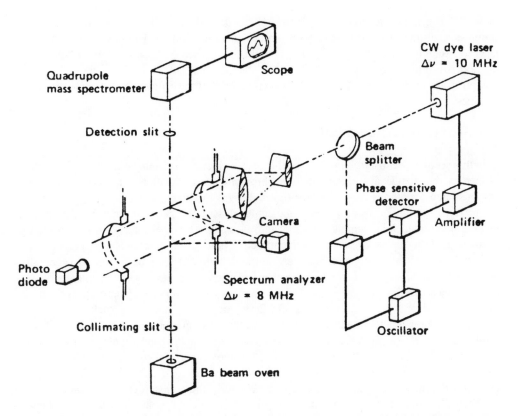

FIGURE 10. Schematic of the Ba phototdeflection experiment. (From Bernhardt, A. F., Duerre, D. E., Simpson, J. R., and Wood, L. L., *Appl. Phys. Lett.*, 25, 617, 1974. With permission.)

which must prevail to maintain the collisionless regime severely limit the net rate of isotope separation. Due to the latter consideration, this scheme has, to date, been restricted to laboratory research applications.

### E. Selective Adsorption

The selective adsorption separation scheme utilizes a cooled surface as a trap for specific istopes in a gas mixture. The process employs selective vibrational excitation of an isotope to lower its capture (sticking) coefficient. This method was first proposed by Gochelashvili et al. in 1975.[67]

The rate of adsorption of molecules in thermodynamic equilibrium with a cold surface is determined by the capture coefficient C:

$$C = 1 - \exp(-E_c/E_T) \tag{15}$$

where $E_c$ is a critical binding energy specific to the molecule and the surface, and $E_T$ is the translational energy of the incident molecule. Theoretical calculations on the interaction of a molecule and a semi-infinite chain of atoms indicate that the atom is adsorbed if $E_T < E_c$ and reflected if $E_T > E_c$.

Experimental data on the variation of capture coefficients with temperature indicate that the effect of vibrational excitation on C may be reasonably explained by replacing $E_T$ by E $= E_t + E_v$, where $E_v$ is the vibrational energy content of the incident molecule. With this model, a different capture coefficient can be associated with a selectively excited isotope.

The first experimental verification of this method for isotope separation was due to Karlov et al.[68,69] The molecule under study was $BCl_3$, with [11]B and [10]B. The experiment used a stainless steel tube reactor with a 160 K wall temperature and the 10.6-μm line of the $CO_2$ laser to excite the $\nu_3$ mode of $BCl_3$. The amount of enriched $BCl_3$ was 2.2 mg in 1 hr. This method is promising because of its wide applicability, but suffers from the requirement of low pressures ($\sim 10^{-2}$ torr). The latter is a consequence of the efficient V-V transfer between isotopes.

### F. Laser-Induced Diffusion (LID)

*1. Overview*

In this section, we discuss separation schemes based on the selective alteration of transport properties due to the absorption of monochromatic laser radiation. This effect has been demonstrated for atomic as well as for molecular gas mixtures.

The transport properties of atoms or molecules in specific quantum states are determined by their intermolecular potentials. When electronic states are excited, the electron distribution changes, and the repulsive as well as the attractive parts of the intermolecular potential are altered. Electronic excitation produces new multipole field distributions, as well as a generally higher polarizability. The latter has the effect of increasing the dispersion interaction. In some species such as acetylene, conformal changes can also occur. These changes occur for small atoms, such as Na, as well as for large conjugated molecules, such as dyes.

The excitation of vibrational modes also changes the intermolecular potentials. This is primarily due to the increase of the molecular bond lengths and dipole moments with increasing vibrational quantum number. These changes in hard sphere radii can be estimated by spectroscopic and low temperature transport and virial coefficient data. Typically, this effect results in, at most, a change of a few percent in the collision cross section.[70]

The two dominant mechanisms resulting in a laser-induced molecular flux of a selected species are elastic collision cross-section changes and velocity direction-changing relaxation. In both cases, the effect can be induced in open as well as closed systems. The development of the theory for laser-induced diffusion (LID) has only recently considered the second mechanism, which is often present simultaneously for certain vibrationally excited molecules.[71] In the next section we discuss the fundamentals of the elastic collision cross-section change effect. Also discussed is recent theory for certain systems, such as the $\nu_3$ excitation of $C_2H_2$, where collisional redirection in the relaxation can cause signficant LID effects.

*2. Elastic Collision LID*

In 1979, Gel'mukhanov and Shalagin reported on the effect of light-induced diffusion (LID) in gases.[72] In a later paper that year, the use of the method for estimating transport coefficient was also discussed.[73] In 1980, a more rigorous theory for the LID effect for an ideal two-level system was presented.[74] This work was based on the solution of the coupled set of density-matrix and kinetic equations. Although this latter work clearly explains the effect, it could not be readily applied to polyatomic systems where several degrees of freedom and many quantum states are involved.

Recent publications dealing with the LID effect in polyatomic systems discuss the significantly more complicated theoretical questions which arise for complex molecules absorbing IR radiation.[75]

The essence of the LID phenomenon is as follows. Let the frequency of a plane monochromatic wave differ slightly from the frequency, $\nu_o$, of the transition of the molecule. Then, due to the Doppler effect, the velocity distribution in the ground and excited states becomes asymmetrical. Depending on the ratio of homogeneous to inhomogeneous widths, the asymmetry can vary in its shape. The most important effect of this asymmetry is that the average velocity in each state is no longer equal to zero. Hence, in each of the states involved there is directed movement which can be characterized as a flux.

At this point it is important to recall that the "size" of an atom differs depending on its state of excitation. This is a hard-sphere description of the state dependence of transport coefficients. When the absorbing atoms or molecules are mixed with another gas, the presence of the second gas produces "friction" with the flows of the absorbing species. However, the friction forces for excited and nonexcited atoms differ due to the difference in transport coefficients. Thus, a net force results with which the second buffer gas acts on the absorbing atoms, causing them to have directed motion. Another view of the process is that the selective excitation and accompanying cross-section change of molecules moving toward or away from the laser result in an anisotropic diffusion process.

The laser-induced diffusion effect in a closed system can be said to reach steady state when the flow due to ordinary Fickian diffusion exactly balances the laser-induced flow. Since the system is closed, the total pressure along the cell will be constant, while the partial pressures of the active gas and the buffer gas will be a function of the position. Designating the flux due to the LID effect as $J_A$ and the density of the active has as $\rho_A$ the equilibrium condition is represented by:

$$J_A = D_{AB} \nabla \rho_A \qquad (16)$$

where $D_{AB}$ is the mutual diffusion coefficient of the gas mixture. This expression is the starting point for the theory of polyatomic LID.

The general form for $J_A$ is derived based on an optical pumping rate equation and a model of a mass moving through a dissipative medium (buffer gas) driven by a constant force (thermal translational bath).[74] This view is based on considering dissipation as any process that changes the direction of an active gas molecule; i.e., a Langevin model. The equation of motion for such a system is given by:

$$m\dot{x} + \beta\dot{x} = F \qquad (17)$$

The solution of this equation yields a limiting velocity

$$v_o = F/\beta$$

where m is the mass of the species, F is the force, and $\beta$ the friction constant. We are, therefore, interested in the limiting velocity for active molecules in the ground vibrational state and those in the excited vibrational state connected by the pump laser.

The macroscopic flow $J_A$ will be a result of velocity distributions summed over the relevant quantum states of the polyatomic system. The derivation for $J_A$ can be given in terms of the molecular velocity group selected by the laser, V, and the ratio $\beta_g/\beta_e$, where $\beta_e$ and $\beta_g$ are the friction constants for the excited vibrational state and the ground state, respectively. Calculations for $CH_3F/He$ mixtures in the Doppler regime result in separation factors of 1.2 for a closed 1-mm diameter, 1-m-long tube at a total pressure of 5 torr, and an initial ratio of $CH_3F$ to He of 0.2.

### 3. Experimental Efforts

Experiments in LID were first reported by Antsignin and Atutov.[76] These experiments represent the first deliberate measurements, while LID effects may have been observed and unexplained by others previously.[77] The work by Antsignin et al.[76] was performed under flowing conditions and utilized sodium vapor as the absorber and helium as the buffer gas. Although these were not closed-system experiments, a two-level model calculation was given for the profile which might exist in such an experiment. The latter indicated that at modest dye-laser intensities, all the sodium could be confined to a length of less than 1 mm. Recently,

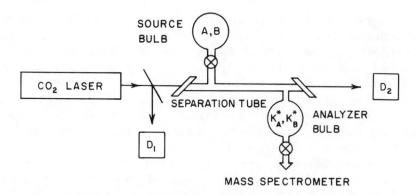

FIGURE 11.    Typical laser-induced diffusion experiment for polyatomic systems ($D_1$ and $D_2$ are detectors for measuring the power absorption). (From Baranov, V. Yu., Velikhov, E. P., Dykhne, A. M., Kazakov, S. A., Mezhevov, V. S., Olov, M. Yu., Pesmennyl, V. D., Starodubtsev, A. I., and Starostin, A., *Zh. Eksp. Teor. Fiz. Pis'ma Red.*, 31, 475, 1980; *JETP Lett.*, 32, 445, 1980. With permission.)

the time evolution of the steady-state profile has been observed for the sodium system.[78] This experimental result led to questions concerning the use of this method as a gas separation scheme for polyatomics which absorb in the infrared. The first such experiment was performed on $SF_6$ in $H_2$ and He mixtures[79] using the $CO_2$ laser 9P(20) excitation of $SF_6$. Measurements were performed by spectrophotometry and mass spectrometry at the end of the cell. Unfortunately, these results were obtained with pulsed excitation which resulted in the excitation of many vibrational modes. Subsequent to this work another experiment was performed in order to more specifically address the issue of gas separation. These experiments were performed on $^{12}CH_3F{:}^{13}CH_3F$ mixtures[80] and resulted in enrichment factors of about 1.1. A typical experimental setup for observing the LID effect is shown in Figure 11.

### 4. Collisional Relaxation and Redirection

Recently, theoretical work on the LID effect has shown that for certain molecular vibrations excited by the pump laser another new mechanism for the generation of a macroscopic flow exists.[81] This effect emerges when nonisotropic scattering is necessary for vibrational relaxation. Such a situation arises when the normal mode excited by the laser is primarily a stretching motion. An example of this is the $\nu_3$ mode $C_2H_2$. This mode is most likely to be de-excited (V-T transfer) by a collinear collision along the bond axis. The collisional relaxation of a vibrating diatomic molecule by a second atomic species under collinear conditions can result in a near-collinearly recoiling (velocity oppositely directed) ground-state molecule if the mass of the atom is large enough. The latter condition is a function of the molecular mass and the natural frequency of the vibration. In the case of $C_2H_2$, the required mass is on the order of 20 amu. Under conditions of continuous optical pumping, this mode of relaxation results in a steady-state flux of molecules in one direction. The theory of this phenomenon shows that when this effect dominates, the direction of the flux is always opposite to the velocity group which absorbs the laser radiation. This collisional mechanism is also expected to vanish for line center pumping. Finally, it should be pointed out that for $C_2H_2{:}Kr$ mixtures, this mechanism is likely to dominate the elastic cross-section change-effected LID for the asymmetric $\nu_3$ stretch mode excitation. This is due to the negligible cross-section change which occurs and the large energy of the vibrational quantum.

### 5. Laser-Assisted Separations in Liquids: Laser-Inhibited Diffusion

The effects of optical excitation on the transport properties of gases were discussed in the

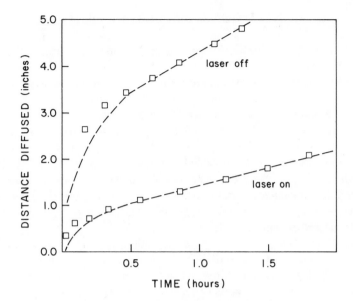

FIGURE 12. Diffusion length of rhodamine 6G in ethanol at 21°C with and without Cd laser excitation. The laser radiation was 17 mW at 441.6 nm. (From Lawandy, N. M., Fuhr, P. L., and Robinson, D. W., *Phys. Lett.*, 84A, 137, 1981. With permission.)

preceding section. The effects of electronic excitation of solutes in a solvent can have even more pronounced effects on the diffusion coefficient. This is primarily due to the large changes in the solvent cage, which, in turn, determine the transport coefficients.

The electronic excitation of large molecules with chromophores usually involves π-electron singlet excitations, resulting in large dipole moment and polarizability changes. For example, the dye *p*-dimethylaminonitrostilbene has a dipole moment of 7.6 and 36.0 D in a ground singlet and excited singlet states, respectively.[82] Such changes, in turn, have pronounced effects on the size of the solvent cage associated with the optically excited solute.

The effects of optical excitation on the diffusion of rhodamine dye on methanol solvent have been studied by Lawandy et al.[83] This work was performed with $10^{-4}$ M solutions and 17 mw from a HeCd laser at 449.6 nm in a diffusion tube experiment. Care was taken to maintain a constant temperature environment to avoid thermal effects. The resultant data are shown in Figure 12 and, as is evident, the effect is dramatic.

The solution to the diffusion equation, with radiative coupling to dyes, has been treated recently in the literature.[84] The theory indicates that the large changes in the effective diffusion coefficient may be due to intersystem singlet-triplet transfer subsequent to optical excitation. This has the effect of trapping a large fraction of molecules in triplet states where radiative relaxation to the ground singlet is spin forbidden. The effect of this mechanism is to produce transport properties with triplet-state characteristics. The work by Lawandy[84] also gives one-dimensional solutions for steady-state concentration profiles when spatially inhomogeneous optical intensity distributions are used to pump the solutions.

The use of this effect to separate species in solution has yet to be exploited. However, the principle is quite straightforward and should easily result in the separations of two species in one solvent, if selective excitation is available. This approach may prove to be extremely useful in biological separations, since many large molecules are thermally labile (thereby eliminating temperature-dependent separation processes from consideration), but also usually exhibit strong UV and visible absorptions which can be used for selective excitation.

## III. CONCLUSIONS AND FUTURE OUTLOOK

Ever since the advent of the laser in 1960, numerous applications of this unique device have been developed and placed into operation. Applications in the area of laser-induced separations are still in their infancy and are restricted primarily to laser isotope separation, most notably uranium enrichment. The reasons for this current state of affairs has been identified as the gross inefficiencies of current isotope enrichment techniques. Large-scale uranium enrichment by LIS is almost a certainty, especially to strip the remaining $^{235}U$ from the tailings of conventional enrichment plants. In addition, the production of highly enriched isotopes via LIS techniques is quite probable for such varied applications as $^{50}Ti$ for structural materials, medical isotopes ($^{10}B$, $^{79}Br$), materials with low neutron absorption cross sections such as $^7Li$, useful radionuclides from irradiation of isotopically pure stable precursor nuclei, etc. These processes are expected to develop concurrently with high-power pulsed dye and excimer lasers and increasingly tunable dye lasers.

Applications of nonisotopic laser-induced separations are not as well developed. The reasons for this are primarily: (1) most of the techniques are still relatively new and, therefore, still in the research stage; (2) competitive schemes are not as grossly inefficient as isotope enrichment processes; and (3) the costs of many existing conventional separation processes do not represent extremely high fractions of the ultimate product value. However, advances in increased laser efficiency and power, and the potentially high separation factors that can be obtained, can conceivably make LIS schemes more attractive for certain applications than current, more conventional separation processes. This is especially true for high value specialty chemicals, such as pharmaceuticals, where many low efficiency separations steps contribute appreciably to product cost. The production of difficult-to-separate, labile bio-chemicals may also become more tractable with the development of high separation factor, low-energy, ambient temperature LIS schemes of the LID type, for example.

The eventual extent of the use of LIS processes is impossible to predict at this stage of their evolution. This will depend entirely on laser development and the detailed comparative economics of specific applications. However, enough is currently known about the subject to indicate that it will be an active area of research for the foreseeable future.

## REFERENCES

1. **Shinskey, F. G.** *Distillation Control,* McGraw-Hill, New York, 1977.
2. **Rhodes, G. W.,** The ERDA laser isotope separation program, in *Quantum Electronics,* Vol 4, Jacobs, S. F., Sargent, M., III, Scully, M. O., and Walker, C. T., Eds., Addison-Wesley, Reading, Mass., 1976.
3. **Denning, R. G.,** Laser isotope separation techniques, *Phys. Technol.,* 9, 242, 1978.
4. **Wilkie, T.,** Laser enrichment — no easy path to proliferation, *Nucl. Eng. Int.,* 25, 13, 1980.
5. **Letokhov, V. S.,** U.S.S.R. Patent No. 65,743, 1977.
6. **James, G. S., Itzkan, I., Pike, C. T., Levy, R. H., and Levin, R. H.,** *IEEE J. Quantum Electron.,* QE-11, 101, 1975.
7. **Tuccio, S. A., Dubrin, J. W., Peterson, O. G., and Snavely, B. B.,** *IEEE J. Quantum Electron.,* QE-10, 790, 1974.
8. **Letokhov, V. S., Mishin, V. J., and Puretzkii, A. A.,** *Prog. Quantum Electron.,* 6, 1977.
9. **Dieke, G. H. and Duncan, A. B. F.,** *Spectroscopic Properties of Uranium Compounds,* McGraw-Hill, New York, 1949, 77.
10. **Landau, L. D. and Lifshitz, E. M.,** *Quantum Mechanics, Non-Relativistic Theory,* Addison-Wesley, Reading, Mass., 1976.
11. **Aldridge, J. P., III, Birely, J. H., Cantrell, C. D., III, and Cartwright, D. C.,** Experimental and theoretical studies of laser isotope separation, in *Physics of Quantum Electronics,* Vol. 4, Jacobs, S. F., Sargent, M., III, Scully, M. O., and Walker, C. T., Eds., Addison-Wesley, Reading, Mass., 1976.

12. **Yeung, E. S. and Moore, C. B.,** *Appl. Phys. Lett.,* 21, 109, 1972.
13. **Marling, J. B.,** *Chem. Phys. Lett.,* 34, 84, 1975.
14. **Leon, S. R. and Moore, C. B.,** *Phys. Rev. Lett.,* 33, 269, 1974.
15. **Dworetsky, S. H. and Hozack, R. S.,** *J. Chem. Phys.,* 59, 3856, 1973.
16. **Hochstrasser, R. M. and King, D. S.,** Proc. 2nd Laser Spectroscopy Conf., Megave, June, 1975.
17. **Hochstrasser, R. M. and King, D. S.,** *J. Am. Chem. Soc.,* 97, 4760, 1975.
18. **Karl, R. R., Jr. and Innes, K. K.,** *Chem. Phys. Lett.,* 36, 275, 1975.
19. **Letokhov, V. S. and Moore, C. B.,** Laser isotope separation (review), *Sov. J. Quantum Electron.,* 6, 129, 1976.
20. **Ambartzumyan, R. V., Letokhov, V. S., Ryabov, E. A., and Chakalin, N. V.,** *JETP Lett.,* 20, 273, 1974.
21. **Chebotaev, V. P., Golgev, A. L., and Letokhov, V. S.,** *Chem. Phys.,* 7, 316, 1975.
22. **Rockwood, S. and Rabideau, S. W.,** *IEEE J. Quantum Electron.,* QE-10, 789, 1974.
23. **Ambartzumyan, R. V. Letokhov, V. S., and Lobko, V. V.,** *Opt. Comm.,* 14, 426, 1975.
24. **Freund, S. M. and Ritter, J. J.,** Paper TC2, 4th Conf. on Chemical and Molecular Lasers, St. Louis, October 21, 1974.
25. **Lyman, J. L. and Rockwood, S. D.,** Proc. of the Electro-Optics Int. Laser '75 Conf., Anaheim, Calif., 1975.
26. **Freund, S. M. and Ritter, J. J.,** *Chem. Phys. Lett.,* 32, 255, 1975.
27. **Ambartzumyan, R. V., Gorokhov, Yu, A., Letokhov, U. S. and Makarov, G. N.,** *JETP Lett.,* 21, 171, 1975.
28. **Houston, P. L. and Steinfeld, J. I.,** *J. Mol. Spectrosc.,* 54, 335, 1975.
29. **Lyman, J. L., Jensen, R. J., Rink, J. P., Robinson, C. P., and Rockwood, S. D.,** Paper 13, B.10, CLEA, Washington, D. C., May 28, 1975.
30. **Ambartzumyan, R. V., Gorokhov, Yu. A., Letokhov, U. S., and Makarov, G. N.,** *JETP Lett.,* 22, 43, 1975.
31. **Lyman, J. L. and Leary, K. L.,** *J. Chem. Phys.,* 69, 1858, 1978.
32. **Cantrell, C. D., Ed.,** Multiple photon excitation and dissociation of polyatomic molecules, in *Topics in Current Physics,* Vol. 35, Springer, Berlin, 1983.
33. **Isenor, N. R., Merchant, V., Hallsworth, R. S., and Richardson, M. C.,** *Can. J. Phys.,* 51, 1281, 1973.
34. **Bloembergen, N.,** *Opt. Comm.,* 15, 416, 1975.
35. **Rice, O. K. and Ramsperger, H. C.,** *J. Am. Chem. Soc.,* 49, 1617, 1927.
36. **Kassel, L. S.,** *J. Phys. Chem.,* 32, 225, 1928.
37. **Marcus, R. A.,** *J. Chem. Phys.,* 20, 259, 1952.
38. **Luria, M., de Pena, R. G., Olszyna, K. J., and Heicklen, J.,** Kinetics of particle growth. III. Particle formation in the photolysis of sulfur dioxide-acetylene mixtures, *J. Phys. Chem.,* 78, 325, 1974.
39. **Knudtson, J. T. and Eyring, E. M.,** Laser-induced chemical reactions, in *Annual Review of Physical Chemistry,* Eyring, H., Christensen, C. J., and Johnston, H. S., Eds., Annual Reviews, Palo Alto, Calif., 1974, 255; **Letokhov, V. S.,** *Nonlinear Laser Chemistry,* Springer-Verlag, New York, 1983.
40. **Ronn, A. M.,** Particulate formation induced by infrared laser dielectric breakdown, *Chem. Phys. Lett.,* 42, 202, 1975.
41. **Ronn, A. M. and Earl, B. L.,** Laser induced dielectric breakdown studies of the reaction $UF_6 + H_2$, *Chem. Phys. Lett.,* 45, 556, 1977.
42. **Lin, S. T. and Ronn, A. M.,** Laser induced sulfur particulate formation, *Chem. Phys. Lett.,* 56, 414, 1978.
43. **Tam, A. C., Moe, G., and Happer, W.,** Particle formation by resonant laser light in alkali-metal vapor, *Phys. Rev. Lett.,* 35, 1630, 1975.
44. **Tam, A. C., Happer, W., and Siano, D.,** Oscillating laser-production of particulates in a $Cs/D_2$ vapor, *Chem. Phys. Lett.,* 49, 320, 1977.
45. **Allegrini, M., Bicchi, P., Dattrino, D., and Moi, L.,** Ultrafine particle formation by resonant laser excitation of potassium molecules, *Opt. Comm.,* 49, 39, 1984.
46. **Kappes, M. M., Kunz, R. W., and Schumacher, E.,** *Chem Phys. Lett.,* 91, 413, 1982.
47. **Calo, J. M.,** Heteromolecular clusters of $H_2O$, $SO_2$, $CO_2$, CO, and NO, *Nature,* 248, 665, 1974.
48. **Gilbert, R.,** *J. Chim. Phys. Chim. Biol.,* 60, 205, 1963.
49. **Hall, R. J. and Pimentel, G. C.,** *J. Chem. Phys.,* 38, 1889, 1963.
50. **Odiorne, T. J., Brooks, P. R., and Kasper, J. V. V.,** *J. Chem. Phys.,* 55, 1980, 1971.
51. **Arnoldi, D. and Wolfrum, J.,** *Chem. Phys. Lett.,* 24, 234, 1974.
52. **Mayer, S. W., Kwok, M. A., Gross, R. W. F., and Spencer, D. J.,** *Appl. Phys. Lett.,* 17, 516, 1970.
53. **Freund, S. M. and Ritter, J. J.,** *Chem. Phys. Lett.,* 32, 255, 1975.
54. **Basov, N. G., Belenov, E. M., Gavrilina, L. D., Isakov, V. A., Markin, E. P., Oraevskii, A. N., Romaneko, V. I., and Ferapontov, N. B.,** *JETP Lett.,* 20, 277, 1974.

55. **Basov, N. G., Belenov, E. M., Gavrilina, L. D., Isakov, V. A., Markin, E. P., Oraevskii, A. N., Romaneko, V. I., and Ferapontov, N. B.,** *Sov. J. Quantum Electron.,* 5, 510, 1975.

56. **Hartley, H., Ponder, A. O., Bowen, E. J., and Merton, T. R.,** *Philos. Mag.,* 43, 430, 1922.

57. **Pertel, R. and Gunning, H. E.,** *Can. J. Chem.,* 37, 35, 1959.

58. **Dunn, O., Harteck, P., and Dondes, S.,** *J. Phys. Chem.,* 77, 878, 1973.

59. **Schmidt, C. F., Reeves, R. R., Jr., and Harteck, P.,** *Ber. Bunsenges. Phys. Chem.,* 72, 129, 1968.

60. **Letokhov, V. S. and Semishen, V. A.,** *Spectrosc. Lett.,* 8, 263, 1975.

61. **Liu, D. D., Datta, S., and Zare, R. N.,** *J. Am. Chem. Soc.,* 97, 2557, 1975.

62. **LaMotte, M., Dewey, H. J., Keller, R. A., and Ritter, J. J.,** *Chem. Phys. Lett.,* 30, 165, 1975.

63. **Pressman,** U.S. Patent No. 2,558,877, 1971.

64. **Ashkin, A.,** U.S. Patent No. 3,710,279, 1973.

65. **Bernhardt, A. F., Duerre, D. E., Simpson, J. R., and Wood, L. L.,** Separation of isotopes by laser deflection of atomic beam. I. Barium, *Appl. Phys. Lett.,* 25, 617, 1974.

66. **Bernhardt, A. F., Duerre, D. E., Simpson, J. R., and Wood, L. L.,** Multifrequency radiation pressure laser isotope separation, *Opt. Comm.,* 16, 169, 1976.

67. **Gochelashvili, K. S., Karlov, N. V., Orlov, A. N., Petrov, R. P., Petrov, Uy. N., and Prokhorov, A. M.,** *JETP Lett.,* 21, 302, 1975.

68. **Karlov, N. V., Korev, Yu. B., and Prokhorov, A. M.,** *JETP Lett.,* 14, 117, 1974.

69. **Ambartsumyan, R. V. and Letokhov, V. S.,** *Appl. Opt.,* 11, 354, 1972.

70. **Lawandy, N. M.,** Total collision cross-sections for laser induced diffusion, *IEEE J. Quantum Electron.,* QE-18, 1258, 1983.

71. **Lawandy, N. M.,** Laser-induced diffusion by collisional redirection and relaxation, in Paper THII3, 13th Int. Quantum Electronics Conf., Anaheim, Calif., June 18, 1984.

72. **Gel'mukhanov, F. Kh. and Shalagin, A. M.,** Diffusive suction and extrusion of atoms by a light field, *Zh. Eksp. Toer. Fiz.,* 77, 461, 1979; *Sov. Phys. JETP,* 50, 234, 1979.

73. **Gel'mukhanov, F. Kh. and Shalagin, A. M.,** Theory of optically induced diffusion of gas, *Zh. Eksp. Teor. Fiz.,* 78, 1674, 1980; *Sov. Phys. JETP,* 51, 839, 1981.

74. **Gel'mukhanov, F. Kh. and Shalagin, A. M.,** Light induced diffusion of gases, *Zh. Eksp. Toer. Fiz. Pis'ma Red.,* 29, 773, 1979; *JETP Lett.,* 29, 711, 1979.

75. **Lawandy, N. M. and Cumming, E. W.,** Analytical solutions for infrared-laser-driven diffusion in polyatomic gas mixtures, *Phys. Rev.,* A27, 2548, 1982.

76. **Antsignin, V. K., Atutov, S. N., Gel'mukhanov, F. Kh., Shalagin, A. M., and Telegin, G. G.,** Gas diffusion induced by resonance light field, *Opt. Commun.,* 32, 237, 1980.

77. **Bjorkholm, J. E., Ashkin, A., and Pearson, D. B.,** Observation of resonance radiation pressure on an atomic vapor, *Appl. Phys. Lett.,* 27, 534, 1975.

78. **Werij, H. G. C., Woerdman, J. P., Beenakker, J. J. M., and Kuščev, I.,** Demonstration of a semi-permeable optical piston, *Phys. Rev. Lett.,* 52, 2237, 1984.

79. **Baranov, V. Yu, Velikhov, E. P., Dykhne, A. M., Kazakov, S. A., Mezhevov, V. S., Olov, M. Yu., Pesmennyl, V. D., Starodubtsev, A. I., and Starostin, A.,** Excitation of drift motion of multiatomic molecules by resonance radiation, *Zh. Eksp. Teor. Fiz. Pis'ma Red.,* 31, 475, 1980; *JETP Lett.,* 32, 445, 1980.

80. **Panifilov, V. N., Stronin, V. P., Chaspovskil, P. L., and Shalagin, A. M.,** *JETP LETT.,* 33, 48, 1981.

81. **Lawandy, N. M., Rabinovich, W. S., and Willner, R. P.,** Laser-induced macroscopic flows by collisional redirection and relaxation, *Phys. Rev. A.,* submitted.

82. **Lippert, E. Z.,** *Electrochemistry,* 61, 692, 1957.

83. **Lawandy, N. M., Fuhr, P. L., and Robinson, D. W.,** Laser-inhibited diffusion in rhodamine-ethanol solution, *Phys. Lett.,* 84A, 137, 1981.

84. **Lawandy, N. M.,** Laser-induced profiles in solution, *IEEE J. Quantum Electron.,* QE-19 (9), 1359, 1983.

Chapter 15

# PROCESS APPLICATIONS OF PEDS TECHNOLOGY: PELTIER EFFECT DIFFUSION SEPARATION (PEDS) CONCEPT

**Milton Meckler and R. W. Farmer**

## TABLE OF CONTENTS

# I. INTRODUCTION

Processes involving the concentration of dilute, aqueous solutions are common throughout the fluid processing industry, including chemical and milk processing operations. Typically, these operations are carried out via some evaporative means, such as a multiple effect evaporation (MEE) or mechanical vapor recompression (MVR). For most of the currently applied methods, the primary energy input is heat, produced by the burning of hydrocarbon fuels.

Unfortunately, the design of evaporator units requires heat energy to be utilized at a very low availability level compared to the hydrocarbon source. Because of this, real second-law efficiencies of such processes are usually lower than the energy efficiencies calculated on the basis of first-law analyses.

Clearly, second-law efficiencies are more valid indicators of the utilization of energy "quality", that is, availability. Given the significant proportion of total energy consumption devoted to concentration processes, it is apparent that development of systems which can perform these tasks at a high second-law efficiency should receive greater attention. Recent expansion in research and development spending[1] by both industry and government sectors is indicative of the drive to implement new, energy-efficient technologies.

This paper describes an innovative system currently being developed by The Meckler Group that can perform aqueous separations at high efficiencies. The system is called Peltier Effect Diffusion Separation (PEDS). The specific application of PEDS to the milk processing industry will be examined through comparisons to state-of-the-art systems currently being used.

A computer model of the PEDS system was recently completed by California State University, Northridge, under contract to The Meckler Group.

This in-depth computer analysis indicated that effective coefficient of performance values of between 40 and 50 were easily achievable with PEDS.

# II. PELTIER EFFECT DIFFUSION SEPARATION (PEDS)

PEDS employs thermoelectric heat pumps configured to provide multieffect levels of separation work input in the form of electricity. By combining the inherently high coefficient of performance (COP) of thermoelectric devices with the concepts of multistage utilization of availability and parallel crossflow of evaporating streams, PEDS demonstrates surprisingly high overall COP. Examination of the process from a second-law viewpoint reveals the reason for this.

A design example is provided at the end of this chapter to illustrate some important parameters to be considered for a specific PEDS application in the chemical industry. Based on second-law criteria, the energy efficiency of the process may be compared to state-of-the-art conventional systems. To provide background, configurations of the conventional systems and the PEDS unit are described. Fundamental mass transfer and hydrodynamic relationships are also given.

## A. Thermoelectric Heat Pump — PEDS Prime Mover

The key factor in achieving high second-law efficiency with PEDS is use of a thermoelectric heat pump (TEHP). This heat pumping phenomenon is a manifestation of quantum-mechanical effects which occur when P- and N-type semiconductor materials are connected electrically and thermally (see Figure 1 for schematic details).

When direct current is passed through such a circuit, a heat flux results parallel to the electron flow. This is the Peltier effect.

A temperature difference proportional to the applied voltage can be maintained between

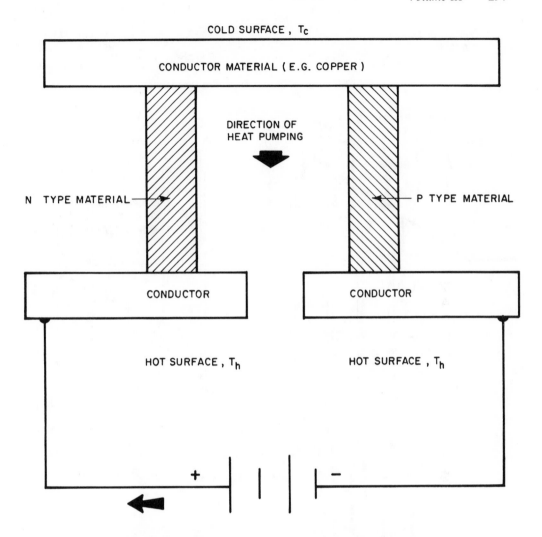

FIGURE 1. Peltier effect heat pump element (simplified diagram).

hot and cold reservoirs (conductor materials). The heat flux, or heat pumping rate, is found to be dependent on the current, as is the COP, since electric current here corresponds to work input. For the TEHP to operate effectively, the P and N materials must be electrical semiconductors and thermal insulators.

A qualitative indicator of the ratio of electrical to thermal conductivity is termed the figure of merit, Z. More rigorously, this parameter is defined as a function of the free-electron densities in the N and P materials. A typical value of this quantity for state-of-the-art materials is $3.2 \times 10^{-3}/1°C$.[2] A theoretical operational description of TEHP devices is presented in more detail, subsequently.

As will be shown, a substantial number of TEHP devices are required for most PEDS applications. For this reason, the economic viability of PEDS technology depends on the cost of suitable thermoelectric devices. Fortunately, the electronics industry is developing new and cheaper production technology at a rapid pace.

Typical of these new techniques are ribbon-type processes for the production of large material volumes at low cost.[3] Automated bonding processes promise an economical means of producing large quantities of semiconductor/conductor interfaces.[4] New semiconductor

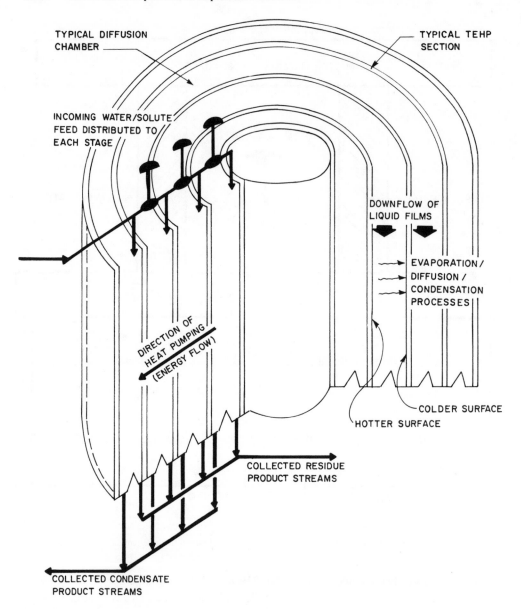

FIGURE 2.   PEDS unit configuration — parallel flow (four stages).

materials are also being explored,[5] which may deliver high Z values. This recent rapid expansion of electronic technology has made PEDS an economically viable alternative.

## B. PEDS Unit Configuration

Instrumental to the high second-law efficiency of PEDS is a staged configuration, a significant departure from conventional evaporating equipment. Essentially, the unit can be envisioned as a series of concentric annuli, or stages, as shown in Figure 2. Mass transfer occurs in diffusion chambers, which alternate with annular sections containing TEHP devices.

As a result of the heat pumping phenomenon described previously, a finite temperature differential is maintained across these TEHP sections. Thus, the opposing diffusion chamber walls are at relatively lower and higher temperatures.

A parallel stage flow scheme is used for aqueous concentrations, as opposed to a series stage arrangement proposed for binary or multicomponent distillations.[6]

As shown, the feed solution is fed as a thin, distributed film to the hotter chamber surface of each stage. Heat and mass are transferred simultaneously to the colder surface via successive evaporation, diffusion, and condensation mechanisms. These will be examined in more detail later. Heat removed from the condensing material is "pumped" by the TEHP to the hotter surface of the adjacent diffusion chamber, driving the transport processes in that stage.

At steady-state operation, condensate and residue falling films flow down the opposing walls of each diffusion chamber. Energy, in the form of vapor enthalpy, and pumped heat flow crosscurrently to these material flows. Each of the TEHP annular sections can be considered as points of work input along the energy flow path. This arrangement is somewhat analogous to a multistage compressor, in that the total work input is applied in small increments. This results in overall improved process efficiency.

At the bottom of the column, the residue and condensate streams are collected from each stage, the condensate being essentially pure water and the residue the concentrated aqueous solution.

## C. Transport Processes in the Diffusion Chambers

In the annular diffusion chambers, simultaneous heat and mass transfer occurs between liquid films falling down the TEHP's heated and cooled walls. To maximize the available film surface area per unit chamber volume, extended fins project from the chamber walls. The relatively hotter and colder fins are juxtaposed (see Figure 3). The effective transfer surface includes not only the primary chamber walls, but also the extended fin surfaces.

The overall transport mechanism taking place in these chambers consists of three successive steps. Water, vaporized from the hotter liquid film, is diffused through stagnant air across the gap separating the opposing surfaces, and condenses on the colder falling film surface. The flow of these films is characterized by a film Reynolds number. Ideally, the temperature and concentration gradients across these films should be negligible, implying that the films are well mixed. However, the falling films must also be stable and nonsplashing.

To ensure this condition, a laminer-turbulent transition Reynolds number range is specified as part of the design criteria. In this way, heat mass transport resistances in the films themselves can be considered negligible. (Therefore, only the three primary transport mechanisms outlined above need be included in the physical description of the process.)

In the design analysis, film temperature is assumed constant over the entire effective surface. In reality, there will be some temperature gradient across the extended fin. For a more rigorous design analysis, an appropriate fin efficiency based on empirical results should be included to reflect this nonisothermality.

## D. Thermoelectric Heat Pump Operation

The prime movers in the PEDS separation process are the thermoelectric heat pump sections. These annular sections each consist of an array of the Peltier effect heat pump devices described previously.

As shown in Figure 4, the thermoelectric devices are distributed throughout the void space between the walls of adjacent diffusion chambers and are isolated from the flowing liquids. The N- and P-type semiconductor substrates are in contact with mounting brackets made of conductive material (e.g., copper). Direct current is delivered to the thermoelectric devices via bus connections along the wall surfaces.

Not shown in Figure 4 are the shock absorbent mountings required between the semiconductor strips and the wall brackets. Such mountings are necessary to insulate the semiconductor substrates from mechanical stresses.

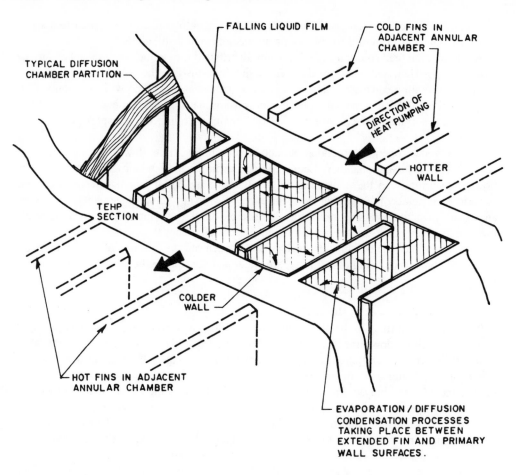

FIGURE 3.    PEDS diffusion chamber detail.

Each TEHP annular section can be viewed as a single heat pump, utilizing electrical work input to remove the vaporization heat from the cooler condensing surface. This heat is then pumped along with internally generated heat (i.e., $I^2R$ losses) to the hot reservoir, to provide the vaporization heat at the hotter surface of the adjacent stage. Process control is effected electronically. The heat pumping rate and temperature differential may be controlled by manipulating the current and voltage, respectively, delivered to the TEHP devices.

## III. THEORETICAL BACKGROUND

### A. Heat and Mass Transfer

The description of simultaneous heat and mass transfer to condensing falling films has been presented by several authors.[7,8] A similar approach has been employed to model the heat and mass transfer in a PEDS diffusion chamber.[9] The mass flux equation for water through the stagnant air in the chamber may be written for the system illustrated in Figure 5.

$$N_T = \frac{C\,D_{AB}}{1 - Y_w} \frac{d\,Y_w}{d\,\ell} \tag{1}$$

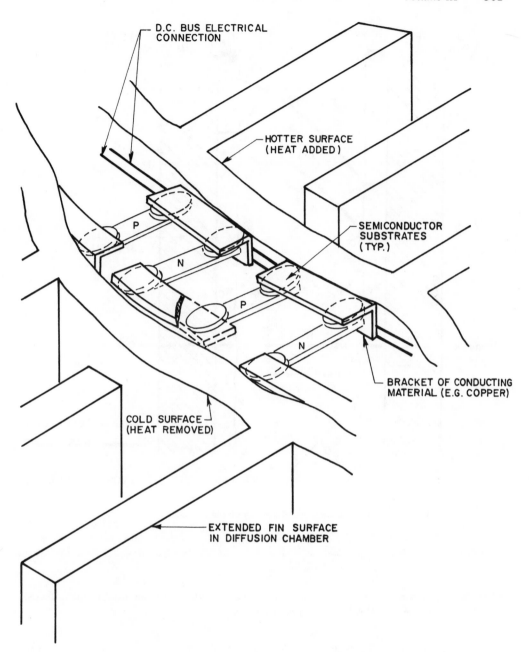

FIGURE 4.    Thermoelectric heat pump (TEHP) section detail.

where z is the distance coordinate perpendicular to the wall surfaces. Recognizing that this flux is constant for all $\ell$, the continuity equation becomes:

$$\frac{d\ N_T}{d\ \ell} = 0 \tag{2}$$

Inserting Equation 1 into Equation 2 and integrating, the concentration profile in terms of the partial pressure of water, $P_w(z)$, is

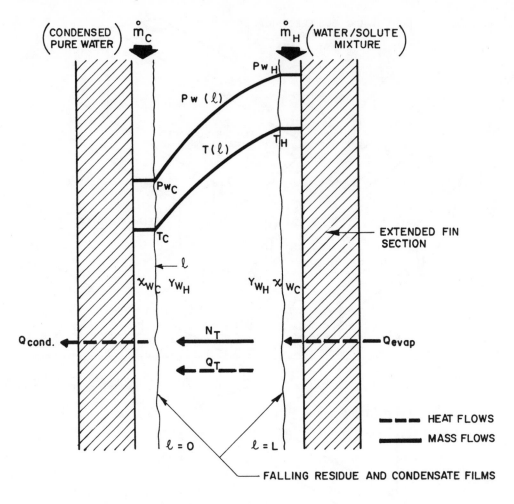

FIGURE 5.    Simultaneous heat and mass transfer in a PEDS diffusion chamber.

$$\left(\frac{1 - P_{W/P}}{1 - P_{WC/P}}\right) = \left(\frac{1 - P_{WH/P}}{1 - P_{WC/P}}\right)^{\ell/L} \tag{3}$$

where boundary conditions $P_{WC}$ and $P_{WH}$ are fixed by equilibrium at the liquid film surfaces. The steady-state energy flux, $E_\ell$, for this system is

$$E_\ell = K \frac{dT}{d\ell} + N_T E_W \tag{4}$$

Radiative flux is not included, since it will be negligible for any practical temperature differences between the films. Similarly, the conductive flux, represented by the first term in Equation 4, also will be relatively small. For the general case, however, the temperature profile may be derived from Equation 4 using film temperatures $T_H$ and $T_C$ as boundary conditions:

$$T(\ell) = T_H + (T_H - T_C)\left[\frac{\exp\left(\frac{A\ell}{L}\right) - 1}{\exp(A) - 1}\right] \tag{5}$$

where

$$A = \frac{(N_T C_{PW})L}{K} \tag{6}$$

The conduction heat flux is given as:

$$Q_K = h(T_H - T_C) \frac{A}{\exp(A) - 1} \tag{7}$$

The enthalpy contribution due to convective mass transfer to the total energy flux similarly is found to be

$$Q_E = h(T_C - T_H) \frac{A}{1 - \exp(A)} \tag{8}$$

For the PEDS application, it has been found that $Q_E \gg Q_K$. Values for the interface convective heat transfer coefficient, h, at the evaporating film surface have been correlated experimentally for a turbulent film by:[10]

$$h_e = 0.0087(Re)^{0.4}(Pr)^{0.344}\left(\frac{v^2}{gk^3}\right)^{-1/3} \tag{9}$$

For the condensing film interface, the following empirical expression for the turbulent heat transfer coefficient has been proposed:[11]

$$h_c = 0.021\left(\frac{k^3 \varphi_W g \Gamma}{\mu}\right)^{1/3} \tag{10}$$

However, the most important design equation, with regard to establishing the size of a PEDS unit, is the integrated mass flux expression obtained from Equation 1:

$$N_T = \frac{-C\, D_{AB}}{L}\, \ell_u\left[\frac{1 - Y_{WH}}{1 - Y_{WC}}\right] \tag{11}$$

Rewritten in terms of partial pressures at the liquid interfaces, Equation 11 becomes:

$$N_T = \frac{-C\, D_{AB}}{L}\, \ell_u\left[\frac{P - P_{WH}}{P - P_{WC}}\right] \tag{12}$$

Since water is the only volatile component of the aqueous solutions considered here, the partial pressures may be approximated by the solution vapor pressures. A further simplification can be made in Equation 12 if it is assumed that the vapor in the chamber is an ideal gas. Then, Equation 12 becomes:

$$N_T = \frac{-P\, D_{AB}}{G\, \overline{T}\, L}\, \ell_u\left[\frac{P - P^o_{WH}}{P - P^o_{WC}}\right] \tag{13}$$

This last equation can be used to calculate the mass transfer flux, which is equal to the condensation rate by continuity. If mass transfer is identified as the controlling mechanism, then Equation 13 is the basis for the unit size determination for a given set of chamber operating conditions.

## B. Falling Film Hydrodynamics

The important parameter in characterizing the falling film flow regime is the film Reynolds number, defined to be

$$R_e = \frac{4\Gamma}{\mu} \tag{14}$$

Resultant film thickness and falling velocity depend not only on the liquid loading, $\Gamma$, but, also, on the density difference between the falling liquid and the vapor phase and bulk liquid viscosity. Expressions for these quantities have been given as Equation 13:

$$m = \left[ \frac{3\Gamma\mu}{g\varphi_w(\varphi_w\varphi_v)} \right]^{1/3} \tag{15}$$

$$v = \frac{\Gamma}{m\varphi_w} = \frac{g(\varphi_w - \varphi_v)m^2}{3\mu} \tag{16}$$

As noted previously, it is necessary that these falling films be stable and nonsplashing. To improve mass transfer, however, it is desirable that the flow not be laminar, so as to prevent significant temperature and concentration gradients within the films. For this reason, an appropriate design criterion is the average film Reynolds number specified within the laminar-turbulent transistion regime, identified as 25 to 200. This fixes the allowable liquid loading along the perimeter of the evaporating surface.

Since the perimeter of the available evaporating surface depends on the number of annular stages and diffusion chamber geometry (e.g., fin spacing), film hydrodynamics also can be employed as a sizing criterion. As will be illustrated by the design example, either film hydrodynamics or mass transfer is the more restrictive constraint. The final unit size will thus be based on whichever criterion dictates the larger size, that is, the greater number of annular stages.

## C. Thermoelectric Design Equations

Perhaps the most critical factor of an effective PEDS design is the evaluation of thermoelectric heat pump performance. As mentioned previously, the total heat pumping rate, $q_c$, and temperature differential, $\Delta T$, are functions of the applied current and voltage, respectively.[13] An energy balance around an individual heat pump device yields the following:

$$q_c = \chi T_c I - \frac{1}{2} I^2 r - K\Delta T \tag{17}$$

In this balance, the first term represents the net heat removed from the cold reservoir and the remaining terms are the internal heat generation and back-conduction heat flows. The heat pump COP is defined as:

$$\phi = q_c/P_{dev} \tag{18}$$

If the positive direction of current is from n to p at the cold junction (as in Figure 1), the applied voltage for a given $\Delta T$ is

$$V = \chi\Delta T + Ir \tag{19}$$

The power, $P_{dev}$, is the product of applied voltage and current:

$$P_{dev} = VI = \chi I \Delta T + I^2 r \tag{20}$$

Substituting Equations 17 and 20 into Equation 18 gives the device COP as a function of reservoir temperatures, thermoelectric properties, and current, I:

$$\phi = \frac{\chi T_c I - \dfrac{1}{2} I^2 r - K \Delta T}{\chi I \Delta T + I^2 r} \tag{21}$$

Solving this equation for I as a function of COP yields:

$$I = \frac{\chi(T_c - O\Delta T) - \left[ (\phi \chi \Delta T - \chi T_c)^2 - 4r\left( \phi - \dfrac{1}{2} \right) K\Delta T \right]^{1/2}}{2r\left( \phi - \dfrac{1}{2} \right)}$$

$$COP = \phi \tag{22}$$

However, the range of $\phi$ is limited by the theoretical maximum COP, defined as a function of thermoelectric figure of merit and reservoir temperatures:

$$\phi_{max} = \frac{\omega - T_H/T_C}{\omega + 1} \tag{23}$$

where

$$\omega = (1 + Z\overline{T})^{1/2} \tag{24}$$

$$\overline{T} = \frac{1}{2} (T_H + T_C) \tag{25}$$

Interactions between the characteristic operating parameters are illustrated in Figure 6. From these relationships it is noted that for fixed $\Delta T$, the current levels for maximum COP and heat pumping rate, qc, do not coincide. Depending on the particular application, either of these maxima may be specified, or some appropriate intermediate condition. For the PEDS unit, it is reasonable to define an effective COP based on the net condensing load removed from the chamber wall.

$$COP_{eff} = q'_C/P_{dev} = \chi T_c I/P_{dev} \tag{26}$$

As will be demonstrated in the design example, this ratio increases as the actual COP of the devices decreases. The reason is that the current is also being reduced, resulting in smaller internal heat losses (i.e., $I^2R$ loss). As a result, the individual TEHP devices operate at a lower heat pumping rate, but at a higher efficiency.

Unfortunately, since many more devices are then required to meet the total condensing load, unit cost increases. A realistic design point for the current would be a value sufficiently low enough to result in minimal $I^2R$ losses, yet provide adequate heat pumping rate, so the number of TEHP devices required is not excessive. The cost trade-offs between operating (energy) and construction (TEHP) costs will dictate the proper optimum.

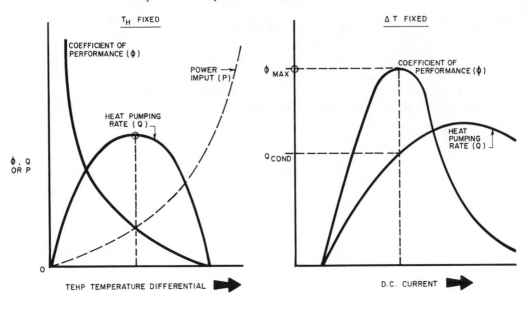

FIGURE 6.    TEHP performance characteristics.

From the fundamental equations presented, it is apparent that TEHP operating characteristics are strongly dependent on reservoir temperatures. For a PEDS application, these are
the temperatures of the evaporating and condensing chamber surfaces. For the parallel flow
scheme, all stages operate at the same temperature and pressure, which are fixed by process
requirements. This greatly simplifies design, since the only remaining degree of freedom is
the heat pumping rate, as determined by the input current.

## D. Procedure for Preliminary Design

A procedure for a preliminary PEDS design can now be outlined. This will permit esti-
mation of PEDS unit size, energy consumption, and the number of TEHP units required for
a particular application. The procedure follows:

- Establish process requirements based on flow rates, stream compositions, temperature,
  and pressure.
- Calculate the mass transfer rate per unit area from Equation 13 to determine the total
  surface area required, $A_T$.
- Using liquid properties and appropriate Reynolds number criteria, determine the allowable liquid film loading, $\Gamma$, from Equation 14.
- The total evaporating surface perimeter is based on the total feed rate and $\Gamma$:

$$L_P = F/\Gamma \tag{27}$$

- Unit height can now be computed from this hydrodynamic constraint:

$$H = A_T/L_P \tag{28}$$

- Falling film stability limits allowable unit height to about 6.1 m (20 ft) for transition
  regime flows. If H is greater than this, mass transfer rather than film hydrodynamics
  is identified as the controlling factor in unit sizing.
- For mass transfer-limited systems, the surface area is increased by increasing the
  number of annular stages until H in Equation 28 is within the film stability limit.

- With the unit size established, TEHP requirements are evaluated by selecting an optimum operating COP based on the following economic criteria:

total power requirement

$$P_T = \frac{Q_C}{COP_{eff}} \tag{29}$$

number of thermoelectric devices

$$N_{TC} = \frac{P_T}{P_{dev}} \tag{30}$$

## E. Second-Law Considerations

The advantages of the PEDS process, from a second-law viewpoint, are readily apparent. As mentioned previously, the multistage work inputs allow greater availability utilization. Also, since proper design focuses on minimizing irreversible heat losses, thermal efficiency is high. Since thermoelectric heat pumps do not involve moving parts, there are none of the friction losses associated with mechanical heat pumps or vapor recompression systems.

In its most fundamental form, the second-law efficiency is defined as:

$$\eta_{II} = \frac{\text{available energy in useful product or work}}{\text{available energy supplied in ``fuels''}} \tag{31}$$

Methods have been established for evaluating this ratio for various mechanical, chemical, and thermal processes.[14] Ultimately, Equation 31 may be expressed as the first law energy ratio, with each term multiplied by a quality factor, C, which reflects the fraction of available energy that can be withdrawn.

$$\eta_{II} = \frac{C_2 \Delta E_2}{C_1 \Delta E_1} = \frac{\text{change in availability required by the task}}{\text{change in availability required by the task}} \tag{32}$$

These quality factors have been calculated for many types of heat and work energy.[15] For electricity and hydrocarbon fuels, the C factor is 1.0; for steam, $C_{steam}$ is a function of pressure.

It should be noted that the $\Delta E$ ratio in Equation 32 is merely the first-law energy efficiency of the process.

Therefore, the second-law efficiency expression can be rewritten:

$$\eta_{II} = \frac{C_2}{C_1} \eta_I \tag{33}$$

which shows that $\eta_{II}$ will always be equal to or less than $\eta_I$, since $C_2/C_1$ is equal to or less than unity. For example, typical steam power plants have been cited to have first-law efficiencies of up to 45%, yet second-law evaluation yields $\eta_{II} = 33\%$. Most of the irreversible losses occur in the steam boiler, for which $\eta_I = 91\%$, while $\eta_{II} = 49\%$ for a typical unit.[16]

This points to another advantage of the PEDS concept. The energy used by PEDS is of higher quality, so irreversible losses are minimal. To compare the two processes from the

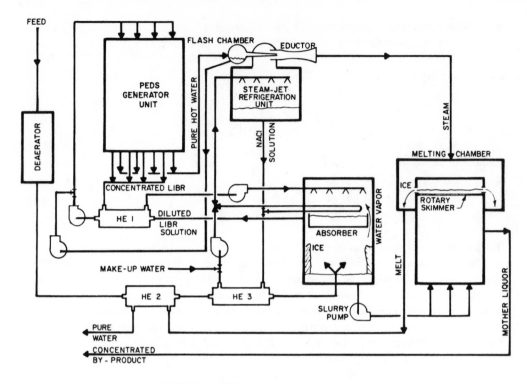

FIGURE 7.    PEDS freeze-concentration system.

standpoint of availability utilization, a second-law efficiency ratio must be multiplied by the ratio of actual energy consumed:

$$R_{II} = \frac{\Delta E_1}{\Delta E_2} \left[ \frac{\eta_{II_2}}{\eta_{II_1}} \right] \tag{34}$$

The second-law efficiencies in Equation 34 serve to normalize the actual energy consumed to the availability of the initial (i.e., hydrocarbon) fuel source. Values of $R_{II}$ greater than unity indicate that process 1 is the more energy efficient.

It should be recognized that in most absorption/refrigeration systems, absorbent regeneration represents a substantial portion of the energy input. As a mass transfer unit, PEDS may be employed as an absorbent regenerator, thus efficiently improving the overall process.

Examples of such applications are freeze-concentration processes for desalination of seawater, dewatering of dilute biological (e.g., whey) solutions, and the concentration of milk. Another variation would be to use the PEDS regenerator in an absorption heat pump for space heating and cooling. This latter case serves as the basis for a design example to demonstrate the high second-law efficiency of the PEDS process.

PEDS technology is not limited to regenerator-type aqueous concentrations. In a series-stage mode, binary and multicomponent distillations can be carried out, as mentioned previously. The physical description and design methodology for these applications are somewhat more complicated than those described here.[17] The PEDS configuration, with the capability to maintain near-isothermal surface conditions across a large effective area, can be employed as a heterogeneous reactor, with catalyst-coated fin surfaces.[18]

### F. Freeze-Concentration Process

As depicted in Figure 7, the PEDS freeze-concentration system utilizes a two-stage absorptive refrigeration mechanism to bring about the formation of ice from the feed solution. For both stages, the main driver is the separation work performed by the PEDS unit.

The first stage involves a steam-jet refrigeration unit where sodium chloride solution is chilled by removal of evaporated vapors by a Venturi ejector pump. This brine circuit removes the heat of solution from the absorber and aids in the precooling of the feed (HE-3).

A concentrated lithium bromide (LiBr) solution produced by the PEDS regenerator provides the second stage of refrigeration by direct contact absorptive cooling in the absorber. The feed is first deaerated and precooled (HE-2 and HE-3) before being sprayed into the lower chamber of the absorber. A sufficient cooling effect is obtained to permit ice crystals to form on the chamber walls. A cooled and diluted lithium-bromide solution is returned to the PEDS regeneration unit while the ice/solution slurry is pumped to the melter/washer unit.

The melter/washer is similar to a flotation tank; a rotary skimmer removes the ice from the mother liquor, which is the concentrated by-product. Steam from the ejector pump enters the upper melting chamber. The outlet melt is the purified water product.

There are two advantages to such a process from a second-law viewpoint. First, all energy inputs are in the form of work (i.e., thermoelectric and pump electrical input), rather than heat. Second, it was found that in this freeze-concentration process 7 kg of ice may be removed for each 1 kg of water evaporated in the PEDS unit. This is because the fusion of water is utilized as the concentrating mechanism, as opposed to evaporation.

In applications involving brackish seawater or delicate biological solutes, problems such as corrosion, scaling, or solute breakdown are minimized, since no heat is directly applied to the feed components.

## G. Absorption Heating/Refrigeration Process

Space heating and cooling represents a major end use of energy, comprising about 20.4% of the total 1972 U.S. energy consumption.[19] In view of this, heating and air conditioning are important applications of PEDS technology. Perhaps the most practical means of employing PEDS in this area is as a regenerator in an absorption cycle, as shown in Figure 8.

As in the freeze-concentration process, a hot pure water product from the PEDS regenerator is flashed to lower pressure steam to provide a cooling effect in a steam-jet refrigeration unit. A converging-diverging nozzle accelerates the flashed vapor stream, simultaneously drawing off vapors from the main chamber, thus, acting as an ejector pump. Evaporation of vapor in the main chamber provides a first stage of cooling for space refrigerant (i.e., water). The vapor effluent from the ejector is condensed and sent on to the evaporator, along with the cooled space refrigerant.

In the evaporator, the space coolant fluid in a closed-tube bundle is further chilled by evaporation of water sprayed over the bundle. Subsequently, the fluid is absorbed by concentrated solvent solution in the absorber section. This cooling effect is sufficient to supply chilled water to the space at about 6.7°C (44°F).

The concentrated absorbent solution regenerated in the PEDS unit is sprayed into the absorbent chamber, absorbing water vapor from the evaporator with generation of heat due to mixing. This heat may be used to effect space heating, if required, or is released to ambient through a cooling tower.

Heat flows in this system are balanced by the PEDS unit feed preheater (HE-1) and by the cooling tower. This latter unit acts as a low-level energy rejection sink, removing mixing heat from the absorber and/or supplementing condenser demands.

Many of the second-law advantages of PEDS as a freeze-concentration process are in evidence in this case. The main driver is the PEDS regenerator, which directly utilizes electrical work input at high effective COP levels. Energy rejection, through the interior spaces or a cooling tower, is at low availability levels, contributing to the overall second-

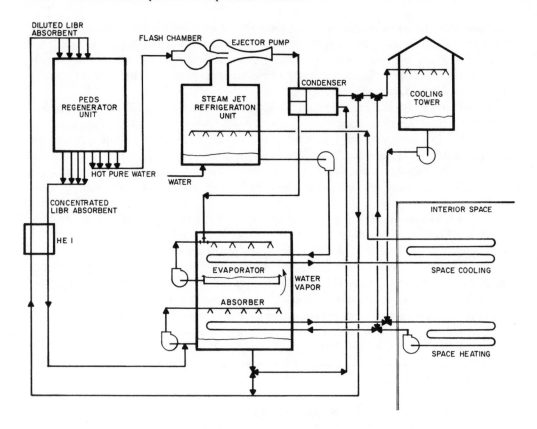

FIGURE 8.   PEDS absorption heating/refrigeration process.

law efficiency of the cycle. Because of these energy conservation benefits, application of such absorption heating/cooling cycles deserves greater attention.

## IV. MILK PROCESSING SYSTEMS

The two most energy-intensive fluid processing areas in the dairy industry today are concentration of solids by evaporation and refrigeration.

The most common method used for concentrating milk solids is multiple effect evaporation (MEE). However, the state-of-the-art technology becoming increasingly popular is mechanical vapor recompression (MVR). Freeze concentration, an alternative method to these evaporation systems, is also being explored by the dairy industry.

The standard refrigeration process used today is vapor recompression with electric drive and a working fluid of ammonia or freon. The high efficiency alternative to this process is the water-lithium bromide absorption cycle.

### A. Multiple Effect Evaporation

The most common method for concentrating milk is multiple effect evaporation (MEE). Typically, raw milk of about 13% solids is concentrated to about 50% solids. The advantage of MEE is that for every pound of steam used, more than 1 lb of water can be evaporated.

This is done by using vapor from one effect at the heating medium for the next effect. Multiple evaporative efficiency is usually categorized by steam economy, i.e., pounds of water evaporated per pound of steam supplied.

In practice, steam economy is always less than the number of effects, because the boiling point rises with higher concentrations. Also, the enthalpy of supplied steam diminishes from one effect to the next. A typical multiple effect evaporator exhibits a steam economy of 2.0 to 2.5.

Although it may appear there is no limit to the number of effects that can be employed, in practice the limit usually is three to eight effects, depending on the material evaporated.

This limitation is imposed because of the large capital cost increases associated with each additional effect. In the dairy industry, evaporation is usually limited to three or four effects.

To compare MEE to PEDS and MVR, steam economy must be translated to COP. The COP value will be about 10% higher than the steam economy value. This is because the latent heat of vaporizing the fluid entering the evaporator is greater than the latent heat of vaporizing entering steam.

In summary, the COP for the milk multiple effect system is less than 3.5 for a quadruple effect system (the most probable configuration). This is significantly less than COP values for MVR or PEDS.

## B. Mechanical Vapor Recompression (MVR)

Mechanical vapor recompression (MVR) is becoming increasingly popular in high volume evaporation applications. It has been used successfully in the desalinization, paper and pulp, distillery, and dairy industries. It is considered a state-of-the art technology, and, therefore, is a good point of reference for evaluating PEDS.

MVR takes its name from the fact that steam leaving the evaporation chamber is compressed mechanically before it is condensed. Compression makes vapor available for use as the heating medium for the stage which produced it.

The primary advantage of the MVR process is its efficient use of input energy. Much less energy is needed to run a mechanical compressor than a steam evaporator of the same capacity.

In addition, the energy may be provided in several forms. The compressor can be driven by an electric motor steam turbine, diesel, or gas engine. With today's cost and availability of fossil fuels, the ability to use alternate energy sources is attractive.

The main disadvantage of the system is a relatively large initial capital outlay in comparison to other present-day evaporative systems. It is apparent, however, that lower operating costs make the system appealing to many users, and an increasing number of manufacturers have invested in the system.

MVR systems are evaluated and compared to other evaporators by calculating a COP. Values of at least 10 are easily obtainable with MVR systems. This is in contrast to COPs of less than 3.5 for practical MEE.

## C. Freeze Concentration

Freeze concentration can be called a frontier technology in the milk industry. This process is now being actively pursued as an alternative to evaporation processes, because it does not require steam to concentrate milk.

Because water is removed by freezing and removal of the resulting ice (instead of removal by evaporation), energy requirements are reduced. A refrigeration cycle is required to achieve freezing of the water.

## D. Refrigeration

Refrigeration is essential to the preservation of milk and milk products. Vapor compression, with electric drive and a working fluid of ammonia or freon, is the most common system used in the dairy industry.

An alternative to vapor compression is the water-lithium bromide absorption cycle. The absorption refrigeration system, however, is not useful for cooling below 4°C, a temperature required in milk refrigeration. This problem can be solved by providing supplementary refrigeration using a conventional refrigeration system.

The water-lithium bromide cycle uses two basic factors in producing a refrigerant effect. First, water boils and flash cools itself at low temperatures when maintained at a high vacuum. Second, certain substances — such as salts — absorb water vapor.

A lithium bromide solution is a hygroscopic (readily absorbed) salt solution. It has the best solubility-vapor pressure relationship for high cycle efficiency.

In the absorption unit, water flashes off to vapor and the remaining water temperature is lowered. The affinity of water to salt is measured by the depression of the water-vapor pressure. This affinity is more pronounced as salt concentrations increase.

### E. PEDS Application

PEDS can be applied to many different areas in milk processing. For example, PEDS can be used in evaporative, freeze concentration, or absorption refrigeration systems. In addition, a PEDS evaporative system could be used to concentrate milk directly on the farm. farm.

### F. Evaporation

Perhaps the most important use for PEDS is an evaporative system. Raw milk containing 12.6% solids can be concentrated to 50% solids at a relatively low temperature of about 21°C. This low temperature operation is ideal in milk concentration, as it sustains milk quality.

More significantly, COPs between 40 and 50 are easily obtained. As a result, PEDS evaporative system operates at a very high efficiency. This, combined with PEDS' relatively low capital cost due to an ever-decreasing thermoelectric cost, makes PEDS an attractive alternative to conventional evaporative systems.

### G. Freeze Concentration (PEDS)

The freeze concentration process involves a refrigeration cycle. PEDS can be employed as a regenerator in an absorption refrigeration cycle using water-lithium bromide as a working fluid.

PEDS operates at high temperatures and pressures (see Figure 7) to use its distillate in steam jet refrigeration. About 121°C water at 206,800 Pa is flashed down to 137,900 Pa to produce 116°C steam, which is fed into the steam jet refrigeration unit.

### H. Absorption Refrigeration

As in the freeze concentration process, PEDS is used as a regenerator in an absorption refrigeration cycle using water-lithium as a working fluid.

For efficient operations, the water-lithium bromide solution enters the absorber at 64% concentration and leaves at 59.9% concentration. However, due to the boiling point rise problem and the high temperature and pressure of PEDS operation, the 59.5% concentration solution is diluted to 10% prior to entering the PEDS unit (see Figure 8).

Inside PEDS, the 10% concentration solution is concentrated to 30% which is then flashed in a vacuum for concentrating to 64%.

The high temperature and pressure distillate product from PEDS is flashed to produce some of the steam required for the steam jet refrigeration unit.

Thus, a 1,265,220-kJ/hr refrigeration cycle is able to produce an additional 316,305 kJ/hr through steam jet refrigeration. On the condenser side, the 59.5% concentration solution from the absorber is used to recover part of the condenser heat prior to entering the PEDS unit.

This results in efficient operation of PEDS at its high temperature and pressure. Therefore,

## Table 1
## ABSORPTION REFRIGERATION COP
## COMPARISON

| | Conventional | | |
| --- | --- | --- | --- |
| | Single-stage | Two-stage | PEDS |
| Equipment COP | 0.48 | 0.73 | 1.77 |

## Table 2
## ECONOMICS

| Costs × 10⁶ | PEDS | MVR | Conv. |
| --- | --- | --- | --- |
| Initial (1985) | A ($2.600) | 0.9A | 0.7A |
| Operational | B ($0.500) | 3.0B | 4.8B |
| Maintenance | C ($0.060) | 5.6C | 1.4C |
| Present     (1985) | D ($8.800) | 2.4D | 6.8D |
| worth      (1990) | 1.4D | 3.8D | 7.1D |

*Note:* PEDS COP = 40; MVR COP = 11; Conv. COP = 4.

using the working fluid to recover heat from the condenser results in additional gain over the cycle.

## I. PEDS vs. State of the Art

PEDS absorption refrigeration process offers a significant improvement in COP values over state-of-the-art systems now being used, due to two factors: additional refrigeration provided by the steam jet effect; and heat recovery through the split condenser by the dilute 59.5 (wt %) solution leaving the absorber.

A comparison between conventional LiBr absorption refrigeration systems and PEDS indicates that PEDS refrigeration system can achieve a COP of 1.77. Conventional single-stage and double-stage absorption refrigeration systems achieve COPs of 0.48 and 0.73, respectively (see Table 1).

A comparison of PEDS evaporative system and conventional evaporative systems (MEE and MVR) indicates the significant advantages of PEDS (see Table 2). The comparison was based on milk concentration systems processing 28,400 kg/hr.

## V. DESIGN EXAMPLE: PEDS REGENERATOR

As demonstrated by the process applications described previously, PEDS can be employed as a regenerator for lithium chloride (LiCl), sodium chloride (NaCl), or lithium bromide (LiBr) refrigerant solutions. Here, a preliminary design of a PEDS regenerator for a nominal 1,265,220-kJ/hr LiBr absorption-refrigeration system (see Figure 9) is described. A qualitative comparison from the standpoint of second-law efficiency, with a dual-effect steam regenerator of equal capacity, then can be carried out.

To provide a realistic design example, as well as a basis for comparison with conventional unit performance, a PEDS unit having inlet and outlet streams identical to a typical 1,265,220-kJ/hr (nominal) regenerator was designed. The process conditons are shown in Figure 9 for both a PEDS regenerator and conventional two-stage steam regenerator. These process

## PEDS REGENERATOR

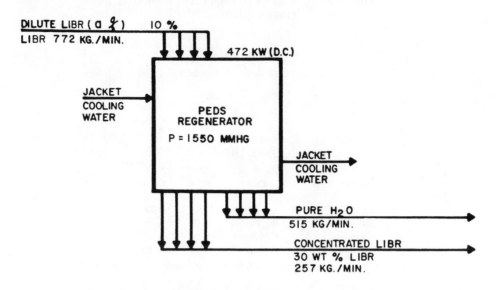

## TWO - STAGE STEAM REGENERATOR

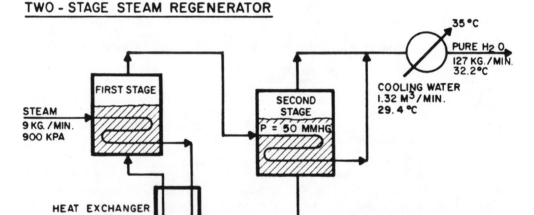

FIGURE 9.   Lithium chloride regenerator design example process conditions (for nominal 100-ton unit).

conditions are typical for absorption refrigeration cycles described in industry literature.[20] The two-stage regenerator is generally regarded as the most energy-efficient, currently applied technology, utilizing high pressure steam as the primary input.

A basic concern in diffusion (or removal) of water from aqueous hygroscopic solutions, such as LiBr, is that with concentrations in excess of, for example, 30 wt % the effect of boiling point elevation (BPE) can be detrimental to TEHP system performance. This is because the temperature difference between TEHP evaporating ($T_H$) and condensing ($T_C$) heat transfer surfaces increases with increasing weight percent LiBr to maintain a constant driving force:

$$\frac{P - P^o_{WH}}{P - P^o_{WC}}$$

Thus, the COP is adversely impacted. Also, referring to Equation 13, notice that the resulting increase to $\bar{\tau}$ will adversely affect the rate of mass transfer term, $N_T$. As a consequence, the PEDS regenerator unit is arranged to operate between a feed stream of 10 wt % LiBr with an effluent of 30 wt % LiBr at the elevated temperature and pressure.

In this way, the PEDS effluent can be flashed to a 64 wt % LiBr concentration prior to heat exchange with incoming feed as shown and prior to returning to absorber section. The weak solution leaving the absorber normally at 59.5 wt % LiBr passes through the split condenser before mixing with by-passed condensate from the same split condenser. This enables the feed stream entering the PEDS regenerator to be reduced by a 10 wt % LiBr concentration.

Thus, by enabling operation at, for example, 1550 mmHg and 120°C (vs. 37.8°C and 50 mmHg), the normally adverse effects of BPE are eliminated and, with a higher TEHP operating temperature, $T_c$, a more favorable device COP is achieved.

Based on the fundamental mass transfer equation discussed previously, charts can be constructed showing mass flux rates as a function of temperature differential and fin spacing for various chamber pressures. Figure 10 shows this relationship for 1550 mmHg.

This will be the PEDS condensing pressure assumed for the hypothetical design case. Note the increasing nonlinearity of the rate curves as pressure decreases. This is indicative of declining returns from incremental increases in temperature differential.

For an aqueous solution, the permissible liquid loading for a given hydrodynamic constraint (Reynolds number) may be related to the fin spacing and average film viscosity. The linearity of these relationships is apparent from the basic definition of the film Reynolds number and unit geometry parameters. Such charts allow a number of design options, such as various temperature differentials, to be more easily evaluated.

Since process requirements fix the distillate and residue temperatures as well as chamber pressure, the mass transfer rate may be directly inferred from Figure 10. The closest practical fin spacing (0.32 cm) is selected, since $N_T$ for this particular system is NT = 2.0 kmol/hr-m². The required surface area is readily calculated based on the known distillate flow rate:

$$AT = \frac{D}{N_T} = \frac{30,868 \text{ kg/hr}}{(2.0 \text{ kg mol/hr m}^2)(18 \text{ kg/kg mol})} = 857 \text{ m}^2$$

Experience indicates that mass transfer rather than film hydrodynamics controls the overall process. Typically, a unit height limitation of about 6.1 m (20 ft) should be observed to insure stable, nonsplashing films. Given this result, the known required surface area and the known fin spacing, one can calculate the number of annuli needed for this operation.

At this point the energy demand for TEHP may be estimated. It is in this calculation that an optimum must be established betweeen electrical demand (i.e., operating cost) and the number of thermoelectric devices required in the device (i.e., capital cost). As discussed earlier, the effective COP, based on the heat of vaporization removed from the condensate, increases as thermoelectric COP decreases.

However, the heat pumping rate per device simultaneously declines, hence the number of thermoelectric devices needed to meet the condensing load will increase. The optimization process must be aimed at minimizing the total owning cost function, consisting of amortized capital and operating costs.

For the preliminary design case considered here, an effective COP of 40 can be used to obtained the total electrical demand. First, the total condensing duty, equaling the heat removal rate from the condensate, must be calculated:

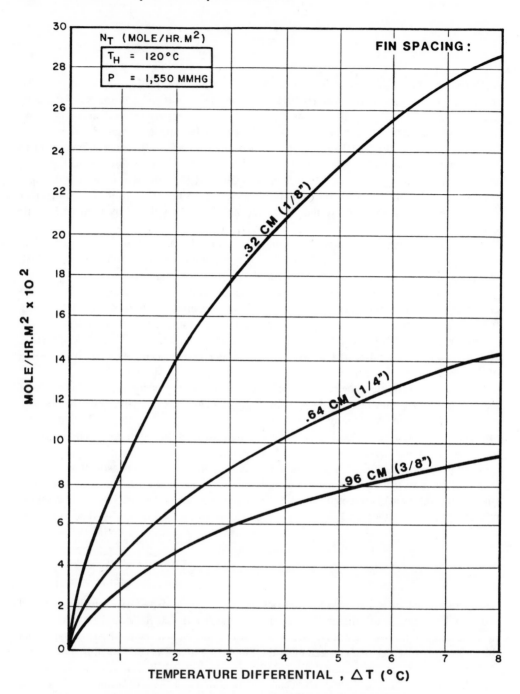

FIGURE 10.    Mass transfer rates as a function of stage $\Delta T$.

$$Q_c = DH_{vap}T_h = 515 \text{ kg/min (2203 k/kg)}$$

$$= 18,909 \text{ kW}$$

Then the total power input is readily determined by:

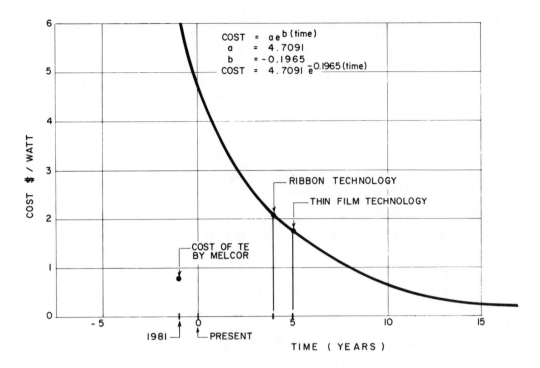

FIGURE 11.   Correlation of projected cost for manufactured TE materials with time.

$$\Delta E_{elec} = \frac{QC}{COP_{eff}} = \frac{18,909 \text{ kW}}{40} = 472 \text{ kW}$$

$$= 2.8 \times 10^4 \text{ kW/min}$$

Total number of thermoelectric devices then can be determined by using power per device, $P_{dev} = 0.0105$ W:

$$N_{TC} = \frac{P_T}{P_{dev}} = \frac{472 \times 10^3 \text{ W}}{0.0105 \text{ W}} = 4.5 \times 10^7$$

Clearly, the cost per device must be extremely low in order to make PEDS economically viable. As discussed previously, solid-state technology will soon provide such low cost devices.

Figure 11 indicates a significant reduction in thermoelectric cost in the future, making PEDS a viable option to conventional systems.

Now it is possible to compare the performance of PEDS on a second-law efficiency basis with the conventional two-stage steam regenerator shown in Figure 9. To normalize these efficiencies to the primary fuel source (hydrocarbon fuels), the second-law efficiencies of the conversion processes delivering energy to the two devices must be known. For the generation of electricity as described before:

$$N_{II \ elec} = 0.33$$

The second-law efficiency of the boiler supplying steam to the dual-effect generator is known to be a function of the steam pressure.[21] For this case, saturated steam at 900 kPA

(130 psig), having an enthalpy of 2773.9 kJ/kg, is fed at a rate of 9 kg/min for a total energy input rate, $\Delta E_{steam}$:

$$\Delta E_{steam} = 9 \text{ kg/min } (2773.9 \text{ kJ/kg})$$

$$= 2.5 \times 10^4 \text{ kJ/min}$$

The energy quality of steam at these conditions is given as 0.312, assuming a first low boiler efficiency of 70%:

$$N_{II\ steam} = \frac{C_{steam}}{C_{final}} (N_{I\ steam}) = \frac{0.312}{1.0} (0.7) = 0.218$$

The second law efficiency ratio of the two processes is calculated to be

$$R_{II} = \frac{E_{dual\ stage}\ \Delta E_{steam}/N_{II\ steam}}{E_{Peds}\ \Delta E_{elec}/N_{II\ elec}}$$

$$= \frac{2.5 \times 10^4 \text{ kJ/min}/0.218}{2.8 \times 10^4 \text{ kJ/min}/0.33} = 1.35$$

Since this ratio is significantly greater than one, it can be concluded that the PEDS process is considerably more efficient than the comparable dual-stage steam regenerator.

## A. Conclusion

A new separation process for concentrating dilute, aqueous solutions has been described. The Peltier Effect Diffusion Separation System (PEDS) has been found to reduce energy consumption required for a given duty and to provide high second-law efficiency.[22,23] This is a result of employing thermoelectric heat pumps (TEHP) as the prime movers,[24,25] in conjunction with a unit geometry permitting multistages of work input.

## NOMENCLATURE

| | |
|---|---|
| A | $N_T C_{pw} L/K$ = convective heat transfer parameter |
| $A_T$ | Total transfer surface area, $m^2$ |
| $C_P$ | Heat capacity, kJ/mol K |
| C | Vapor concentration ($mol/m^3$) or energy quality factor |
| $D_{AB}$ | Diffusivity of water in air, $cm^2/sec$ |
| D | Distillate flow rate, kg/min |
| E | Enthalpy (kJ/kg mol), energy availability utilized, kJ/min |
| $E_1$ | Energy flux, $kJ/m^2$ sec |
| E | Process energy consumption, kJ/sec |
| F | Feed flow rate, kg/min |
| G | Gas law constant, $m^3 kPa/mol$ K |
| g | Gravitational acceleration, $m/sec^2$ |
| H | Unit height, m |
| h | Convective heat transfer coefficient, $kJ/m^2 sec$ K |
| I | Electric current, amps |
| K | Thermal conductivity of thermoelectric couple, KJ/sec K |
| k | Thermal conductivity of vapor, kJ/sec Km |
| L | Fin to fin spacing, cm |
| l | Coordinate perpendicular to fin surface |
| $L_p$ | Total wall perimeter, m |

| m | Film thickness, cm |
|---|---|
| m | Mass flow rate, kg/min |
| $N_T$ | Mass transfer rate, mol/m²sec |
| $N_{TC}$ | Number of thermoelectric devices |
| n, N | n-Type semiconductor materials |
| p, P | p-Type semiconductor materials |
| P | Power (w), partial pressure, stage total pressure, (kPa) (mmHg) |
| $P_T$ | Total unit power, w |
| $P_{dev}$ | Power to thermoelectric device, w |
| P° | Vapor pressure, mmHg |
| $Q_k$ | Conductive heat flux, kJ/sm² K |
| $Q_E$ | Convective heat flux, kJ/sm² K |
| $Q_C$ | Total condensing load, kJ/sec |
| Q | Heat flux, kJ/sm² |
| $q_P$, $q_c$ | Thermoelectric heat pumping rate, kJ/sec |
| $q_c$ | Effective thermoelectric heat pumping rate, kJ/sec |
| R | Average annular radius, m |
| r | Thermoelectric resistance, ohms |
| $R_{II}$ | Second law efficiency ratio |
| T | Temperature, K |
| T | Temperature differential, K |
| T | Average stage temperature = $\frac{1}{2}$ ($T_H$ + $T_C$) K |
| V | Voltage, volts |
| W | Thermoelectric parameter = $(1 + ZT)^{1/2}$ |
| X | Fin to fin spacing (cm), liquid phase mole fraction |
| Y | Vapor phase mole fraction |
| Z | Thermoelectric figure of merit, °C$^{-1}$ |
| Re | Reynolds number = $4\Gamma/_Y$ |
| Pr | PrandltH number = $c_p y/k$ |
| COP | Coefficient of performance |
| $COP_{eff}$ | Effective PEDS COP |
| χ | Seebeck coefficient for thermoelectric couple, v/K |
| φ | COP |
| $\phi_{max}$ | Maximum theoretical COP |
| Γ | Perimeter liquid loading, kg/mmin |
| μ | Liquid dynamic viscosity, CP |
| ϑ | Liquid kinematic viscosity, cm²/sec |
| φ | Density, kg/m³ |
| $\eta_I$ | First-law efficiency |
| $\eta_{II}$ | Second-law efficiency |

## Subscripts

| c, cond | Condensing or colder chamber wall |
|---|---|
| H, e, evap | Evaporating or hotter chamber wall |
| w | Water, aqueous solution |
| v | Vapor phase |
| 1, 2 | Process designations |
| wc | Water at hotter film interface |
| wh | Water at colder film interface |

# REFERENCES

1. Special report: facts and figures for chemical R & D, *Chem. Eng. News*, p. 46, July 27, 1981.
2. **Meckler, M.,** Peltier Effect Heat Pump System, Paper #73-WA/PID-3, 94th Winter Meet., American Society of Mechanical Engineers, Detroit, 1973.
3. New silicon ribbon process promises low cost, high volume capabilities, *Semicond. Int.*, 3(11), 10, 1980.
4. **O'Neill, T. G.,** The status of tape automated bonding, *Semicond. Int.*, 4(2), 33, 1981.
5. **Stauffer, R. N.,** The changing scene in semiconductors, *Manuf. Eng.*, July, 61, 1981.
6. **Meckler, M. and Farmer, R. W.,** Peltier effect heat pump/diffusion system utilizes availability potentials in chemical separation, at AIChE Natl. Meet., Houston, April 1981.
7. **Bird, R. B., Stewart, W. E., and Lightfoot, E. N.,** *Transport Phenomena*, John Wiley & Sons, New York, 1960, 572.
8. **Sherwood, T. K., Pigford, R. L., and Wilke, C. R.,** *Mass Transfer*, McGraw-Hill, New York, 1975, 257.
9. **Meckler, M.,** Peltier Effect Heat/Mass Transfer Systems, Paper #CT1.7, Proc. 5th Int. Heat Transfer Conf., Tokyo, Japan, September 1974.
10. **Chun, K. R. and Seban, R. A.,** Heat transfer to evaporating liquid films, *J. Heat Transfer — Trans. ASME*, November, 391, 1971.
11. **Bird, R. B., Stewart, W. E., and Lightfoot, E. N.,** *Transport Phenomena*, John Wiley & Sons, New York, 1960, 418.
12. **Perry, R. H. and Chilton, C. H., Eds.,** *Chemical Engineers Handbook*, 5th ed., McGraw-Hill, New York, 1973, 5—57.
13. **Heikes, R. R. and Ure, R. W.,** *Thermoelectricity: Science and Engineering*, Interscience, New York, 1961, 463.
14. **Petit, P. J. and Gaggioli, R. A.,** Second law procedures for evaluating processes, in *Thermodynamics: Second Law Analysis*, Gaggioli, R. A., Ed., ACS Symp. Ser. 122, American Chemical Society, Washington, D.C., 1980.
15. **Gyftopoulos, E. P. and Widmer, T. F.,** Availability analysis: the combined energy and entropy balance, in *Thermodynamics: Second Law Analysis*, Gaggioli, R. A., Ed., ACS Symp. Ser. American Chemical Society, Washington, D.C., 1980.
16. **Reistad, G. M.,** Available-energy utilization to the United States, in *Thermodynamics: Second Law Analysis*, Gaggioli, R. A., Ed., ACS Symp. Ser. 122, American Chemical Society, Washington, D.C., 1980.
17. **Meckler, M. and Farmer, R. W.,** Peltier effect heat pump/diffusion system utilizes availability potentials in chemical separation, at AIChE Natl. Meet., Houston, April, 1981.
18. U.S. Patent 3,393,130; 3,671,404; 3,801,284; 4,290,273; and patent pending.
19. **Fowler, J. M.,** *Energy and Environment*, McGraw-Hill, New York, 1975, 81.
20. *Equipment Handbook*, American Society of Heating, Refrigerating, and Air Conditioning Engineers, New York, 1979, 14.1.
21. **Gyfropoulos, E. P. and Widmer, T. F.,** Availability analysis: the combined energy and entropy balance, in *Thermodynamics: Second Law Analysis*, Gagglioli, R. A., Ed., ACS Symp. Ser. 122, American Chemical Society, Washington, D.C., 1980.
22. **Meckler, M. and Farmer, R. W.,** PEDS concept permits utilization of availability potentials, in AIChE Natl. Meet., Orlando, 1982.
23. **Meckler, M.,** Reducing industrial energy use with T.E. diffusion heat pumps, in 17th Intersociety Energy Conversion Engineering Conf., Session E4, Los Angeles, 1982.
24. **Meckler, M.,** Use of Peltier heat pumps to improve process separation availability, in 14th Intersociety Energy Conversion Engineering Conf., Boston, 1979.
25. **Meckler, M.,** Thermoelectric diffusion chemical systems, in 18th Natl. Symp. and Exhibition, Society of Aerospace Material and Process Engineering, Los Angeles, 1973.

# AUTHOR INDEX

## A

Aberth, E. R., 158 (ref. 24)
Adamson, A. W., 72 (ref. 2, 12), 73 (ref. 48), 74 (ref. 70)
Adduci, A. J., 13 (ref. 3)
Adler, R. J., 42, 43 (ref. 8)
Aida, T., 104 (ref. 105)
Albano, V. G., 13 (ref. 20)
Aldridge, J. P., III, 292 (ref. 11)
Alexander, A. E., 73 (ref. 29)
Allegrini, M., 282, 293 (ref. 45)
Allen, G., 102 (ref. 30, 32)
Altman, M., 158 (ref. 24)
Alty, J., 60, 73 (ref. 23)
Ambartzumyan, R. V., 293 (ref. 20, 23, 27, 30), 294 (ref. 69)
Ammerman, K. K., 43 (ref. 12)
Anderson, P. J., 73 (ref. 28)
Anolick, C., 104 (ref. 97, 99)
Antonini, E., 14 (ref. 33)
Antonson, C. R., 250, 263 (ref. 9)
Antsignin, V. K., 289, 294 (ref. 76)
Aresta, M., 13 (ref. 19—21)
Arnoldi, D., 293 (ref. 51)
Ascenzi, P., 14 (ref. 33)
Ashkin, A., 294 (ref. 64, 77)
Atutov, S. N., 289, 294 (ref. 76)
Ayarza, J., 102 (ref. 38)

## B

Bailes, P. J., 67, 74 (ref. 58)
Baker, C. H., 102 (ref. 30, 32)
Balch, A. L., 13 (ref. 30, 31), 14 (ref. 32)
Balder, J. R., 103 (ref. 72)
Baldwin, J. E., 14 (ref. 35)
Balogh-Nair, V., 171 (ref. 8)
Baranov, V.Yu., 294 (ref. 79)
Barkelew, C. H., 13 (ref. 5)
Basmadjian, D., 158 (ref. 27)
Basolo, F., 13 (ref. 12), 14 (ref. 35)
Basov, N. G., 293 (ref. 54), 294 (ref. 55)
Bdzil, J., 176, 195 (ref. 20)
Beavon, D. K., 42
Belenov, E. M., 293 (ref. 54), 294 (ref. 55)
Bell, L. G., 14 (ref. 39)
Benakker, J. J. M., 294 (ref. 78)
Benner, L. S., 13 (ref. 30, 31)
Benton, A. F., 137, 144, 148, 157 (ref. 16)
Bercaw, E., 14 (ref. 39)
Berg, C., 157 (ref. 1)
Bernhardt, A. F., 285, 294 (ref. 65, 66)
Beverwijk, C. D. M., 176, 195 (ref. 29)
Bhattacharyya, D., 108, 133 (ref. 4—10)
Bibome, K., 74 (ref. 67)
Bicchi, P., 293 (ref. 45)

Bier, M., 72
Biles, R. H., 13 (ref. 4)
Billmeyer, F. W., Jr., 195 (ref. 24)
Bird, R. B., 320 (ref. 7, 11)
Birely, J. H., 292 (ref. 11)
Biros, J., 102 (ref. 33)
Bishop, A. A., 103 (ref. 68)
Bjorkholm, J. E., 294 (ref. 77)
Bloembergen, N., 278, 293 (ref. 34)
Bockris, J. O'M., 72, 73 (ref. 41)
Bohen, M. C., 103 (ref. 48)
Bojarski, J. T., 226 (ref. 17)
Booth, F., 72 (ref. 15)
Bowen, E. J., 294 (ref. 56)
Bradley, W. E., 157 (ref. 1)
Brauman, J. I., 14 (ref. 36)
Breitenbach, J. W., 103 (ref. 70)
Breiter, M. W., 195 (ref. 20)
Brian, P. L. T., 43 (ref. 11)
Brintzinger, H. H., 14 (ref. 39)
Brody, S., 171 (ref. 11)
Brooks, P. R., 293 (ref. 50)
Brosilow, C. B., 42, 43 (ref. 8)
Brown, D., 53 (ref. 6)
Brown, M. G., 72, 73 (ref. 43)
Brown, W. R., 42, 43 (ref. 8)
Brunner, G., 101 (ref. 13), 104 (ref. 96)
Brunori, M., 14 (ref. 33)
Bull, H. B., 73 (ref. 19)

## C

Cadogan, W. P., 158 (ref. 23, 32)
Cahn, R. P., 225 (ref. 8)
Calabrese, J. C., 13 (ref. 24)
Calberson, O. L., 103 (ref. 56)
Calderazzo, F., 13 (ref. 28)
Calleja, G., 158 (ref. 19)
Calo, J. M., 293 (ref. 47)
Calvin, M., 13 (ref. 4, 5)
Cantrell, C. D., III, 292 (ref. 11), 293 (ref. 32)
Carfagno, J. A., 103 (ref. 66)
Carlier, C. C., 195 (ref. 20)
Carraher, C. E., Jr., 72
Carruthers, J. C., 59, 60, 73 (ref. 24)
Carter, J. W., 150, 159 (ref. 46)
Cartwright, D. C., 292 (ref. 11)
Cassidy, R. T., 159 (ref. 53)
Caywood, S. W., 104 (ref. 98)
Cenini, S., 14 (ref. 43)
Chai, C. P., 92, 102 (ref. 39), 104 (ref. 79)
Chan, Y. N., 150, 158 (ref. 43, 44)
Changyin, J., 208 (ref. 3)
Chao, K. C., 152, 159 (ref. 50)
Chapman, T. W., 66, 73 (ref. 53)
Chappelear, D. C., 103 (ref. 52)
Chaspovskil, P. L., 294 (ref. 80)

Robinson, D. B., 91, 92, 103 (ref. 74), 104 (ref. 76), 141, 143, 158 (ref. 30)
Robinson, D. W., 294 (ref. 83)
Robinson, S. D., 14 (ref. 49)
Rockwood, S. D., 293 (ref. 22, 25, 29)
Roddy, J. W., 56, 72 (ref. 5)
Rogers, K. A., 158 (ref. 28)
Rolinick, P. D., 53 (ref. 3)
Romaneko, V. I., 293 (ref. 54), 294 (ref. 55)
Ronn, A. M., 280, 281, 293 (ref. 40—42)
Rony, P. R., 251, 263 (ref. 10, 11)
Roselius, W., 104 (ref. 94, 95)
Rosi, G., 14 (ref. 33)
Rosseau, R. W., 42
Roughton, F. J. W., 175, 194 (ref. 10)
Rowlinson, J. S., 102 (ref. 25, 26, 31)
Rudy, C. E., 53 (ref. 8)
Rupp, W., 103 (ref. 73)
Ruthven, D. M., 136, 157 (ref. 13)
Ryabov, E. A., 293 (ref. 20)
Ryan, J. M., 32, 43 (ref. 2)
Ryan, R. C., 13 (ref. 27)
Ryohei, H., 226 (ref. 14)
Ryoji, K., 226 (ref. 14)

# S

Sanad, W. A. A., 74 (ref. 56)
Sandler, S. I., 103 (ref. 75)
Santangelo, J. G., 13 (ref. 1)
Sarafim, A. F., 43 (ref. 11)
Sato, T., 74 (ref. 68)
Saunders, J. T., 158 (ref. 37)
Schaaf, D. P., 133 (ref. 7)
Schedock, J. P., 53 (ref. 14)
Scheffer, F. E. C., 101 (ref. 2), 102 (ref. 21, 36), 104 (ref. 103)
Schell, W. J., 244 (ref. 6)
Schily, W., 104 (ref. 92)
Schmidt, C. F., 294 (ref. 59)
Schneider, A., 53 (ref. 14)
Schneider, G. M., 101 (ref. 1), 102 (ref. 19, 20, 29), 104 (ref. 78)
Scholander, P. F., 175, 194 (ref. 1)
Schubert, F. H., 194 (ref. 9)
Schugerl, R., 175, 194 (ref. 6)
Schuit, G. C. A., 53 (ref. 9)
Schultz, J. S., 175, 178, 194 (ref. 2, 3)
Schumacher, E., 293 (ref. 46)
Schutz, E., 104 (ref. 92)
Schwedock, J. P., 53 (ref. 12)
Schwimmer, S., 170 (ref. 4)
Scibona, G., 73 (ref. 51), 74 (ref. 64)
Scott, R. L., 80, 93, 102 (ref. 27, 28)
Scott, R. P. W., 167, 171 (ref. 16)
Seban, R. A., 320 (ref. 10)
Sebastian, J. J. S., 13 (ref. 8)
Sefcik, M. O., 244 (ref. 5)
Semishen, V. A., 294 (ref. 60)
Senftl, H., 103 (ref. 70)

Seno, M., 133 (ref. 13)
Senoff, C. V., 14 (ref. 47)
Sesler, J. L., 14 (ref. 36 )
Shalagin, A. M., 288, 294 (ref. 72—74, 76, 80)
Shaw, D. J., 72 (ref. 9, 14)
Shaw, J. N., 72 (ref. 18)
Shendalman, L. H., 150, 158 (ref. 40)
Sherwood, T. K., 43 (ref. 11), 320 (ref. 8)
Sheth, A. C., 159 (ref. 49)
Shilov, A. E., 13 (ref. 14)
Shimshick, E. J., 104 (ref. 102)
Shinskey, F. G., 266, 292 (ref. 1)
Shor, A. J., 133 (ref. 18)
Shulman, J. H., 73 (ref. 35)
Siano, D., 293 (ref. 44)
Sidgwick, N. V., 72, 73 (ref. 42, 44)
Silver, B. L., 13 (ref. 6)
Simmons, J. H., 170 (ref. 3)
Simpson, J. R., 65, 294 (ref. 65, 66)
Siow, K. S., 102 (ref. 35)
Sips, R., 144, 158 (ref. 36)
Sircar, S., 150, 157 (ref. 14), 159 (ref. 47)
Sirkar, K. K., 246, 251, 253, 255, 261, 263 (ref. 4, 6, 12)
Skarstron, C. W., 158 (ref. 38)
Skoog, D. A., 74 (ref. 65), 133 (ref. 20)
Sloan, E. D., 194 (ref. 7)
Slocum, E. W., 104 (ref. 97)
Smith, D. R., 175, 176, 194 (ref. 4), 195 (ref. 16)
Smith, R. H., 53 (ref. 16)
Smith, S. B., 158 (ref. 25)
Smith, W. T., 133 (ref. 18)
Snavely, B. B., 292 (ref. 7)
Sneeden, R. P. A., 13 (ref. 26)
Snyder, R. E., 53 (ref. 17)
Soave, G., 91, 104 (ref. 77)
Sobocinski, D. P., 43 (ref. 5)
Solash, J., 53 (ref. 13)
Sorrie, A. J. S., 176, 195 (ref. 28)
Sotelo, J. L., 158 (ref. 19)
Sourirajan, S., 133 (ref. 3)
Speck, G., 53 (ref. 13)
Spencer, D. J., 293 (ref. 52)
Squires, T. G., 104 (ref. 105 )
Stahl, E., 97, 101 (ref. 1), 104 (ref. 91, 92)
Stammett, V. T., 103 (ref. 49)
Starodubtsev, A. I., 294 (ref. 79)
Starostin, A., 294 (ref. 79)
Stauffer, H. C., 53 (ref. 7)
Stauffer, R. N., 320 (ref. 5)
Steigelmann, E. F., 176, 195 (ref. 17)
Steinfeld, J. L., 293 (ref. 28)
Sternberg, J. C., 165, 171 (ref. 15)
Stewart, R. F., 13 (ref. 8)
Stewart, W. E., 320 (ref. 7, 11)
Sticher, O., 171 (ref. 10)
Stookey, D. J., 244 (ref. 3)
Stover, C. S., 53 (ref. 15)
Strain, S. M., 170 (ref. 3)
Strangeland, B. E., 53 (ref. 8)
Strathman, H., 133 (ref. 2)

Wong, Y. W., 158 (ref. 43, 44)
Wood, F. W., 73 (ref. 31)
Wood, L. J., 64, 73 (ref. 27)
Wood, L. L., 294 (ref. 65, 66)
Wood, R. W., 274
Wyszynski, M. L., 150, 159 (ref. 46)

# Y

Yagodin, G. A., 56, 72 (ref. 7)
Yakatan, G. T., 226 (ref. 17)
Yamabe, T., 133 (ref. 13)
Yamamoto, K., 170 (ref. 3)
Yang, R. T., 159 (ref. 48, 51, 54)
Yang, T. T., 210, 225 (ref. 9)
Yanik, S. J., 53 (ref. 7)
Yogan, T. J., 102 (ref. 38, 41)
Yon, C. M., 144, 158 (ref. 34)

Yost, D. M., 195 (ref. 21)
Young, C. L., 102 (ref. 23)
Young, D. M., 136, 157 (ref. 8)
Young, E. S., 293 (ref. 12)
Yuen, K. H., 252, 263 (ref. 2, 13)

# Z

Zahka, G. J., 195 (ref. 33)
Zanazzi, P. F., 13 (ref. 16—18)
Zanzari, A. R., 13 (ref. 17)
Zare, R. N., 294 (ref. 61)
Zeman, L., 102 (ref. 33, 34)
Zhata, S., 226 (ref. 15)
Zieger, D. H., 104 (ref. 86)
Ziegler, W. T., 158 (ref. 28)
Zink, D. L., 162, 167, 170 (ref. 1)
Zosel, K., 104 (ref. 93)

# SUBJECT INDEX

## T